WORLD WATER '86

Water technology for the developing world

WORLD WATER '86

Water technology for the developing world

Proceedings of an international conference organized by the Institution of Civil Engineers and held in London on 14–16 July 1986

Thomas Telford, London

Conference Organizer: The Institution of Civil Engineers

Co-sponsors:
Association of Closed Circuit Television Surveyors
Association of Consulting Engineers
Association for the Instrumentation, Control and
 Automation Industry
British Consultants Bureau
British Effluent and Water Association
British Electrical and Allied Manufacturers Association
British Foundries Association—Ductile Iron Pipe
 Committee
British Plastics Federation
British Pump Manufacturers Association
British Valve Manufacturers Association
British Water Industries Group
British Water International
Clay Pipe Development Association
Export Group for the Constructional Industries
Institute of Water Pollution Control
Institution of Public Health Engineers
Institution of Water Engineers and Scientists
International Commission on Irrigation and Drainage
 (British Section)
International Water Resources Association
Overseas Development Administration
Ross Institute of Tropical Hygiene
United Nations Development Programme
Water Companies Association
Water Research Centre
World Health Organisation
World Water Magazine

Conference Organizing Committee: G. Edwards (Chairman), L. Bays, N. J. Dawes, Dr R. G. Feachem, M. A. Hartley, J. Hennessy, B. A. O. Hewett, P. Lowes, M. J. Lowther, H. R. Oakley, H. Phillips, T. D. Pike, W. R. Rangeley, K. F. Roberts, Mrs J. Segal, R. West, R. Wiseman, D. Wood

British Library Cataloguing in Publication Data
 World Water '86 : water technology for the developing world : proceedings of an international conference organised by the Institution of Civil Engineers and held in London on 14–16 July 1986
 1. Water-supply — Developing countries
 2. Sewage disposal — Developing countries
 I. Institution of Civil Engineers
 628.1'09172'4   TD201

ISBN 0 7277 0360 9

First published 1987

© The Institution of Civil Engineers, 1986, 1987, unless otherwise stated

All rights, including translation, reserved. Except for fair copying, no part of this publication may be reproduced, stored in a retrieval system, or transmitted in any form or by any means electronic, mechanical, photocopying, recording or otherwise, without the prior written permission of the publisher. Requests should be directed to the Publications Manager, Thomas Telford Ltd, Telford House, 1 Heron Quay, London E14

Papers or other contributions and the statements made or opinions expressed therein are published on the understanding that the author of the contribution is solely responsible for the opinions expressed in it and that its publication does not necessarily imply that such statements and or opinions are or reflect the views or opinions of the ICE Council or ICE committees

Published for the Institution of Civil Engineers by Thomas Telford Ltd, Telford House, 1 Heron Quay, London E14

Printed and bound in Great Britain by Robert Hartnoll (1985) Ltd, Bodmin, Cornwall

Contents

Opening address. H.R.H. THE PRINCESS ANNE		1
Opening address. J. PATTEN		3
Keynote address. Water resources: the decade issue of the nineties. M. S. Z. KILANI		5
Keynote address. The International Drinking Water Supply and Sanitation Decade (1981–90): the situation at mid-decade. G. A. BROWN		9
Discussion on Keynote addresses		17
1.	Water resources management. H. FISH and CHEN SHENYI	19
2.	Safeguarding water quality. R. HELMER and G. OZOLINS	25
Discussion on Papers 1 and 2		31
TN1.	Application of artificial groundwater recharge for rural water supply. E. H. HOFKES, J. T. VISSCHER and A. VAN DAM	39
TN2.	Contamination of potable water in rural areas: should it be systematically monitored as part of impact management of various interventions? P. O. LEHMSULUOTO	41
TN3.	Drilling in disaster areas. A. HAYES	43
TN4.	Water: supply and demand forecasting. B. DENNESS	45
TN5.	Occurrence of iron and thionic bacteria in groundwater of Quaternary formations. K. M. OLAŃCZUK-NEYMAN	47
TN6.	Water and salt balance in the Fayoum, Egypt. E. F. RAMADAN	49
3.	Distribution and collection systems. D. B. FIELD, G. F. MOSS and S. H. WHIPP	51
4.	Non-sewered sanitation in developing countries. A. M. WRIGHT	57
Discussion on Papers 3 and 4		63
TN7.	Rural water supply in Rukwa region, Tanzania. G. MOVIK	69
TN8.	Low-cost sanitation in Mozambique. M. MULLER	71
TN9.	Research and demonstration needs related to on-site waste water treatment and management. T. VIRARAGHAVAN	73
TN10.	Aqualis technologies to solve drinking water distribution problems. D. VILLESSOT	75
5.	Trends and issues in rural water supply: a case study. F. F. PADERNAL	77

6.	Treatment and disposal of industrial waste waters. A. L. DOWNING, G. R. GROVES and C. J. APPLEYARD	81

Discussion on Papers 5 and 6 — 87

TN11. The influence of surface exploitation of lignite on water pollution. A. K. DJELI, A. H. IBAR and M. M. BERISHA — 91

TN12. Appropriate least cost reliable water supply technologies for developing countries. C. RADHAKRISHNAN — 93

TN13. The East Java water supply project. R. D. H. TWIGG — 97

TN14. Appropriate technology for rural water supplies. P. G. WILLIAMS and L. R. J. VAN VUUREN — 99

TN15. Recent developments in refugee water supplies. E. G. THOMAS — 101

TN16. Trends and issues in rural water supply in the Sahel area. H. C. BALFOUR and J. M. WENN — 103

7. Recent developments in wastewater treatment. P. J. HUISWAARD and J. G. BRUINS — 105

8. The activated sludge process and energy efficiency for the 80s. G. A. THOMAS and A. J. RACHWAL — 111

Discussion on Papers 7 and 8 — 123

TN17. Energy saving in the activated sludge process. C. F. SKELLETT — 127

TN18. The use of an upward flow sludge blanket process for sewage treatment. W. G. DAVIES and D. E. SMITH — 129

9. Water resources for irrigation. B. S. PIPER — 131

10. Trends in irrigation. G. LE MOIGNE — 137

Discussion on Papers 9 and 10 — 143

TN19. The Sukhothai groundwater development in Thailand. G. R. THORPE and B. PANTANAVIBUL — 145

TN20. Water resources management: its role in effective utilization of water resources in developing countries — a case study of Gujarat State, India. G. S. PARTHSARTHY, P. N. SUTARIA, N. M. BHATT and P. M. MODI — 147

TN21. Slingram as a water exploration method. M. A. JAEMTLID — 149

TN22. A combined geophysical method for borehole siting, Kano State, Nigeria. C. R. C. JONES and S. BEESON — 151

11. New developments in surface irrigation. W. R. RANGELEY and R. BARNSLEY — 153

12. Recent developments in sprinkler and drip irrigation. M. D. SQUIRE — 161

Discussion on Papers 11 and 12 — 171

TN23. New developments in water distribution with hard hose travellers. J. STEINER — 175

13. Recent developments in horizontal drainage. W. C. HULSBOS — 177

14. Mechanisms in watertable control. G. R. HOFFMAN, R. F. STONER and J. H. PERRY — 183

Discussion on Papers 13 and 14 — 191

TN24.	The decay and rehabilitation of irrigation systems. M. J. SNELL and R. J. WELLS	195
15.	Health aspects of water supply and sanitation. A. A. GADDAL, A. FENWICK and O. TAMEEM	197
16.	Irrigation and health. R. BAHAR	201
17.	Social aspects of water supply and sanitation. G. WANCHAI	211

Discussion on Papers 15-17   215

TN25.	The natural disinfection of rural water supply by solar irradiation. S. KOOTTATEP, S. KARNCHANAWONG, S. WATTANAJIRA and V. KITJANAPANICH	219
TN26.	Investigation of the effectiveness of engineering methods for schistosomiasis control in a small holder irrigation scheme in Zimbabwe. P. BOLTON and A. J. DRAPER	221
TN27.	Hydroponic use of treated sewage effluent and waste waters for crop irrigation. R. F. LOVERIDGE, J. E. BUTLER and D. A. BONE	223
18.	Financing development projects. J. M. HEALEY and J. B. WILMSHURST	225
19.	Project casework based training for better planning and management of water pollution control in developing countries. G. THARUN	231

Discussion on Papers 18 and 19   241

TN28.	Meter reading and billing with portable microcomputers. D. H. TOMPSETT	247
TN29.	Beira water and sanitation — the aid component. M. MULLER	249
TN30.	Appropriate technology? L. O. WILD	251

General discussion   253

Summary of Conference. M. J. HAMLIN   255

Closing address. A. CHURCHILL   259

H.R.H. The Princess Anne

Opening address

This Conference concentrates on water and sanitation of less developed countries, and has both technical and commercial thrust.

I am not a civil engineer or in business, but I have been shown ample examples of the benefits of water and sanitation to developing countries. I am also aware that to the experts a little knowledge can be a very dangerous thing and that it can be easy to over-simplify.

The founder of the Save the Children Fund, Eglantyne Jebb, even in 1919 when she started the Fund, was very keenly aware of the importance of clean water in relation to infant mortality and to the health of mother and child; the importance of clean water at birth and during the growing-up process with health and hygiene, washing and especially simple things like washing the eyes frequently; and the many diseases caused by just lack of hygiene, not only in the cooking and preparation of food, but also because of lack of attention to the boiling of water and, in recent times, the misuse of powdered milk with unboiled water and bottles just being left around to get dirty and covered in flies. These things have in the past probably been the greatest problems to children and the actual cause of death.

In health and hygiene, clean water is still important, but teaching the importance of clean water to mothers is one thing and having that water available is quite another.

I suppose there are really two major problems: one is a genuine lack of water and the other is that of having too much. Lack of water can be real and apparent - real in the sense that there is just very little rain and virtually no surface water around, and apparent in the sense that in countries like Nepal, where people tend to live on the top of 3000 ft hills, the water is very much at the bottom. The people frequently boil the water, when it is available, but not the goats' milk which is there all the time, and that also causes problems. There is too much water in the areas of high monsoon rains and low-lying land like estuarial countries. Bangladesh is the prime example where the dangers of stagnant water, no drainage and disease are greatly enhanced.

In areas of very little water, there are other problems - problems of erosion brought on by a number of different cultural and traditional problems such as the search for firewood. It seems a major contradiction in terms to be saying on one hand that you must boil your water, and on the other hand that you must not cut down trees. This, of course, is a major dilemma all over the world. Available resources, either wood or perhaps dung, may have to be used to boil water. Fodder for livestock and problems of tradition in culture, in the sense that your livestock is often your wealth, produce difficulties in storage and the use of your wealth and the livestock available to you.

In rural areas, water is fundamental for nutrition; in urban areas it is fundamental for health, lack of sanitation being probably the greatest danger. The Save the Children Fund works in some of the more distant and some of the harshest parts of the world. The availability of water is one of the major priorities in any new project and so the setting up of water schemes, such as well drilling and pipe laying, are often part of the work that the Fund's teams have to do.

Making use of some of the traditional knowledge and methods in a particular area is no bad thing, if they are adequate for the community - for two reasons. First, people can then understand and maintain systems and, second, it probably does less harm ecologically than some large-scale projects. There are many arguments for and against, for instance, the big dams that are being built. They do retain valuable silt, depriving land downstream, and, if they are there to provide power, they become inefficient because of the build-up of silt. In some cases irrigation ditches get clogged with weed, because the clean water coming out of the dam encourages the growth of weed.

The pros and cons of such developments give a very different equation, but a look back through history gives hints and guide-lines that might well be usable today. In fact, I have met a student at the London School of Economics who was studying, with the use of computer graphics and modelling, the efficiency of a low technology water-harvesting system for desert agriculture. The system was started in 1500 BC in the Negev Desert and was still going in AD 600, and it was using an elevation of a mere 50 m down the side of the valley. Aerial photographs still show that the system was amazingly sophisticated, but simple to manage. The rain-water did not rush down the steep sides of the valleys, so there was less erosion and the silt was more likely to be swept down into the flatter area of the valley which was enclosed as an irrigation area.

Such projects must be good. Modern technical capabilities have led us to make some probably

OPENING ADDRESS: PRINCESS ANNE

basic errors and serious imbalances, and we must learn from those mistakes. I have seen one or two examples that I suppose with hindsight were not quite what was called for. In the Gambia there was quite a large-scale irrigation system which, after two years, had completely silted up really through lack of maintenance. There is also a knock-on effect in human terms of some projects. For instance, large numbers of people have to be moved in order to make way for a lake created as a result of damming. In Zimbabwe, people were moved from the side of a river where they were fishermen and rice-growers to a semi-arid area where they simply had not the knowledge or capability to withstand such a different existence, and they are suffering considerably as a result.

Despite that, however, the type of aid now available is encouraging. Britain's own water industry, through WaterAid, is making a real contribution to the requirement, the knowledge, expertise and practical application of water and sanitation schemes. The Register of Engineers for Disaster Relief has shown itself to be a very valuable friend to many agencies requiring engineering expertise for small projects - it is a way in which individual engineers can contribute their knowledge. Company support has, for example, provided the Save the Children Fund with 100 t of pipe which has been used in Kampala, Uganda, in Ethiopia for piping springwater, in Somalia for main connection and in the camps in the Sudan. The Fund is very grateful for that assistance - not just the piping but also the manpower assistance that has come with it.

The Fund is usually involved with small projects, whereas many engineers are involved in considerably larger projects. However, in distant rural areas small is beautiful - from installation to management and maintenance. Irrespective of the size of a water or sanitation project, whatever its aims, the importance of training, management and maintenance is paramount if all that money, effort and goodwill involved is not to be wasted. Waste is perhaps one of the things that one sees most often in one's travels. I met a retired army major in Bangladesh who was running an apprentice school. He managed to get apprentices from all the far-flung corners of Bangladesh and he ran a small engineering apprentice scheme specifically designed to build and maintain a type of well apparatus and irrigation equipment that they would not only use simply but that they would be able to maintain themselves on a long-term basis.

It is worth remembering that Britain spends a great deal of time and energy looking to help developing countries, but it must also be prepared to look at its own water supplies before its own highly industrialized societies find that they too have their own water problems.

Water is a global necessity - an element which must be conserved carefully and shared wherever possible. That makes water engineers very responsible people.

JOHN PATTEN, Minister for Housing,
Urban Affairs and Construction, UK

Opening address

The first World Water conference, held in 1983, concentrated on the aims of the United Nations Drinking Water Supply and Sanitation Decade. It discussed the scale of the world problem, including the resources of finance, manpower and the materials needed to solve it. This second conference has two basic themes. First, it continues to address the aims of the present decade, but also it looks ahead to examine the issues likely to be important in the next decade.

At the half-way point through the first world water decade it is appropriate to review progress and perhaps to forecast how many of the initial aims will be achieved by the end of the second five-year period, as is done in Dr Brown's keynote address.

WaterAid was the charity set up five years ago by the water authorities and water companies in England and Wales to provide practical help to the smaller and poorer settlements in areas without adequate water supply or sanitation. It is supported by donations from employees in the UK water industry, from the trade unions, from firms and consultants associated with the industry and from the wider public in the form of local associations, clubs, churches, schools and so on. There is a practical input by volunteer UK engineers who provide first-hand training in self-help to those in need to enable them more efficiently to mobilize their own resources and to maintain and make best use of the simplest of hardware.

The total value of this effort is modest in relation to the world-wide scale of need – perhaps about £1 million in 1986, but I see this contribution as leading to a much greater benefit over the longer term through the reverse flow of information and knowledge which these volunteers bring back to the UK.

They have absorbed at first hand not only what it is really like to lack the facilities which we take for granted, but also how best to provide them, having regard to local circumstances. In time this awareness will be spread throughout the UK water industry, and will make it easier to appreciate the true nature of the problem and how best to overcome it and to achieve the broader aims of this decade of effort. I feel this is very important, and that it should also encourage those who may feel that progress towards these aims has been far too slow.

Five years is a very short time for progress to be properly gauged. As experience is gained the rate of improvement will accelerate – and of course the end of the eighties will not mean the end of our efforts.

On the broader subject of managing resources for water supply and sanitation, I think the UK can make a substantial contribution to the needs of the developing nations.

The water authorities are unique in the way in which they are responsible for all aspects of the water cycle within regions based on river catchments. They oversee pollution prevention in the water environment as well as providing and distributing clean water and collecting and treating waste waters. This regional management system has been operated in England and Wales for 12 years. Mistakes have been made, but they have been overcome and the experience has been beneficial.

This wealth of knowledge will prove a useful source of advice which can be made available to those in the developing countries who carry the responsibility for planning water supply and sanitation facilities.

This theme of overall management of resources is not something which can be incorporated quickly into early programmes for the first-time provision of water services, but I urge that its long-term benefits be borne in mind when administrative and financial structures for their future maintenance and development are being set up.

Although this conference is concerned essentially with health and sanitation in respect of the water cycle, I should like to emphasize one further important aspect. In the drive to ensure that rivers, lakes and underground waters are kept safe from contamination it must not be forgotten that some parts of them are capable of absorbing a substantial proportion of the waste products of urban developments without suffering permanent harm. This facility must be used effectively; it is as important as the complementary benefits of maintaining high quality for water supply purposes.

Wastes must have some final resting place. If too many restrictions are placed on the separate sectors of the available environment there may be a danger of having no disposal options at all.

M. S. Z. KILANI, BScEE, President, Water Authority, Amman, Jordan

Keynote address. Water resources: the decade issue of the nineties

Some of the issues that engineers face and will continue to face until the end of the century are the problems of too much and too little water, the management of our available resources and the protection of water from contamination. To focus attention and funds on these and other global issues, the 1990s should be declared the International Water Resources Decade.

1. Good morning, fellow water engineers, ladies and gentlemen:

2. I am honoured to be invited to be one of the keynote speakers at the World Water '86 Conference. It is always a pleasure to come to London, which is one of my favourite cities.

3. The other keynote speaker, Dr. Brown, will talk to you on issues of the International Drinking Water Supply and Sanitation Decade. I have been asked to speak on non-decade issues. Before I do, I would like to say a few words about the decade in my country, Jordan. I fully expect that by 1990 more than 95 percent of all Jordanians will have piped water in their houses, all communities with populations of more than 5,000 people will have sewers and sewage treatment and the rest will have improved on-site disposal of waste water. We owe this progress to the dedicated efforts of our own engineers but realise that it could not have been achieved without generous financial contributions from the World Bank, United Nations agencies, various Arab funds, the European Community and bilateral assistance from many countries including the United Kingdom. We are indeed grateful for their help.

4. Some of the more important non-decade issues that we as water engineers are facing and will continue to face are the problems of too much water and too little water, the management of our available resources and the protection of water from contamination. Developed countries can solve most of these problems nationally, but we in the developing world will need both technical and financial international assistance.

5. Let us look first at the problem of the contamination of water. For centuries people have just dumped their wastes into the nearest water course. In many countries rivers have become open sewers, not fit for fish. Surface waters can be cleaned up in a relatively short time by building wastewater treatment plants and controlling non-point sources of pollution. This was dramatically proved in England on the Thames river, where I understand that fishing is possible again. I was pleased to read that the Thames Water Authority will help the Indian government clean up the Ganges river.

6. Then there is the contamination of the oceans and seas from nuclear and industrial wastes. Because of the large amount of dilution, the oceans and seas will recover once the discharges are stopped.

7. The contamination of ground water is a different story. For years many of us believed that passing water through soil would purify it. We have since learned that, although soil will remove bacteria, some dissolved chemicals will persist almost forever, because water moves so slowly through the ground. The United States is spending billions of dollars in an effort to clean up landfills and chemical dumps. In Jordan we are concerned about this problem because most of our drinking water comes from the ground; but fortunately, with only a few industries, we can still do something about it.

8. In the decades ahead, in addition to protecting our ground water, we must conserve it and use it carefully. Over 30 percent of the world's land surface has been classified as arid or semi-arid and 19 percent of the world's population live in these areas. Because rainfall is uncertain, surface supplies are seasonal and arid areas must depend largely on ground water.

9. One of the driest areas is sub-Saharan Africa. And yet I am told that there are vast underground reservoirs that extend across national boundaries from the Arabian peninsula throughout most of North Africa. They were filled by rain thousands of years ago. Some people feel that tapping these water sources could help solve some of the problems caused by drought in the countries involved.

10. It is not known whether these aquifers will be as productive as these people believe because they have not been systematically studied transnationally. I understand that findings based on investigations at national levels have been incomplete or misleading because of limited hydrologic or hydrogeologic data. Separate development by each country of a common water source will never reach its

World Water '86. Thomas Telford Limited, London, 1986

ultimate usefulness without the full knowledge of the shared basin's main parameters, which include climate, geomorphology, hydrology, hydrogeology, soils, vegetation and the socio-economic conditions prevailing in the whole basin.

11. A comprehensive study of these parameters will take a long time, will require close cooperation of national governments that share the basins and will require support from international organizations. In the next decade, we engineers as well as our fellow-scientists in other disciplines may be asked to study these basins under internationational auspices. As we provide the information on which to base irrigated agriculture for these areas, we must learn how to design projects that avoid the problems associated with irrigation, such as water logging, salinity and disease.

12. Rain in the geological past that created the giant underground aquifers referred to above also left behind areas of arable land. In an article which appeared in the Jordan Times in December 1985 but which was credited to The Guardian newspaper, Dr. Farouk El-Baz, an Egyptian scientist, reported that "In November 1981 a shuttle-borne radar instrument unraveled the terrain beneath the sands of the western desert of Egypt. In an area that is now bone-dry without a single blade of grass, the radar revealed ancient river courses as wide as the Nile valley. Nearby a region was selected to drill for water. Eight wells were dug and all brought fresh water from depths between 25 and 250 feet (7.6 and 76 metres). Today there is an experimental farm that may be the nucleus for a vast agricultural settlement in this parched land." Elsewhere in the article, Dr. El-Baz reports that photographs of Ethiopia taken from a shuttle with a large-format camera showed areas with potential groundwater resources.

13. I have quoted these items to illustrate the tremendous potential uses of the satellites and other devices that are being put into space. As one of the two representatives from Jordan to the World Meteorological Organization (WMO), I have followed with great interest how its programmes have benefited from space technology and from the development of powerful computers. The World Climate Research Programme (WCRP) may be able to predict seasonal and annual variations in climate much sooner than we had expected a few years ago. If we know more about the cyclic patterns that produce rain, we may be able to plan better for the years of plenty and the years of famine. For example, the years of drought that caused such misery in 1984 were followed by dangerous floods in the Nile valley in 1985.

14. One of the most interesting and important parts of the WCRP is the Tropical Ocean and Global Atmosphere Programme (TOGA), sponsored by WMO and other international scientific organizations. It is a ten-year study of the large-scale climatic variations of the tropical oceans and the global atmosphere (for example, El Nino, monsoons in Asia and Africa, droughts in Africa and South America and other special events).

15. The TOGA programme resulted from the dramatic climatic changes in 1982-83 that significantly affected the lives of nearly half the worlds's population. The shift in the Southern Oscillation of sea level pressure in the Pacific ocean appeared as an El Nino event in September 1982 when the temperature of the ocean off the coast of Peru rose 4°C in 24 hours and ended abruptly in June 1983. The devastation to the people, villages, and economy of several South American countries caused by excessive rainfall and loss of fisheries was unprecedented. The widespread effects elsewhere included heavy snows in North America, colder weather in China and New Zealand, floods in some parts of Brazil and drought in others, and a weak monsoon on the Indian subcontinent. The most tragic impact, however, was the enhanced drought in all of Africa.

16. These climatic changes are the results of global interactions of the atmosphere and the oceans. If we can understand the causes of these changes and predict their occurrence, it would allow us some time to adjust and prepare for the impact of such changes. For us as water engineers it is particularly important to understand the variability of the monsoons and the occurrences of floods and droughts. Too much water and too little water affect the design and operation of projects.

17. I believe that it is realistic to expect that in a few years, as a result of these programmes, we shall be able to _predict_ the climate and to use the predictions in our planning and operations. In the more distant future, using satellites and the more powerful computers that certainly will become available, we may be able to _influence_ the climate to help increase rainfall in areas that need water and to help decrease flooding. Engineers and scientists from all over the world working together can probably achieve this goal. Far more difficult will be the task of the political leaders who will have to decide, for example, which areas will receive limited rainfall or the areas to which storms must be diverted to protect major cities.

18. A problem for all of us in charge of water organizations is the effective management, operation and maintenance of our projects. This is particularly important for us in the developing world and can only be solved by training. One of my highest priorities is the training of my staff at all levels from mechanics and operators to top managers. We are setting up training facilities in Jordan that we hope will ultimately serve as a regional training center. We still need academic and practical training opportunities outside of Jordan for our middle-level staff. We are sending as many people as we can afford to such training programmes. We need help for training now and will continue to need help for the next decade. I appeal for more bilateral and international assistance for training. I am

sure that my colleagues from other developing countries will support me in this appeal.

19. I have touched briefly on only a few of the problems of world water that must be addressed before the end of the century. With the new developments in science and technology, we have a chance to move more quickly towards solving them. We have seen how the International Drinking Water Supply and Sanitation Decade has focused world attention and financing on the problems of that sector. I propose that the United Nations declare the decade of the nineties as the International Water Resources Decade. The cost will be high but probably less than one percent of the amount that is being spent on armaments. I am sure that it will be worth it.

G. A. BROWN, Chairman, Steering Committee for Co-operative Action of the International Drinking Water Supply and Sanitation Decade; and Associate Administrator, United Nations Development Programme, Switzerland

Keynote address. The International Drinking Water Supply and Sanitation Decade (1981-90): the situation at mid-decade

We are now separated by three years - and some 3 miles - from our last World Water Congress. Since then a lot of water has flowed under the bridges of the Thames and past our World Water '83 South Bank Site. But, I am happy to say that even more water has been cleaned and channelled into new pipes, and pumps, and wells as a direct result of the International Drinking Water Supply and Sanitation Decade. A Decade which, as we agreed in 1983, is not just an impersonal international effort, but our Decade - yours and mine, and everyone's.

So, it is a great pleasure to be back in London for this triennial event, organized by the Institution of Civil Engineers. We are all grateful for the excellent conference organization and for the professionalism of the interesting and varied programmes we are to enjoy. This time we not only have a more extensive exhibition, but also will address the theme of irrigation as did, incidentally the Decade founding Conference at Mar del Plata in 1977. Even more appropriately, our Congress will run in tandem with a conference on Drought and Famine. Thus, "World Water '86" will highlight the symbiotic links between different water sub-sectors. All too often, these have hitherto been approached as separate self-contained problems, leading inevitably to waste and to inefficiency.

Firstly, it is my pleasant duty, as chairman of the Steering Committee of the International Drinking Water Supply and Sanitation Decade, to bring you up to date on our Decade - its constraints and progress - then to review with you where we are in this mid-Decade year, and finally to suggest how we may profit in the second half of the Decade from our experiences in the first.

INTRODUCTION

Background

1. Water means life and no one gets by without it. So, the Decade addresses one of Society's and the individual's most basic needs. There is nothing new there, but what the founding Conference at Mar del Plata in 1977 did do, that was new, was to emphasize drinking water and sanitation as issues by themselves. So often before, they had been submerged in higher priority and more investment-attractive programmes. The Action Plan of Mar del Plata drew attention to the need for governments to set up realistic goals and to concentrate on the underserved - those really in need. Those two goals realistic targets and the underserved - both quite new - remain the twin pillars of the Decade, but now in the second half to achieve them we must introduce a third element, a shift in emphasis to the software areas.

2. I propose, rather than again to review the history of the Decade, to focus on its meaning by asking the essential question: is the DECADE just a new way of speaking about water supply and sanitation? My answer to this question is no. The need to provide better drinking water and better sanitation was recognized before the Decade and the efforts to satisfy these needs will continue way beyond the year 1990, probably for ever.

3. Of what, therefore, does the DECADE consist? In my opinion, the Decade implies that during the ten years from 1981 to 1990, there must be a special effort in both quantity and quality. Because quantity must not be developed without reference to quality and this will lead us towards the software areas and the so-called Decade approaches, themes on which I shall be particularly focussing my remarks.

4. To achieve this, we hoped - and we still hope - that each country will establish goals for 1990 which match, as far as possible, global expectations. These were indeed that all people, having the right to access to drinking water in quantities and, of a quality equal to their basic needs, will in fact, enjoy such rights. Similar considerations apply equally to all that concerns the disposal of wastes.

5. There have been many Decades before. Water supply and sanitation were integral parts of both the UN's first two development Decades of the 1960's and 1970's, and Latin America has even already had its first Drinking Water Decade decided in Punta del Este in 1961. So this Decade of ours will not see the end of the problem because the level of services can always be improved, but we can set, and meet targets, and we can concentrate on the underserved, which is what the Decade is about. So, let us now look

World Water '86. Thomas Telford Limited, London, 1986

at the challenge and how we have responded.

The challenge

6. In economic and political terms, 1981-1990 was not a good Decade in which to launch this particular effort in water supply and sanitation. So we may, according to our taste, deplore the foolhardiness or applaud the courage of our founding fathers. They met at Mar del Plata well aware that in the disadvantaged world, 80 per cent of all sicknesses was attributed to unsafe and inadequate water supply and sanitation and half of the hospital beds were unnecessarily occupied by people with water-related diseases.

7. With sanitation - the less attractive twin sister of the Decade - the drive encountered vastly different religions, cultures, attitudes, habits, economic conditions, and political ideologies among regions, nations and provinces within nations. Simple formulas for gaining improvements fell by the wayside, while frequent changes in approach became necessary.

8. Disparities among people have caused uneven progress. Varieties in accomplishment are due to various controlling factors. I shall mention three of them:

(a) Average per capita incomes. These dramatically illustrate the constraints upon the programme. They range from some $ 23 000 in some oil rich States to somewhat less than $ 300 for nearly half of the rest of the world.

(b) Infant mortality. This is a valuable index for measuring water supply and sanitation impacts. Some 34 000 to 40 000 children under 5 years old die each day, while in far too many countries, the mortality per 1000 births may be 10 times as high as in the 'western' world. A sanitation programme could possibly reduce these losses by about half.

(c) The urban drift. In most countries, at the end of World War II before the march to the city and its fringes got really underway, the rural population exceeded by far the urban numbers. In the interval the global population has more than doubled to around 5 billion. Over ten urban conglomerates have risen to 20 million each, and they continue to grow. Predictions are that, by the year 2000, for the first time in history, more than half of the world's population will be living in urban areas. The largest of them will be in the less developed countries - projections list Mexico City as the most populous with an anticipated 31 million people - if its infrastructure holds up.

THE DECADE AT MID-POINT

Data collection

9. The Mar del Plata Water Conference in 1977 recommended in its Action Plan that through international cooperation, the ongoing activities of the UN family in maintaining and reporting on the status of community water supply and sanitation, be intensified. For this reason, we in the UN System and particularly its Steering Committee attach high priority, not only to the collection and analysis of data produced by countries, but also to the dissemination of this information as a promotion tool, and as a support to governments in their efforts to stimulate the sector.

10. Accordingly, a monitoring exercise was conducted by the World Health Organization (WHO) in 1980-1981 at the beginning of the Decade to provide baseline data provided by the countries themselves. This exercise was repeated for the end of 1983 and since then a continuing process has yielded vital statistics, and identified constraints and solutions.

Constraints

11. On a general note, the Decade system is handicapped by traditional approaches which require development activities to be carried out through time-worn channels, when what is really needed is a more flexible approach. Multi-sectoriality should be the order of the day - strong ministries must collaborate with weaker ones, and professionals recognize the interdependence of engineers and doctors, of social scientists and administrators. The efforts of Non Governmental Organizations and of UNICEF have been particularly successful in exploiting these needs by finding new ways to effect what are essentially old programmes.

12. But traditional intransigence apart, our monitoring exercises have consistently shown that the countries themselves identify the same major constraints in the same order of importance.

(a) Human resources. Shortage and maldistribution of skilled manpower (particularly trained sub-professional and professional manpower) has been identified by the countries as their single most severe and persistent constraint. Huge training programmes are planned, but correct priorities are needed. Inadequate management of human resources has led in some cases to overstaffing at certain levels which has increased bureaucracy.

(b) Operation and Maintenance inadequacies and the need for rehabilitation. Broken down and badly performing systems demonstrate the failure of operation and maintenance arrangements. There is a need to stress the quick and economic benefits to be achieved by rehabilitation and maintenance improvements.

(c) Institutional constraints. Multi-sectoral approaches are needed because combined programmes are more attractive to finance, and fragmented approaches to water supply and sanitation development limit impact and waste resources. Institutional failures most frequently occurring are faulty management, lack of policies and budgets and inadequacies of training.

(d) Failure to attract more external finance. Shortage of funds for accelerated Decade programmes was reported as particularly crucial among the least developed countries where these funds are needed to prime their own financial pumps.

(e) <u>Water resources management</u>. Competition over the use of water resources causes problems. Countries report difficulties in preserving water quality and quantity, and that environmental and water rights issues can seriously delay projects.

(f) <u>Imbalance in coverage in urban and rural areas</u>. A disproportionate distribution of facilities occurs between urban and rural areas, and even within urban areas themselves – in particular at the urban poor fringes. Sanitation services usually lag far behind water which is more popular and politically sensitive.

(g) <u>Ill-considered community participation and choice of technology</u>. Community participation has been established as a crucial element in water supply and sanitation development, but its implementation calls for careful consideration and proper planning.

(h) <u>Low government priority for the sector</u>. There appears to be a lack of influence of agencies and ministries concerned with the water and sanitation sector on governmental policy and resource allocations. The latter tend to go more readily to more productive sectors.

Software
13. The consistency and persistence of the same major constraints underscore clearly the need to put more effort into the 'software' elements of programmes to support its capital investment activities. This conforms to my main theme – the need for new Decade approaches for the remainder of the Decade, because investment in the sector without the existence of the necessary absorptive capacity will not be effective.

Achievements
14. Latest returns received in February 1985 and based on 94 developing countries representing 87% of the developing world (but not including China) are shown in Table 1 along with changes in levels of service coverage in ten years preceding the Decade and its first five years.

15. Put simply, these results show that so far during the Decade we are holding our own with urban water and sanitation (in spite of the urban population drift), that we are losing ground in rural sanitation but making some good progress in the rural water sector.

Population served and yet to be served
16. It is estimated that during the first half of the Decade 270 million additional people received water services and around 180 million received access to sanitation facilities. At the end of 1985, however this left still about 1 200 million people in the developing countries (excluding China) without access to a water supply and slightly over 1 600 million without access to a sanitation facility. It is clear that most of the enormous task of providing water supply and sanitation services to the unserved still lies ahead.

17. So much for the macro statistics of the Decade progress, there are three other levels from where we can now usefully glance at the Decade: country level, the external support community and the global level.

Country level:
Activities
18. Whatever country you to go in the Third World, you will find Decade-generated activities. In Lesotho, a successful consultative meeting last year was based on the positive continuing work of the National Steering Committee. This brings together not only widely separated government departments, but also bilaterals, Non Governmental Organizations and multilaterals. In Santiago de Chile, an effective National Action Committee has already been instrumental in achieving nearly 100% coverage in urban water and sanitation and rural water. While in Zambia, as in other countries, the 'home grown' National Decade Plan has provided a vital tool for pulling local and international efforts together.

Coordination
19. Progress has been made with regard to the co-ordination of United Nations activities at the country level. There, the organizations of the United Nations system have agreed that UNDP resident representative acts as focal point for the Decade support provided by the United Nations system to the country. In several instances this coordination role of the resident representatives has been expanded to include others in the international development community. In any event the ResReps' offices will have general Decade information, so that, if any of the participants here wants to know what is going on in the Decade in any country, information should be available in the ResReps' office. Technical support teams have also been organized in a number of countries to lend technical support to the National Action Committees or equivalent bodies established by governments. The teams function under the chairmanship of the UNDP resident representative and comprise country-based agency representatives and technical staff.

Decade Consultative meetings
20. In Zaire, Morocco, Lesotho, and Rwanda, governments with the support of WHO and UNDP have organized special consultations on the Decade which have succeeded in mobilizing substantial increased interest and resources for sector development. Several other governments, notably Zimbabwe, Peru, Indonesia, Burkina Faso, and Nepal, are expected to plan similar consultations in the coming years as part of their 'round table' processes.

Table 1: Changes in levels of service coverage in 10 years preceding the IDWSSD and its first five years (1981 - 1985) (Population in millions)

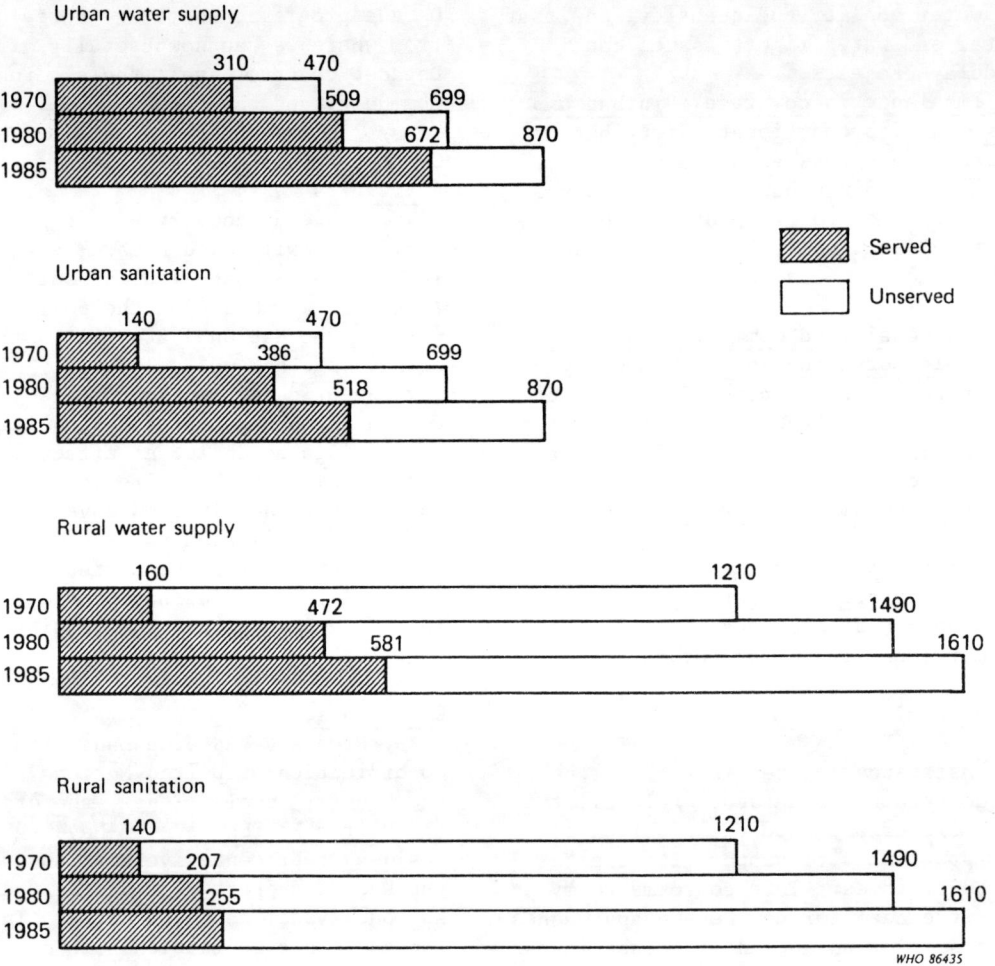

Donor level:

Water Aid

21. There is one particular 'success' story I would like to salute at this point as it involves a home grown British initiative currently providing over £ 1 million per annum of Decade assistance to the developing world. I refer to "Water Aid" the charity born of the Decade and the British water industry. More detailed reference will doubtless be made to Water Aid's outstanding successful initiatives elsewhere in this congress - Here I would just like to express the hope that by its example "Water Aid" will encourage other national water industries to emulate its success.

Aid Coordination

22. In the field of aid-coordination the Decade is catalyzing sectoral collaboration. This progress is of special interest to this congress, where so many diverse interests are represented. The sector in general, and the Decade in particular, have now been singled out by the OECD's Development Assistance Committee (DAC) - the leading donor group - as model efforts of international collaboration. It is hoped the Decade's efforts will stimulate by their example the less successful, but higher profile and economically more vital sectors, like agriculture and transport.

23. However, among the bilaterals, it is recognized that, even more importantly what must be attacked successfully is coordination at the country-level. It will be not so much by the water and sanitation systems, as by the viable ongoing institutions with trained manpower, that we leave behind there in the country itself, that our Decade will be judged.

Consulting Engineers

24. These donor level issues have a direct impact on those attending this congress: for consulting engineers, the DAC's activities are leading to interesting conclusions which are now shared among the donor agencies. Our system is seeking to coordinate mechanisms among donors and the consulting engineer should know about this because in his competition for contracts he is likely to encounter a more uniform perception among his clients in the donor community.

25. More important may be the strong emphasis which donors are now likely to give to more software component in projects. Is the

consulting engineer prepared for this? More particularly, does the consulting engineer know what is wanted? Has he expertise in software areas such as manpower development, water quality, community-based methods for Operation and Maintenance, self-help projects, local labour, etc.? Has the consulting engineer made arrangements to have access in the developing countries to professional associations, universities and similar institutions, etc. if he himself has no in-house competence in these areas? Has the consulting engineering association dealt with this matter adequately? Has information transfer been organized within the professions to respond to this new approach?

26. The consulting engineer should also know that more emphasis is being placed on the use of "local money". Will this imply changes in project design technically or institutionally? But remember that local money is likely to be short and therefore economies in design must be attempted both by bringing down unit cost and possibly by lowering planning and design horizons.

Software investment

27. World Water '86 is a forum where one would more logically choose to put the case for hardware rather than for software investment, because often the traditional engineering approach of the past still persists. This has been to put the pipes in the ground, throw the switches and proceed to the next project which would be handled in the like manner without adequate attention being given to what we might call the software, e.g. human resources development, the strengthening of institutional performance, the mobilization of the community through health education and other support communication and, by assuring that programmes are linked to other relevant development programmes, e.g. those dealing with primary health care. Our information is that, today, perhaps only one percent of the total investment is made available to software development. Now even if this should be raised to, say, five percent it would still be much less than the commercial sector would normally spend on marketing and research.

Manufacturers

28. For manufacturers, the essential message is that in spite of the financial difficulties much money continues to be invested into water supply and sanitation even if the squeeze is on; therefore, lower unit costs are important and where manufacturers provide credits, local term conditions will be sought. Local assembly or production will clearly be coming increasingly important.

29. Scepticism about "Water for All" goals implies lower expectations for the manufacturers but this should not discourage them because the need to expand water and sanitation will continue and will not be limited to the 1980's.

Global level:

Steering Committee

30. An important response of the United Nations system to the Founding Conference at Mar del Plata's recommendations was the establishment, in 1978, of an inter-agency Steering Committee for Co-operative Action for the Decade. This Steering Committee of which I am the chairman promotes closer co-operation among United Nations agencies in support of the Decade at both the global and country levels. The Committee which now meets once a year functions as a focal point within the system to exchange information on members' activities related to the Decade and as a mechanism for harmonizing policies and taking joint action in areas considered to be of special importance to the progress of the Decade.

31. An important positive spin off of the Decade has been the recognition that we must all, countries themselves, bilaterals and multilaterals, NGO's and private sector share our experiences, information and technology. This sounds simple but earlier development decades have wasted many resources on unnecessary rivalries. We have also learned that progress must be from the bottom up with countries themselves, and preferably with the local communities, involved in general development from as early a stage as possible.

Least developed countries

32. As a general rule, the poorest countries, especially those in Africa, are most dependent on external support for financing the development of drinking water supply and sanitation. The poorest countries, however, frequently have greater difficulties in servicing access to loans and credits from development banks for new investment projects. In fact, loans have tended to flow towards the largest and the relatively more affluent developing countries of Asia and Latin America.

33. This situation is undoubtedly related to countries' absorptive capacity, their repayment capability, the tendency to give priority to the financing of income-generating projects, and, as noted above, the availability of well-prepared projects for financing. These factors are to some extent offset by policies of bilateral and other international organizations which, particularly in recent times, are making deliberate efforts to address themselves to the need of the poorest countries. In the case of the United Nations Development Programme (UNDP), for instance, approximately 33 per cent of the funding for drinking water supply and sanitation projects flows to the African region.

Urban versus rural

34. Projects in urban areas and water supply projects continue to absorb the largest amounts of external assistance flowing into the sector. It is estimated that in 1983 bilateral agencies allocated from 84 to 95 per cent of their sector

assistance to water supply and, with one exception, larger proportions of assistance to urban water supply projects than to those in rural areas. The same situation exists with regard to the financing of urban and rural projects by development banks, although the proportion of World Bank funds allocated to rural projects has increased from an average of 8 per cent between 1974 and 1980 to an average of 13.6 per cent between 1981 and 1984.

35. There are signs that policies relating to official development assistance for the sector are changing to correct the imbalances between water supply and sanitation and between rural and urban projects. In fact, for 1982 and 1983 a number of bilateral agencies report a concentration of activities in rural areas. At the moment, however, the shift from water supply to sanitation is less marked than the shift in emphasis from urban to rural water supply programmes.

New Approaches

36. However, an even more important change is that external funding agencies are also shifting funds into programmes to eliminate the main non-financing constraints identified by governments. Those funds are supporting human resources development and public health education programmes, as well as programmes to improve operation and maintenance, promote community participation, and improve institutional arrangements.

37. Several bilateral development agencies have indicated that they have accepted Decade approaches i.e. new thrust towards software activities, in their funding to the sector. These include Belgium, Canada, Germany, the Federal Republic of, Italy, New Zealand, Norway, Sweden, Switzerland, the United Kingdom of Great Britain and Northern Ireland and the United States of America. In fact, some of these countries have already worked out strategies and policy statements for the Decade, or are in the process of doing so, notably the Federal Republic of Germany, Sweden, Switzerland and the United States of America. The need for increased attention to community participation, health education and the link to sanitation has been stressed by bilateral agencies from Finland, France and Norway.

38. Governments and communities used to ask where is the money? Now they ask where can we find the expertise, the training and the technologies? But so much for the Decade's background, its constraints and achievements. Let us now turn to the second half of the Decade, look at its issues, and then finally make some proposals.

ISSUES AT MID-DECADE

Country priorities

39. It is encouraging that in spite of the difficulties faced by the Developing Member Countries at this troubled time of the world economy, the investment for water supply and sanitation has not declined but rather increased and that, similarly, contributions from the Official Donor Agencies, though variable, are on the increase. This is an important positive indicator of a shift in country priorities.

40. As regards the qualitative aspect of the Decade, we all remember well that priority was to be accorded to the poor and less privileged and to water-scarce areas. This implies that priority be accorded not just to the poor and less privileged but almost invariably to those whose health is threatened most by the conditions in the environment in which they live. We all know that this is a particularly difficult challenge of the DECADE because of the apparent dilemma, on the one hand, of the right of the people to safe drinking water supply and adequate sanitation and, on the other hand, the need to pay for these services - in part at least. Some people have termed this seeming contradiction the "cost recovery dilemma".

PROPOSALS

41. But it is not enough just to identify mid-Decade issues; we need new proposals for the second half of the Decade like those suggested by the UN Geneva Assembly last December in its mid-Decade review. Here are five:

(a) <u>Better use of human resources</u>. The Decade's biggest resource need is trained people. We have already noted that shortage of skilled manpower consistently tops the list of constraints identified by developing countries as hampering more rapid sector development. A much more integrated approach is needed in human resources development, linking training with institutional changes resource mobilization, career development, and planning and management systems.

(b) <u>Rehabilitation, operation and maintenance</u>. The developing world is littered with broken down or badly functioning water and sanitation facilities. Failure to carry out proper operation and maintenance increases the hazard to health, as water quality deteriorates, or people are forced to return to alternative unsafe sources. Defective water systems have a double impact. Investment is wasted, because potential beneficiaries are deprived of intended services but equally important, the non-functioning system has a strong demotivating effect, discouraging future investment.

(c) <u>Mobilizing local money</u>. However, much priority governments may assign to water supply and sanitation, the world economic climate means that resources are limited. It would be unduly optimistic to presume that any substantial increase in funding can be achieved quickly, either from national development budgets or through external financing institutions. We must therefore concentrate on alternative sources of finance for water supply and sanitation projects, for instance by developing and implementing cost recovery policies to minimize government subsidies through appropriate tariff structures or contributions from users for construction and operation of facilities. The funds so generated should be used to replicate successful systems.

(d) *The need for Decade campaigns.* Positive news about Decade progress exists, but is not being adequately reported. National newspapers are an important medium by which government ministers may be able to sense public attitudes and respond with appropriate priorities, thus the positive reporting of IDWSSD activities should have a beneficial effect on future programmes. By increasing public awareness, project reports can also help to mobilize support for water supply and sanitation activities. The over-riding message involves us all. Engineers, manufacturers, planners, owners and operators or projects, should see themselves as promoters of the Decade, and recognize the publicity value of success.

(e) *Reemphasizing of Decade concepts.* The first four proposals above imply necessary changes. To tackle the constraints revealed during the first five years of the IDWSSD, countries and external support agencies need to take new initiatives. My fifth proposal is different. It reflects the identification of a number of Decade concepts which, in some cases, are being neglected or misunderstood with the result that even countries with a strong commitment to the sector may not be obtaining maximum benefit from that investment.

42. Accordingly, I would like to see special emphasis now given to Water Quality and Health, Community Participation, and choice of Technology, Water Resources Management and to seed money for activities which will be eventually taken over and financed nationally.

CONCLUSION

43. I hope my remarks have helped you to put the Decade into perspective: to understand better some of the constraints, and to reflect on new ways by which we can speed the real progress that, despite formidable odds, we have undoubtedly made since "World Water '83" and intend to make before "World Water '89"

44. On our side we in the United Nations, will keep the pressure up on governments and institutions, and as our partner conference here bears witness we shall emphasize the close link between our Decade and Drought and Famine. We shall also follow up the need to internationalize and integrate relief and development.

45. The variety in the world ensures that there is not just one International Drinking Water Supply and Sanitation Decade, but 150 National ones. So inevitably much more goes on than we 'know or need to know'. Our task is to maintain the climate in which a thousand decade flowers can bloom.

Discussion on Keynote addresses

MR T. J. WOODS-BALLARD, Sir Alexander Gibb & Partners, UK
Mr Kilani's interest in satellites is sure to stimulate discussion. The idea of understanding and controlling the weather has enormous appeal; such control could be used for the benefit of mankind, not only impacting on water resources but the whole question of energy supply. Further research in this field is urgently required. Energy, like water, is unevenly distributed over the face of the world and is in ever-increasing demand. The problems of energy and water supply pass over national borders and therefore a solution can be achieved only by peaceful co-operation among nations and will always require the allocation of funds from the wealthy to those in need.

Jordan is a country whose water resources are a critical factor in its future development. It is essentially a desert kingdom sitting astride the escarpment which falls 1200 m between the high plateau of the great Nefud Desert to the east and the base of the Rift Valley to the west. The rift runs along the entire western boundary of Jordan from Lake Tiberius in the north through the Dead Sea to the Gulf of Aqaba to the south. The edge of the escarpment catches a modest amount of winter rain moving in from the Mediterranean to the west.

The country has more than 2.5 million people, who live mostly in and around Amman in the north. Drinking water is supplied to the city essentially from boreholes sunk into the limestone aquifer in and around the built-up areas. Amman has been expanding rapidly in the past two decades and provides safe homes in a pleasant climate for not only resident Jordanians but also those from Palestine to the west of the Jordan river. Refugees from local trouble-torn areas and Jordanians working in the Gulf returning to their homeland have contributed to the population increase. The water demand has thus been increasing rapidly and indeed exceeds available safe supplies from the aquifer. It has been necessary to look further afield for additional resources.

The Water Authority of Jordan was established in 1984 as the single body in Jordan charged with the identification and development of water resources throughout the country. Some alternative supplies to Amman have already been commissioned and others are being studied. One option is the utilization of resources traditionally allocated to agriculture, primarily in the Jordan Valley. This possibility has led to the need to reassess the whole concept of water resource planning in the Jordan Valley. Another option is to develop fossil groundwater resources from well fields as far south as Disi, but the wisdom of mining such resources has to be carefully evaluated.

Jordan's main surface water resource is the Yarmouk river which supplies the requirements for the huge Jordan Valley Authority irrigation system. International considerations have put a severe brake on the development of this major river to meet Jordan's needs. Rainfall - the main supply base for the recharge of the aquifer that provides drinking water for the principal towns in the north of the country - has been meagre over recent years. However, Jordan is a country used to facing problems and the challenge of finding additional water resources is being tackled with the usual determination. Programmes for the conservation of water resources, including improvement in training and water management, are under way. Methods of controlling the water supply systems are continuously under review and a comprehensive study of telemetry is in hand to ensure that maximum use is made of the available resources. Developments of surface water storage, improvement in irrigation techniques, reduction in waste and the treatment and reuse of waste water are all proceeding to meet these objectives. The return flows in the Zerqa and Wadi Arab from treatment plants in the urban areas will in future represent an increasingly significant proportion of the total water resources available for irrigated agriculture. These are the measures that are being taken in Jordan and are the sort of measures that will have to be taken in many countries sooner or later.

By 1990 Jordan will have substantially achieved the objectives of the world water decade. The development of water resources in that country is now the key issue being tackled. I would like to second Mr Kilani's proposal that the United Nations declare the decade of the nineties as the International Water Resources Decade.

H. FISH, BSc, CBE, CChem, PPIWES, FIPHE, FIWPC, Natural Environment Research Council, UK, and CHEN SHENYI, Beijing Municipal Commission for Science and Technology, China

1 Water resources management

Water resources management is an endeavour of water conservation. The degree of management required, locally or regionally, depends primarily on climatic, geological and geographic circumstances. The principles of this management have been clearly explained in scientific and engineering terms, but by no means universally applied. Often where these principles are understood they are not properly applied for a variety of reasons which ignore the fundamental water conservation considerations involved. This paper examines these principles, and the water conservation factors which underpin them, as a contribution to the better pursuit of sensible and economic management of water resources worldwide.

INTRODUCTION

1. One of the frustrations of modern times is that there is, in the technologically-developed parts of the world, a very substantial understanding of how water resources should be developed and managed while there is only limited opportunity for ensuring that this know-how is applied world-wide. However it is not the intention of this paper to examine the causes of this situation and the ways and means of correcting it, notwithstanding that water is the second most important commodity to man. Instead the authors intend to set out what they consider to be the most important principles in water resources management in the hope that this will give some useful guidance on the subject of the pursuit of sensible and economic management of water resources.

2. In the broad context of this paper, which uses the global environment as background, all water, fresh, salt, naturally clean, or polluted by man is considered to be a "water resource", and its management is regarded as an endeavour of water conservation. Nature itself, on a global scale, does marvellously well in water conservation through its hydrological cycle. Indeed the global stock of water is to all practical purposes fixed in total quantity, although the burning of fossil fuels temporarily frees new water into the biosphere. In contrast the regional and local availability of water is very variable in location and time, and the essential objective of water resources management is to improve on the natural provision of water resources to ensure that this is adequate for human needs (both utilitarian and environmental) and neither grossly in excess in the flood damage sense, nor in deficit against demands for water.

3. While man's requirements for water resources vary according to his local circumstances, in substantial degree those local circumstances are determined by two factors of geography - climate and the geology of the upper layers of the ground, which factors determine the frequency and intensity of rainfall and the fate of that rainfall after reaching the ground. In any locality, for so long as the minimum local availability of water resources is comfortably in excess of demands for water, that water resource needs no management in the quantitative sense. When this favourable balance shows signs of being lost, water resources management must begin, and the primary requirement of this is to observe, record and understand rainfall patterns and the fate of rainfall run-off. This principle is the same, except in the scale and complexity of its application, no matter whether the local water resource be an accumulation of local rainfall, or a lake, river or underground aquifer which derives most of its water from distant sources. Further management in developing new water resources, often distant, and in controlling different demands for water may well be, and usually is, also required.

4. However this quantitative management of water resources is adequate only where significant man-made pollution of the water cannot occur. Where water pollution can occur, it will occur and where it does occur it will do no good and may do great harm to human health. In the underdeveloped areas of the world the main danger is of contamination of water resources with excremental sources of human disease, and the controlling management required is to protect the water resources from pollution or to give the abstracted water simple hygenic purification. In the densely-populated areas of the world, where the provision of urban, piped water services is the norm, both pathogenic pollution and chemical pollution of water resources occurs. Then extensive management of the quality of water resources, and provision for the comprehensive purification of abstracted water resources, becomes a major requirement.

5. Further, the modern recognition of the need to minimise environmental pollution of all kinds leads to the sophistication of setting standards of quality for surface waters, and

World Water '86. Thomas Telford Limited, London, 1986

for the purification of wastewaters before disposal to rivers. These are the essentials of modern water pollution control. The further step, of seeking economy in the use of water and of making large-scale reuse of wastewaters for all purposes, is being taken in an increasing number of locations in the built-up areas of the world. These water conservation measures are seemingly at the most advanced stage in the United Kingdom, especially in the River Thames basin.

6. In the light of this broad introductory review of water resources management, this paper will proceed to consider the subject in its two main aspects - namely in the understanding of water resources, in the control of water resources including pollution control and the pursuit of economy in water use. Additionally a short look at water resources in the global context is warranted in view of the concerns arising from the changes which are probably taking place in the chemistry and physics of those vital parts of the hydrological cycle - the atmosphere and the oceans.

Understanding Water Resources

7. There are three essential parts in the process of understanding water resources. The first part relates to establishing the physical, chemical and biological characteristics of the water at any location or time. The second relates to understanding the natural phenomena and processes which determine the volume, level, and quality of the water involved; and the third relates to an understanding of how the present characteristics of the water have been affected, if at all, by the activities of man.

8. In generating this understanding, clearly the scale of the management problems involved and the level of scientific capability and technology available are of great relevance. In respect of a water resource in a stream with a small catchment area, it may not be necessary to do anything more than use commonsense, textbooks and a reasonable but by no means extensive knowledge of natural science, augmented as necessary with a little external assistance, to develop an adequate understanding of the water resource. A thorough visible inspection of the stream and its catchment area would yield evidence of pollution levels and pollution risks. Significant pollution sources can be readily seen. Abnormal abundance or lack of expected plant and animal species are reliable pointers to water quality problems. The volume of flow can be established reasonably well by simple weir construction and discharge measurement. The biology of a water is a function of the physical and chemical characteristics of the water, and if the biology of the water, its flow and depth, and the shading of the water are approximately observed, a good approximation to its chemistry can be deduced. Such observations and deductions, and the use of simple field kits can produce remarkably valuable information. The microbiological suitability of the water as a source of drinking water can only be established satisfactorily at present by using laboratory facilities, but these can be effectively very simple. Provided the water can be sampled properly and adequate laboratory facilities can be reached fairly quickly, a few sanitary analyses coupled with the field observations referred to above will suffice to give an adequate understanding of most of the smaller local sources of water, surface or underground. The difficulty of too little science and technology being available is well recognised. The waste of using too much science and technology often occurs and is seldom recognised.

9. The process of establishing the physical, chemical and biological characteristics of water resources is essentially the collection of facts. From these facts, approximately or accurately obtained as the case demands, it is possible to begin to establish the variability of the quantity and quality of the resource, and from this the reliability of the resource and the quantity of water which may be taken continuously for use without major problems in drought periods. Understanding of the natural phenomena and processes which determine the volume, level and quality of a water, and how these are affected by man's activities - for example in impounding water, abstracting water, returning used water or otherwise polluting the water - is a complex matter. Where such understanding is necessary its achievement involves the use of past general research findings and usually applied research on the resource in question. Given this understanding and adequate observations of the physical, chemical and biological facts relating to the water, it is possible to create computerised mathematical models to describe the particular water resource system and to use these models as major tools in managing the system. A well-known example of successful river management utilising computerised river-models is the cleaning-up of the River Thames over the period of 1965 to 1983. A more modern example is the river modelling used to plan the clean-up of the Huang-Pu river in Shanghai.

10. Great advances have been made over the last quarter-century in research into water resource systems and in development of mathematical management models of a variety of kinds. The British laboratories of the Water Research Centre at Stevenage and Medmenham, of the Hydraulics Research Station at Wallingford, and of the Natural Environment Research Council at Wallingford (Institute of Hydrology) and at Windermere and Wareham (Freshwater Biological Association) have played a major role in this progress. Yet there are still important areas where detailed knowledge is insufficient and where research of an increasing multi-disciplinary nature is proceeding - for example in explaining what causes the changes of rainfall quality on passing through the surface and sub-surface soil in different catchment areas; and in assessing the role of phytoplankton in river systems (as distinct from lakes and reservoirs which are now reasonably well understood).

11. A very difficult facet of the process of understanding water resources arises in the context of the consequences of man's actions in

altering natural water resources systems physically (by development), and chemically (by discharges of used water). While to many people such considerations are regarded as environmental matters and to be managed as such, in truth these matters are not only environmental but crucial ultimately to water resources management. A close understanding of water resources is tremendously assisted by a close understanding of how water resources are, or can be, controlled and used and vice-versa. Very many examples can be quoted to support this statement but only one will be referred to here. During the great drought of 1976 in the UK almost every drop of available water resource was harnessed by the Thames Water Authority to maintain water supply to Oxford and to London. Flow of the River Thames was being lost to water supply at these towns because of navigation use of locks, and it was suggested that the Thames navigation be closed to conserve the river flow. However, a close understanding of river science, that is of the physical, chemical and biological processes proceeding in the river showed clearly that if the navigation were closed, and the water saved thereby was put into supply, very quickly stagnation in the river would render the water very difficult to purify and obnoxious for drinking. Consequently instead of taking the simple, but technically-disastrous, decision to close the river to navigation the navigation was kept in use, the extra water supply was taken, and the water loss through locks was made good by pumped recirculation of the river water in the critical reaches at Oxford and London.

12. This very important synergistic effect of integrating an understanding of water resources with an understanding of use and control of water resources has been achieved in very high degree in the United Kingdom, particularly in England and Wales where the ten river-basin authorities undertake all aspects of water management on an integrated basis. More will be said on this theme later.

Controlling Water Resources
13. Given a reasonable understanding of the characteristics of a water resource, it is then possible if necessary to control it. Such control will involve taking one or more of the following courses of action:
 a. carrying out works to control, or to mitigate the consequences of, an excess of the water resource - ie flooding;
 b. sharing out the available water resource between demands for use of it;
 c. carrying out works to increase the availability of the water resource;
 d. regulating the return of waste used waters to the water resource system, and other sources of pollution, so that the water resource remains of a quality appropriate to the use to be made of it;
 e. in all these actions ensuring that the general quality of the aquatic environment is maintained (or restored if not maintained in the past).
It is not intended that these courses of action should each be considered in detail. Instead an overall view of the control situation will be taken for the purpose of identifying the important general principles involved. Before doing so it is necessary to emphasise that whatever action is taken in controlling water resources, it should not only be sensible but economic. It is all too easy, within a public service, to waste public money in unnecessary, and often unnecessarily bureaucratic, regulatory and engineering provision. Unfortunately it is also too easy, within a commercial undertaking, to minimise the relevance of any activity which does not enhance profit. Between the two lies the ideal, which like all ideals is seldom achieved, but can be approximated to successfully.

14. It must be observed that the broad courses of action listed above divide into regulatory or engineering action, but these are very closely interrelated. Furthermore the different actions are also closely related. For example, if a flood control reservoir is to be built on a river, should not serious consideration be given to constructing it as a flow-regulating reservoir, achieving both flood control and means of augmenting the downstream dry weather flow of the river? And if the latter were achieved it would enhance the availability of water resources for use - including the possibility of its use, where possibilities of better use do not arise, to increase the dilution of discharges of used water to the river and thereby producing a cleaner river and improving its fishery value. That is a fairly obvious example of interrelationships in water resources management. Less obvious examples have been demonstrated. For example, there can be some advantages in gross river pollution arising from interrelationships in river management - engineers concerned with critical control of water levels in rivers have been known to complain that the removal of gross pollution of certain rivers has resulted in massive, natural resurgence of weed growth which greatly increased expenditure on weed cutting and removal. Another example of interrelationships comes from the freshwater River Thames. There, the total cost of sewage purification to very high standards of quality is wholly recouped by the value secured from the subsequent re-use of the reclaimed water for water supply along the river system. It cannot be a bad thing that a clean river can be secured at no cost in this way; of course the river system must be subject to a high degree of careful management.

15. In addition to sensible regulatory and engineering control of water resources, sound financial control of management systems is essential to give the whole sensible and economic system required. Undoubtedly the most important aspect of such financial control is the production and maintenance of income, for services provided, to sustain the whole operation. How this income is levied and collected may not be a matter of the highest importance; certainly not on the international scene where a variety of fundamentally different schemes seem to work satisfactorily, or at least all seem capable of being made to work satisfactorily. Nevertheless whatever the

scale of water conservation endeavour involved, one human factor seems universally true. It is that whatever commodity is provided adequately to people, at low apparent cost to them, this is given low value by those people until supply of that commodity is withdrawn. On the water resources scene this means that people generally will not conserve their use of water unless they pay significantly and proportionately for their use of it, or their use of it is rationed at a level considerably less than they desire.

16. Now of the five courses of controlling action set out above, the main (but not all) elements of regulatory action to be taken are in respect of sharing out the available water resources between competing demands and in pollution control. Both courses of action are closely interlinked and have their implications for the quality of the aquatic environment. Returns of used waste water to rivers and underground, done properly, adds greatly to the available water resources through water re-use (see paragraph 14 above); while inadequate control of abstractions of water from a river system, including connected groundwaters, like inadequate control of water pollution, will adversely affect the quality of the river water in the amenity, recreational and environmental contexts and for use in potable water supply, industry and agriculture. Overall, therefore, in controlling water resources it is of paramount importance that, where effective and economic conservation of water resources is required in any country enjoying or pursuing a high level of technological development, a sound system of water management is created and operated.

17. Throughout the world people are very good at setting up management systems which in theory should work well. We are much less successful in making these systems work properly and in keeping them working properly. In the UK, which arguably created in 1974 the best water management system in the world, a new and close attention to the working of the system is required, particularly in the proposed process of privatising that management system, especially the regulatory aspects. This can be done effectively and economically if those arguing for or against the privatisation will only clear their vision of the darkness of theoretical dogma. It is worthwhile considering some of the common practical shortcomings of water resources management systems on the worldwide scene.

18. The commonest shortcoming is that many systems have too many different authorities managing parts of the system. This does not occur in the system in England and Wales where each of the ten water authorities are responsible direct to the Secretary of State for the Environment, and they are in sole charge of all water management functions in their areas. Where there are separate organisations responsible for water resources management, for water supply, and for wastewater treatment and disposal, it is demonstrably possible to achieve effective, if not efficient, overall management. The minimum requirement is that the executive management of <u>water resources</u> in both quantitative, qualitative and environmental terms should be organised on a regional scale and be placed in the hands of one body. As far as practicable the regional unit should be based on entire catchment areas of rivers, but it is not necessary, even if desirable, that interstate or international rivers should be put under a single management body. Provided agreed objectives are set for the minimum volume and level of river flow and for the minimum quality of the river water at interstate or international boundaries, sound management becomes possible. Even where interstate or international boundaries run along the courses of rivers, a sensible common system of management along each bank can be worked out.

19. Certainly where regional management of water resources is not in the executive hand of one body, or not in the hands of two bodies by agreement as indicated in the special cases set out in paragraph 18 above, many sensible and economic water management practices become very difficult to follow. For example, flood alleviation, river-flow regulation, the conjunctive use of surface and groundwater resources, pollution control, conservation of water resources and of the aquatic environment are generally less effective and efficient in multi-management than in single-body management of water resources. Also, standards of quality for river and groundwater resources and for discharges of used water tend to be set too high, or too low; and even if they are set at the correct level it is not unusual for the enforcement of standards to be neglected in practice. And the duplication of effort which often occurs as each separate management body carefully checks its interface with the other management bodies, the consultations which have to proceed before any significant action for change can be taken by any one body, and the exercise of regulatory power without responsibility for all the consequences thereof, can lead to monumental waste, delay and other inefficiencies.

20. A number of comments have already been made in this paper on the subject of water re-use, and other papers at this conference will deal with economy in water supply use and in industry. It is, however, important to stress the importance of economy in water use to the management of water resources. Obviously, reduction of excessive leakage from water supply distribution systems, and the use of schemes levying charges for water supply which discourage waste of water, are important. But even more important in many cases in industrialised countries, it is economy in the use of water in industry which overall can make the largest contribution to the conservation of water resources. There are in many countries today glaring examples of expensive schemes for urban sewerage and sewage purification being provided while little, if any, provision is being made to deal with industrial pollution of water resources. There is no doubt at all that control of industrial pollution of water, to be economically effective, must be based on economy in industrial water use and re-use. This need becomes even more important where

governmental control of industrial development does not properly control the type of factories which may be located along the courses of rivers drawn on for water supply, or for agricultural irrigation, or which are important fisheries. In general the effective pursuit of economy of water use in industry is a pollution control operation which industry should undertake, and it is seen as a remarkable deficiency that in so many countries the pollution control authorities do not apparently press industry to take this action.

The Global Scene

21. On the international and global scale, the worrying problems of acid rain and of possible future climatic changes arising from air pollution need to be taken into account. The damage to forests and freshwater fisheries caused by acid rain, particularly where the sources of the atmospheric pollution contributing to the acidification are in countries different to those in which the acid precipitation occurs, presents political problems as well as technical ones. A satisfactory solution of these problems can only be defined as a result of extensive and detailed scientific research. This is now proceeding rapidly in various parts of the world but there are many questions yet to be answered. Essentially the problems are environmental because acid rain is not yet, nor seems likely to become, a phenomenon which will interfere significantly with the major uses of water, eg for water supply, hydro-power, and irrigation.

22. The possibilities of climatic change occurring as a result of the "greenhouse effect", or earth-warming, arising from increasing emissions of carbon dioxide to the atmosphere from the burning of fossil fuels, are quite alarming. Unfortunately we know so very little about the great global physical processes which govern climate that there is no early prospect of knowing with reasonable certainty whether there will be major problems in the future. The present conjectures, and deductions from large-scale and very crude mathematical models of the driving forces of climate, are only likely to be of any value if they stimulate the need for extensive and intensive scientific research on a collaborative international scale. What is reasonably certain is that if climatic change is likely to occur, even if only slowly, the impact of this change on the airborne part of the hydrological cycle will be likely to have tremendous implications in terms of major changes in patterns of world agriculture and of water resources.

23. The matter of pollution of the seas and oceans, in its global consequences as distinct from more local concerns, does not appear to be a major problem. It is difficult to imagine that man-made, non-degradable, pollutions of the oceans are likely to be significant when compared to that naturally occurring from land run-off. Local difficulties which can be controlled there may well be, as happens already in the confined seas of the Mediterranean and the Baltic, but suggestions of global-scale difficulties seem very far-fetched. And of course it must not be forgotten that there are major interactions between the oceans and the atmosphere that may balance the increased emission of carbon dioxide to the atmosphere. The increasing solution of atmospheric carbon dioxide into seawater, accelerated by the increased growth of phytoplankton which in turn is stimulated by increased man-made discharge of phosphates and other plant nutrients to the sea, may well prove to be a tremendous sink for atmospheric carbon dioxide as calcium and magnesium carbonates.

24. There are many possibilities, some favourable and others unfavourable, for the future of the plant's hydrological cycle. This growing uncertainty must be clarified by basic multi-disciplinary research, globally planned and executed on a priority timescale, with a minimum of duplication in different countries. This will require the creation and execution of a major and well-structured international programme, in which the United Nations Organisation could play a vital sponsoring role.

Conclusions

25. It is reasonable to conclude:-

(1) Water resources management is essentially water conservation, and the need for it varies greatly according to local and regional circumstances.

(2) Water resources and the natural processes involved in the terrestrial element of the hydrological cycle are well, but by no means completely, understood.

(3) The various aspects of water resources management are closely interrelated, and the more a particular water management system reflects these multi-disciplinary, and multi-functional links, the more effecient and effective the system will be in operation.

(4) The creation of sound financial arrangements, particularly charges for services given, within water management systems is fundamental to effective water conservation.

(5) Countries enjoying or pursuing high levels of technological development must have sound systems of water resources management.

(6) While integrated river-basin management of the style employed in the UK is desirable wherever it can be arranged, this is not as essential as it is to have water resources in all respects managed by a single body.

(7) Water resources systems managed in parts by different bodies are wasteful of water resources and other resources.

(8) The pursuit of economy in industrial water use, essentially as a pollution control measure, is an important activity in industrialised countries and is widely neglected.

(9) On the global scene, the effects of atmospheric pollution in contributing to acid rain and to possible climatic change are worrying. New collaborative international research, to secure a better understanding of the driving forces of climate and of possible effects of climatic changes in the future on the global distribution of water resources, is urgently required. The encouragement of such a programme seems to be an appropriate task for the United Nations Organisation.

R. HELMER, PhD, and G. OZOLINS, Prevention of Environmental Pollution, World Health Organisation, Switzerland

2 Safeguarding water quality

Drinking-water quality is threatened mostly by microbiological agents and some chemical constituents. Guidelines have been issued by WHO which describe and quantify different drinking-water quality problems and their remedies. Guideline values are recomended which provide the basis for setting national standards. Special attention is given to rural areas of developing countries for which simple preventive measures and surveillance and control methods are advocated. Current WHO activities concentrate on the application of these guidelines and on the conduct of pilot projects which also serve as demonstration sites for training.

INTRODUCTION
1. The importance of water as a vehicle for the spread of disease has long been recognized and there are numerous studies and reports demonstrating the relation between ill health and deficiencies in water supply and sanitation. The health effects range from outbreaks of water-borne diseases to potentially increased risks of cancer incidence. Situations where there is prevalence of diarrhoea, cholera etc. are commonly found in many developing countries. In addition, the aquatic environment often plays host to a variety of disease vectors such as mosquitos or snails resulting in malaria, schistosomiasis and other severe health impairments. On the other hand, scientific journals over the past years have published many papers concerning the chemical threats that are carried by municipal water supplies, and much has been written, for example, about volatile organics, lead and nitrates. Thus the question arises as to which health aspects of drinking-water concerns whom, or in other words, against what, exactly, should water be safeguarded?

2. A review of typical situations and water quality issues are provided in the following to further illustrate the severity and ubiquity of different health problems associated with drinking-water. The present paper limits itself deliberately to drinking-water alone, since health aspects of sanitation and irrigation are covered under another agenda item of the conference programme.

WATER QUALITY: BACTERIA VERSUS CHEMICALS
3. By and large, it seems common practice to associate developing countries with bacteriological and biological water quality problems, and to associate industrialized countries with problems related to chemical contaminants. In terms of morbidity and mortality rates causatively linked to water quality this is certainly true and the number of infant deaths attributed to diarrhoeas associated with contaminated water is a convincing argument. However, the association of chemicals with the rich and of bacteria with the poor countries might prove to be too simple a formula.

4. In a number of instances, the reverse could also happen. Most of the industrialized nations' health statistics periodically report water-borne outbreaks of communicable diseases, particularly in small-community supplies. Recent studies have shown that thousands of small rural community supplies in industrialized countries are not considered hygienically safe (refs 1 & 2). In particular, the presence of Giardia lamblia has become a rather wide-spread phenomenon in many rural water supplies (ref. 3).

5. On the other hand, many areas in developing countries are faced with chemical water quality problems which are both natural or man-made in origin. There are, for example, geologically confined areas in South America, Eastern Africa (Rift Valley) and Central Asia, where fluorides in groundwater are at such high levels that mottled teeth and even skeletal fluorosis are being found (ref. 4). Groundwaters are also often found more rich in nitrates than considered acceptable due to sewage disposal on land or excessive fertilization of agricultural areas (ref. 5). Chemical contaminants are meanwhile found in all environmental compartments, and drinking-water resources in developing countries are particularly affected due to heavy use of insecticides, herbicides etc. in agriculture.

6. Industrialized nations focus more on the environmental impact of their industrial activities, including organic trace contaminants being found in rivers, lakes and groundwaters - in addition to the heavy metals whose presence has been a subject of concern for some time (ref. 1). Remobilization of toxic

metals due to acid precipitation has emerged as another new problem of potential significance to human health. In recent years, the detection of trichloroethylene and other chlorinated compounds in important groundwater resources caused considerable concern in the populations affected. Municipal waterworks along some major rivers reported the presence of trihalomethanes, particularly chloroform. The fear that carcinogens were being distributed to water consumers stimulated much research on their occurrence, potential health risks and suitable control techniques.

7. The latter case of trihalomethane formation as a by-product of chlorination serves as a valuable example to illustrate the different importance attributed to various aspects of drinking-water quality. In countries where disinfection has limited the water-borne diseases to rare episodic events, public concern over water quality is raised by the health risk that may be posed by minute amounts of potential carcinogens. In the developing countries, however, the immediate danger of water-borne disease outbreaks and their wide-spread endemic prevalence is so much greater that compromising chlorination for the marginal decrease of possible cancer risk cannot be considered acceptable. Risk-benefit studies would demonstrate the difference in importance between thousands and millions of people affected by communicable diseases versus a potential incremental risk of cancer in the population.

8. Severity and ubiquity of the bacteriological health hazards mandate that drinking-water disinfection must be dealt with first or that other preventive measures have to be taken against biological hazards. There are, of course, exceptions, as for example high fluoride and nitrate levels which also cause rather severe health effects, or where water supplies are threatened by industrial and agricultural wastes.

GUIDELINES FOR INTERVENTION

9. Before determining what type of treatment should be given to drinking-water, some basic questions on the actual definition of "safe" drinking-water have to be resolved. The first question concerns the knowledge of risks to human health associated with the different biological agents, physical properties and chemical substances of drinking-water. This question addresses several aspects, including:

- routes of human exposure and their relative significance;
- evidence of health effects and their biological significance;
- identification of the most sensitive population groups at risk;
- expected dose/effect relationships in exposed populations.

10. Based on answers to the above, which can be difficult to obtain for the different water constituents, one can proceed to the second question which deals with the "right" intervention levels. How to determine a quantitative level for water constituents which will ensure the safety of a drinking-water supply? In a generic approach to determine these critical levels, the World Health Organization developed "guideline values" as the centrepiece of the new Guidelines for Drinking-Water Quality (ref. 6). The nature of these guideline values is characterized by the following main features:

(a) A guideline value represents the level of a constituent that ensures an aesthetically pleasing water and does not result in any significant risk to the health of the consumer.

(b) The quality of the water is such that it is suitable for human consumption and for all usual domestic purposes, including personal hygiene.

(c) The guideline values have been derived to safeguard health on the basis of lifelong consumption; short-term exposures to higher levels of chemical constituents may be tolerated for certain substances.

11. Depending on the nature of the water constituent and its predominant health impact, the rationale followed in determining a certain guideline value varied a great deal (ref. 7). In the case of bacterial pathogens, resort to indicator organisms was taken, and as a support measure for implementation, the conditions of chlorine disinfection and residual chlorine levels were specified. This aspect of drinking-water quality is naturally accorded the highest, and in most situations overriding, priority. Its application in small-community supplies was even the subject of a separate volume of the WHO Guidelines for Drinking-Water Quality (ref. 8).

12. Safeguarding of microbiological water quality as a precondition for safeguarding human health would be more effective if the other important routes of human exposure to the hazardous agent were also controlled effectively. Simultaneous interventions in all routes of transmission - drinking-water, excreta disposal, food, soil and personal hygiene - are considered mandatory to achieve a lasting reduction of pathogens in the human environment (ref. 9). Consequently, accompanying control measures in these other areas have to be borne in mind when introducing programmes for the control of drinking-water quality.

13. The health risk due to toxic chemicals in drinking-water differs markedly from that caused by microbiological contaminants. It is not very likely that any one chemical could result in a widespread and acute health

problem, except, of course, in cases of largely accidental and massive contamination of the supply. The problem of chemicals is mainly one of cumulative effects brought about by long-term low-level ingestion. To assess and quantify potential effects on human health associated with the intake of chemicals, heavy reliance on toxicological laboratory animal studies is inevitable, since the available epidemiological and clinical evidence is limited. Inference from animal test results to human health effects requires high-to-low dose and animal-to-man extrapolations followed by the application of safety factors to set guideline values for acceptable drinking-water quality. Such laboratory animal studies provide for some estimate of the risk to human health due to exposure to one particular chemical, whereas in reality man is exposed to complex mixtures of chemicals through drinking-water.

14. Carcinogens require a risk assessment approach which implies the determination of a socially acceptable risk level to derive a "safe" concentration in drinking-water. For the purpose of setting the guideline values, WHO opted for an acceptable risk of 1 additional case of cancer per 100,000 population over a lifetime of exposure (ref. 6). Some national standards follow a similar rationale. Whenever feasible, human exposure via other routes (air, food) have been taken into consideration when deciding upon a guideline value. In the case of pesticides, the acceptable daily intake (ADI) for man was used for recommending guideline values with the assumption that not more than 1% of the ADI be allocated to drinking-water (ref. 6).

15. As the above cases illustrate, the rationale underlying the derivation of guideline values is quite different depending on the nature of the water constituent. Genotoxic substances were assumed not to have a "safe" threshold limit and have been treated differently from non-genotoxic substances which can be tolerated below certain doses. Microbiological and biological agents have to be evaluated in light of their different interference mechanisms with human health. Due to a lack of suitable monitoring techniques no guideline values were feasible for the latter group and other measures for safeguarding water quality were proposed (ref. 6).

16. Guidelines to safeguard drinking-water quality are, therefore, by necessity complex and difficult to formulate and to apply. Consequently, compromises became inevitable when deciding on the "right" intervention levels. Despite all signs of caution and scientific reservations still pending discussion by more expert reunions, the third question is the most pressing one and does not permit further delay: what needs to be done to safeguard drinking-water quality? This final and crucial question cannot await further scientific deliberations and verdicts, particularly where the risks to human health are high and people are suffering. Some fresh approaches to this last question have been initiated recently and are briefly outlined in the following.

WATER QUALITY STANDARDS

17. WHO programme activities on drinking-water quality emphasize two areas, namely (i) the development, adoption and implementation of suitable drinking-water quality standards in Member States and (ii) establishment of appropriate programmes for the control of drinking-water quality particularly in rural areas. The development of standards is meaningful primarily in those countries which have already built, or are committed to building, the necessary infrastructure which guarantees implementation and enforcement. In predominantly rural areas, however, where adequate and safe drinking-water services are still lacking, prevention, surveillance and control programmes with heavy reliance on community involvement would be more appropriate.

18. The development of national drinking-water quality standards must carefully consider a number of different aspects. The standards should be practical and attainable within a reasonable period of time and with the resources which are available. Incorporated within the standards must be measures for checking that the standards are, in fact, complied with - otherwise they have no utility. The establishment of standards without considering the practical measures which will need to be taken with respect to finding new sources of water supply, instituting certain types of treatment, and to providing for adequate surveillance and enforcement, will not yield the desired results. These considerations speak very strongly against the "blind adoption" of a whole array of standards borrowed from other countries or from recommendations made by international groups of experts. In defining the WHO guideline values it is explicitly stated that "in developing national drinking-water standards based on these guidelines, it will be necessary to take account of a variety of local geographical, socioeconomic, dietary, and industrial conditions. This may lead to national standards that differ appreciably from the guideline values" (ref. 6).

19. In view of the economic constraints which many of the countries face, it might be useful in some cases to consider a slower, step-wise approach to standards. Such an approach might be taken with respect to the number of contaminants which are to be covered by the standards. Since the attainment of and compliance with standards requires resources for treatment as well as surveillance and enforcement, it might be better in some situations to concentrate on the priority problems leaving the less pressing ones until later on. This, of course, implies that there is a good understanding of the existing situation of water quality in order that the correct

choices are made. While there are merits to adopting step-wise approaches to drinking-water quality standards, it must be remembered that standards do serve as a deterrant to adding to the contamination load of the water supply.

20. Whether a step-wise approach should also be taken with respect to the stringency of the standards is debatable. The practical considerations might, in some situations, indicate the selection of a standard or standards which are less stringent than those required or desired on a health basis. It can be argued that such standards provide for a realistic intermediate step, with more stringent standards being established later on. There is the danger, however, that such less-than-desired standards become permanent, resulting in the acceptance of water of inferior quality and a false sense of security.

21. The WHO guidelines clearly recognize that the possibilities for providing safe water differ greatly among the different areas of the world. There are vast differences between areas of plentiful water and areas of water scarcity. There are differences between areas as to the level of contamination as well as to the ability to provide treatment or finding more suitable sources of supply. There are differences between the water supplies for large metropolitan areas and those for villages and rural areas. Each case requires separate consideration if the resulting strategies, including the enactment of standards, are to be meaningful and achievable.

SPECIAL FOCUS: RURAL AREAS

22. For small communities in rural areas WHO has opted to strive for a cost-effective application of its guidelines, with additional emphasis on simple preventive and remedial measures. Advice on what ought to be done in such cases is described in volume three of the WHO guidelines (ref. 8). Taking into account the often severe constraints and likely difficulties communities are facing in rural areas, main emphasis is placed first and foremost on the microbiological safety of the drinking-water supplies. The information provided should allow for the selection of methods and techniques in accordance with local conditions, and likely limitations of the available resources and manpower. Where possible, simple well-tried techniques are being advocated.

23. WHO's strategy in cooperating with its Member States for safeguarding drinking-water quality in rural areas is geared towards the prevention of microbiological contamination of drinking-water resources through a series of protective measures as follows:

- Careful planning for surveillance and control, including the design of a workable organizational structure, the assessment of local conditions, and the proper handling and use of information. Special attention is needed for those communities where no surveillance is yet in place and where future possibilities seem to be limited.

- Sanitary inspection is advocated as a crucial part of any surveillance programme which needs to be carried out by trained people at the community or regional level. In many poverty-stricken areas this activity may constitute the only feasible form of surveillance for the time being.

- Field and laboratory methods are made available for microbiological analysis of drinking-water, including the membrane filtration and the multiple tube method for both total and faecal coliforms. Attention has to be given to arrangements for transport, where necessary, of samples to the nearest laboratory.

- In the case of chlorinated water supplies the routine checking of residual chlorine is regarded essential and field methods are provided. This determination is amenable to application at the community level with a minimum of training.

- Immediate remedial measures as well as long and short-term preventive action are considered absolutely essential for achieving the desired control of drinking-water quality. Only if sanitary deficiencies identified during surveillance are remedied will the effort put into the programme render its full benefits.

- Community education and involvement are another mandatory component upon which an effective surveillance programme must be built. The active participation of the members of the community is considered an essential prerequisite in safeguarding drinking-water quality, particularly in remote areas with small communities that are widely scattered. Much of the relevant local health eduction is to be implemented within the framework of primary health care.

24. Surveillance responsibilities have to be shared and coordinated between the water supply and health authorities. The latter, in particular, have the overall responsiblity for ensuring that all drinking-water under their jurisdiction is free from health hazards. All too often, however, the health authorities have neither the necessary programmes nor the qualified staff to implement them. Fulfilling their functions involves usually both sanitary inspections and sampling/analysis of public supplies. Ideally, this is based on legislation, regulatory standards and codes of practice, trained staff, laboratory installations, regular health inspection services,

educated community health workers, etc. However, a realistic programme may have to be started with only a fraction of these requirements fulfilled initially.

WHO PROGRAMME ACTIVITIES

25. At the United Nations Water Conference, Mar del Plata, 1977, it was already agreed that "all peoples, whatever their stage of development and their social and economic conditions, have the right to have access to drinking-water in quantities and of a quality equal to their basic needs." This demand subsequently translated into WHO's strategy for Decade implementation as an objective "to cooperate with governments in the establishment of appropriate quality standards for drinking-water, in the organization of national drinking-water quality surveillance programmes and the protection of drinking-water sources" (ref. 10).

26. The WHO efforts to cooperate with developing countries in the safeguarding of drinking-water quality has centred largely around the new Guidelines for Drinking-Water Quality. Thus a series of regional and national workshops were conducted on the application of the guidelines under the prevailing socio-economic conditions. Some of the subjects of concern emerging from those discussions were the development of standards appropriate for a country's water resource situation, primary water contamination problems requiring control, trained manpower availability for implementation and enforcement of standards, laboratory and transport capacity for surveillance, and the coordination of public waterworks authorities and health authorities at different levels.

27. The series of training courses and workshops will certainly be continued thanks to the assistance of bilateral donors such as DANIDA. If our programme were to end here, we would fail at the crucial point; the ultimate and prime goal being field implementation, particularly aimed at the vast rural areas where surveillance and control of drinking-water supplies is required as a basic and continuing service. To apply and test the approaches advocated in the WHO guidelines they have to be put to work in and by the communities in the critical areas. During the last year a programme has been initiated in cooperation with UNEP where in selected countries the responsible health authorities and water agencies were brought together in working sessions during which approaches and methodology were introduced, and technical as well as manpower requirements for the implementation of surveillance schemes discussed. In each case the health requirements were delineated, and specific constraints, socio-economic conditions, and cultural aspects prevailing within the country discussed. Thus a common basis for active participation of staff at all levels was generated. Also, on this occasion, a rural pilot area was designated and a detailed surveillance programme elaborated.

28. As concerns the appropriate technology for such surveillance programmes, the necessary test methods, instruments, supplies, and information schemes are available. Their reliable functioning in different, more or less difficult, situations requires demonstration under actual field conditions. Differences in climate, distances between the communities and the laboratory, existing communication and transport infrastructure, and availability of personnel are among the factors that determine the surveillance scheme in each case. The incorporation of village health workers, sanitary inspectors, public health officers, district hospitals, etc. in varying geographic, demographic and administrative settings has to be put to the scrutiny of trials under field conditions.

29. Technical support is being provided to each demonstration area under the current programme in the form of:

- equipment provision for bacteriological tests (membrane filtration units, incubators, field kits, etc.);

- on-the-job training of technicians and sanitarians involved in sanitary inspections, sampling, analysis and advisory services on remedial and preventive measures;

- development of local language support communication and audio-visual aids for related health education and specific technical instructions;

- training of trainers in health education through courses provided at the demonstration sites;

- promotion of community education and involvement by local trainers throughout each demonstration area.

30. As of the beginning of 1986, surveillance results and water quality data are being collected and evaluated. During this year we will also jointly review the experience made in each of the pilot areas in order to improve our approaches and methods and to avoid mistakes or inefficiencies in similar situations. The pilot projects will also serve, starting with Zambia, as demonstration areas where water supply staff and health workers from neighbouring countries will be trained.

31. It is envisaged that experience and results of these pilot exercises will serve as valuable examples from which better and more effective guidance can be drawn. In particular, a comprehensive and field-tested package for organizational and technical alternatives for drinking-water quality control in rural areas should emerge as a major output suitable for other countries in comparable situations.

32. It is hoped that the entirety of this technology package will be used in the countries needing it most, and that national action committees for the Decade as well as the international donor community will consider and include an integral component on safeguarding the quality of the water supplied in their programmes. This is the only way in which the full health benefits of the investments in community water supply systems will be achieved.

REFERENCES

1. COUNCIL ON ENVIRONMENTAL QUALITY. Environmental Quality - 1980, the eleventh annual report of the council on environmental quality, chapter 3: water quality, 1980, 81-145

2. MCKAY T.M. EC directive on the quality of water for human consumption - the scale of the problem. Environmental Health Scotland, winter 1986, vol. 1, no. 12, 9-11

3. KIRNER J.C. ET AL. A waterborne outbreak of giardiasis in Camas, Washington. Journal of the American Water Works Association, vol. 70, 1978, 35-40

4. WORLD HEALTH ORGANIZATION. Fluorine and Fluorides. Environmental Health Criteria No. 36, WHO, Geneva, 1984

5. WORLD HEALTH ORGANIZATION/REGIONAL OFFICE FOR EUROPE. Summary report of working group on health hazards from nitrates in drinking-water, Copenhagen, 5-9 March 1984. WHO/EURO doc. ICP/CWS/002/m05(S), 24 May 1984

6. WORLD HEALTH ORGANIZATION. Guidelines for drinking-water quality, vol. 1. Recommendations. WHO, Geneva, 1984

7. WORLD HEALTH ORGANIZATION. Guidelines for drinking-water quality, vol. 2. Health criteria and other supporting information. WHO, Geneva, 1984

8. WORLD HEALTH ORGANIZATION. Guidelines for drinking-water quality, vol. 3. Drinking-water quality control in small-community supplies. WHO, Geneva, 1985

9. BRISCOE J. The role of water supply in improving health in poor countries (with special reference to Bangla Desh). American Journal cf Clinical Nutrition, 1978, vol. 31, 2100-2113

10. WORLD HEALTH ORGANIZATION. Strategy for WHO's participation in the international drinking-water supply and sanitation decade. WHO doc. EHE/82.29 Rev. 1

Discussion on Papers 1 and 2

MR H. FISH
Firstly I should like to comment on the conclusions of Paper 1, as set out in paragraph 25, and then to comment on the water situation in China.

On conclusion (2) there are probably some people, particularly at the higher levels of water management, who would agree that water resources and natural processes are not completely understood, but then conclude that further understanding will only make their job more expensive and less profitable than it is now, and accordingly that expenditure on the pursuit of further understanding should be avoided. This is one reason why events catch mankind unprepared.

Conclusion (4), that the creation of financial arrangements is fundamental, is based on the premiss that most people will not take the trouble to conserve water, or to eliminate its waste, unless there is an advantage to them in taking that trouble. The advantage that most people recognize in this context is a cash saving.

That many developed and developing countries do not have sound systems of water resources management (conclusion (5)), yet believe that they have adequate systems of water resources management, does not invalidate this conclusion. All such countries will have different views on how sound a sound system of water management should be in their particular circumstances and even where a country knows and admits that its system of water management is unsound it may not be politically possible to correct the situation until a major water resources crisis precipitates the taking of corrective action.

Conclusion (6) needs to be considered in conjunction with conclusion (7) (see paragraphs 16-19) In particular I refer to the opening sentences of paragraph 17, which state 'Throughout the world people are very good at setting up management systems which in theory should work well. We are much less successful in making these systems work properly and in keeping them working properly'. Properly in this context is intended to mean 'efficiently and effectively'. Recently, in the view of many people, the UK appeared to be on the brink of making a grave mistake in proving these words by making their water authorities commercial companies. The intention was that in such companies efficiency would be measured precisely by actual profit levels and effectiveness much less precisely by the ratio of performance against target levels of service. However, this major step was postponed.

Conclusion (8), that economy in water use is neglected, particularly in its reference to the pollution control importance of economy in industrial water use, may seem surprising. However, it is a fact of operational water management in the newly developed countries. It is not easy to achieve such water economy by administrative means; it requires considerable knowledge and skill, but it most certainly pays in terms of reduced overall costs and a cleaner aquatic environment. Conclusion (9) relating to the global scene is one which the Authors view with great concern. It deals with global questions to which global and local answers are required. These call for world-wide collaboration and a world-wide organization should take the lead.

In dealing with the subject of water resources management in China, the first thing to be pointed out is the size of the situation in China compared with that in England and Wales. From Table 1 it can be seen that the land area of China is about 65 times that of England and Wales, the average annual run-off per capita in China is about twice that of England and Wales, while the average annual water demand in China is about 40 times that of England and Wales.

However, Table 2 gives an insight into the differences in water uses between China and

Table 1. Water resources data for China and England and Wales, 1980

	China	England and Wales
Land area: km^2	9.6×10^6	0.15×10^6
Population: $\times 10^6$	1000	48.5
Average annual run-off per capita: m^3	2500	1430
Average annual water demand: km^3	500	13

England and Wales. This shows how water use in China is dominated by agricultural demand (very nearly all for irrigation) while the scene in England and Wales is dominated by industrial demand.

Table 2. Percentages of water resources used for various purposes in China and England and Wales, 1980

	China	England and Wales
Use in agriculture	88	2
Use in industry	11	73
Domestic use	1	25

For the future, the outlook for water demand in the two areas is different. The total water demand in England and Wales in the year 2000 is not likely to be greatly different from the 1980 total demand. An expected increase in domestic demand of 20% is likely to be matched by a decrease in industrial demand of 20%. In contrast China's total water demand in the year 2000 is likely to be about 40% greater than in 1980, representing a doubling of both industrial and domestic demand, while the agricultural demand is expected to increase by one-third.

Accordingly on average in England and Wales the future water resources management problems should not be very difficult, at least until the turn of the century. However, there remain considerable problems in improving the general condition of water distribution and sewerage systems and in cleaning up the remaining polluted rivers.

In China, there are major problems of deficiencies in the availability of water resources to meet demands. This arises essentially because of uneven geographical distribution of resources and considerable seasonal variability in rainfall. These effects are most pronounced in the large conurbations of North China, particularly in the Beijing-Tianjin area.

The problems here arise from a succession of dry years coupled with rapidly increasing demands for water for domestic and industrial use, together with a general lack of treatment of waste water discharges to high quality levels before return to rivers. Not only does China intend to protect as much of its aquatic environment as it can afford, it also intends to conserve its water resources to accommodate new water demands. The traditional approach to meeting new water demands has been, world-wide, to carry out engineering works for the conservation of natural water resources usually without much regard to the conservation of the aquatic environment. It is quite feasible that major new water resources could be made available to meet new water demands in the Beijing-Tianjin area by transferring raw water from the Yellow River and the Yangtze River northwards. Such major transfers, parts of which are already in operation to serve Tianjin, will use existing river channels, canals and lake systems as far as is practicable, with the use of pumps and tunnels as appropriate. These are very large and very expensive developments, presenting problems of leakage from channels and severe siltation problems. For example the water in the Yellow River in the summer flood season contains about 4% silt from the great area of loess clay cover in the Yellow River basin.

However, to China's credit, it is now being recognized that the execution of such major works in the near future will not alone offer the optimum solution of the wider water management problems. It is clear that the demand side of water management needs urgent attention to permit reasonable control of the rising water pollution problem and to reduce the considerable wastage of water resources which occurs in the domestic, industrial and agricultural uses of water.

One fundamental problem is that the price of water is so low that much unnecessary use and wastage of water occurs. Major adjustments of the price structure are required to encourage economy in the use of water. With respect to the distribution of potable supplies much also needs to be done to reduce leakage from mains. With respect to industrial water use a major new campaign to combine economy in water use with an effective reduction in the degree of pollution of water used in industry is required. This should be based on a new pricing structure which encourages 'good housekeeping' in industrial uses of water. Such good housekeeping will embrace not only the avoidance of waste but also the separation of clean cooling waters for recycling. Additionally the adjustment of industrial water-using processes to minimize water contamination will be required.

Other endeavours will be required. Losses in water transfer through open channels need to be reduced. The agricultural usage of water by surface irrigation, which has already been shown here to be by far the major usage of water resources, needs tighter control without causing loss of agricultural output.

Action on these lines will require the optimum mix of engineering endeavour, the introduction of new pricing structures for water abstraction and potable supply use, and measures for the enhancement of effective and efficient pollution control operations. To facilitate the implementation of these measures a revision of the existing organizational structure for water management will be required.

At present in China there are many organizations that are responsible for water management operations within the municipal and provincial governments. The creation of new arrangements for deciding policy on water management and for monitoring the execution of this policy, at municipal and provincial level, merits serious consideration at least. In the water resources context the control of abstraction of water from surface and underground sources needs to be unified to ensure that important groundwater resources are properly managed and that the great advantages of the combined use of surface waters and groundwaters can be realized.

These are formidable, but not insoluble, problems which are not unusual globally. They are not by any means the only major problems facing China in its transformation to a major industrial nation. However, it is manifest that major efforts are being made to bring the water resources and broad water management problems under control at as fast a rate as economic and other factors permit. There is no doubt that China will apply, as far as practicably possible, those main principles of good water management which have been outlined in Paper 1.

DR R. HELMER
Paper 2 largely represents the current philosophy which the World Health Organisation has adopted in dealing with problems of drinking water quality. There is a basic dichotomy between microbial and chemical contamination of drinking water. Although gastro-intestinal morbidity and mortality are still the predominant concern of developing countries, these countries also face chemical exposures such as excessive fluoride in groundwater and pesticides in rivers and canals. Rural communities in many industrialized countries have to live with bacteriologically inadequate supplies whereas the big cities are mostly concerned about organic micropollutants and the by-products of chlorination.

The drastic impact of improvements in water supply in urban and subsequently in rural areas in Costa Rica (Fig. 1) demonstrates the potential benefits to human health which could be derived from such national programmes. However, such dramatic impacts cannot always be expected from interventions on drinking water quality only. According to the threshold saturation theory (Fig. 2) there are countries at the low end of socio-economic development with a correspondingly low health status of the population in which solitary programmes on drinking water quality would not translate immediately into measurable health improvements. Further improvements of drinking water supplies in highly developed countries are not expected to produce a measurable incremental increases in the health of the people, however. In between these two extremes lie the vast number of developing countries at various intermediate levels of socio-economic progress which constitute the prime target of drinking water quality programmes of the World Health Organisation.

It is my conviction that during the second half of the water decade, and even more so during the post-decade period, quality aspects of water supplies will have higher priority in the considerations and actions taken by national decade committees as well as the international and bilateral donor community. Thus, the massive investments made into providing drinking water during the decade will generate the much awaited benefits to the health of the consumers.

MR C. J. BINNIE, W. S. Atkins & Partners, UK
I would like to illustrate the importance of the need for one organization to be responsible for the control of all water resources.

In Oman the rainfall is limited and so the chief water resource is groundwater. The rain falls on the hills and charges the local alluvial aquifer.

In the Solar region of Oman traditional groundwater development was by falaj and by hand-dug wells. Recently there has been a great increase in the population and urban development. Associated with this has been the development of irrigated agriculture using motor-driven pumps and tube-wells abstracting groundwater. Much of this is by private farmers.

The quantities of groundwater abstracted for irrigation are very large compared with water supply. The aquifer is already being overpumped. In the medium term the water needed for public water supply can be made available only by reducing established irrigation development.

However, the control of the drilling of boreholes in rural areas for private sector irrigation and the abstraction of water from them is difficult. The borehole/well can have a farm building constructed over it. The pumps can be run by diesel engines or from the farm electricity supply. The water can be piped away under ground. In this area of Oman it could be linked to a traditional falaj irrigation system. All these aspects will make it very difficult in the future to provide proper water resource control.

This may seem to be an extreme example, but there are many other areas of the world where the conflict for scarce water resources occurs, especially between urban potable use and agricultural use by privately owned farms.

A satisfactory balance can be achieved only if one organization is in control of all water resources. However, it is most important that it is given the technical and financial resources and the legal powers for it to be effective. It is never too soon to apply the philosophy of unified control.

MR N. V. P. SARATHI, Central Water Authority, Mauritius
Measures to conserve water in Mauritius have been taken by the Central Water Authority through billing systems.

Domestic water consumption before 1985 was 42 m^3/consumer per month as a result of billing by a water point system. This has been reduced to 28 m^3/consumer per month as a result of a hybrid system, with minimum charges based on water points and metered systems for consumption in excess of an allowed quantity. Another consequence was that revenues went up from Rs 140 M to Rs 190 M (20 Rs $\approx$ £1 sterling).

There is no need to change the corporate structure towards privatization, but simply to privatize such activities as can be done more efficiently and economically by the private sector, e.g. the billing system, cash collection systems, new supplies, contracting out of developments, works, consulting, training and so on.

PAPERS 1 AND 2: DISCUSSION

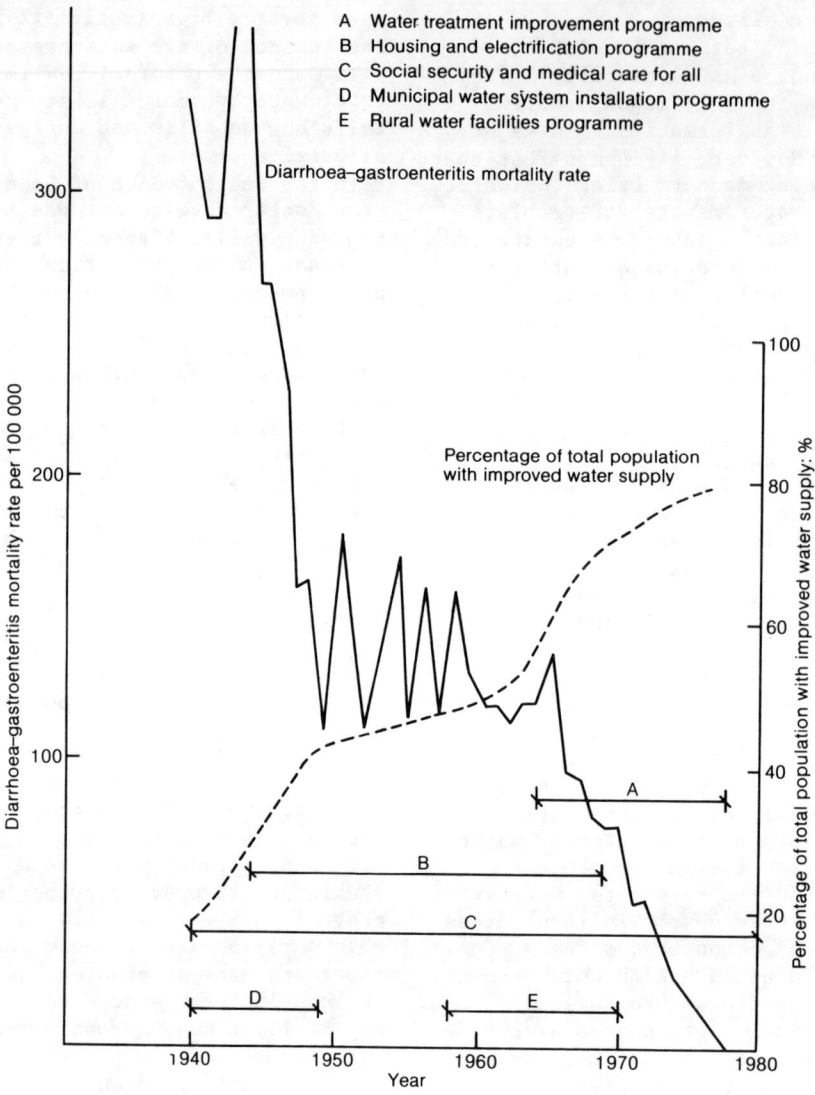

Fig. 1. Diarrhoea-gastroenteritis mortality rates versus time and the percentage of the total population with an improved water supply versus time for Costa Rica (taken from ref. 1)

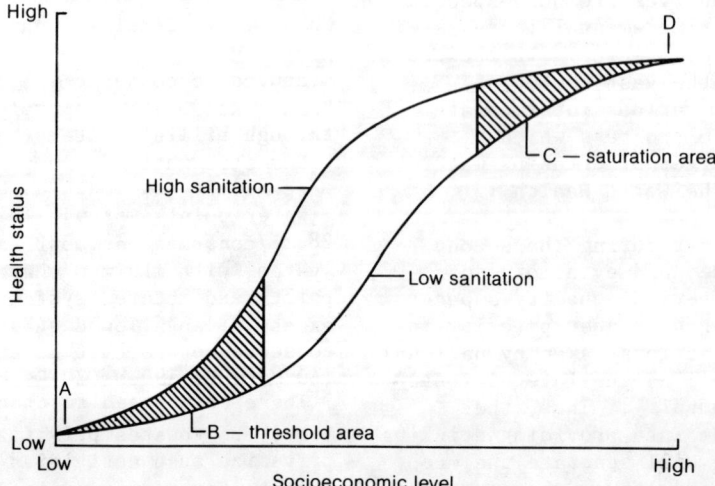

Fig. 2. Threshold saturation theory: effect of sanitation on health status at varying socio-economic levels (taken from ref. 1)

PROFESSOR J. A. PICKFORD, Loughborough
University of Technology, UK
Paper 1 suggests that it is undesirable for
developing countries to adopt high environmental
objectives for pollution control. However, it
may be equally maintained that developing
countries should set high standards before
industrial and municipal effluents cause serious
deterioration in water quality. In this way
they would save the higher cost and greater
effort required to reinstate grossly polluted
water.

Some countries such as India have adopted
legislation authorizing stringent quality
control. However, it is difficult for one state
or district to impose high effluent standards on
industry where there is a possibility of the
industry's moving to another, move permissive,
state or district.

The World Health Organisation has published
various guide-lines, but these are not the same
as standards.

Dr Helmer has related the significant risk to
the emphasis on microbiological criteria in
developing countries, as opposed to chemical
criteria in the industrialized countries. Of
particular concern is the importance attached to
nitrates. It seems that this is given a great
deal of emphasis, for example in the pollution
of groundwater due to pit latrines. The main
danger seems to be blue baby disease, which
affects only infants whose bottle-fed milk is
made up with water containing high nitrates.
This seems to be an insignificant risk where
mothers are encouraged to breast-feed their
babies.

MR H. FISH
I agree that where the environmental
quality is already high the application of high
environmental quality objectives is a sensible
preventative measure. However, where heavy
pollution already exists, few if any developing
countries can afford to set and properly to
enforce immediately high environmental quality
objectives. The setting of high standards which
manifestly cannot be met for many years is
derisory and dangerous in the sense of making
mockery of the control system.

MR W. M. H. RICHARDSON, American Water Works
Association, and Alvord Burdick & Howson, USA
The American Water Works Association has had the
same experience as the World Health Organisation
in establishing water quality guide-lines rather
than finite numerical water quality standards.
This is due to the variation in source of supply
and economics of treatment. In arid areas a
poor source of supply not meeting a numerical
standard would be considered an excellent source
if it is the only water available.

DR A. H. ABOUZAID, National Drinking Water
Office, Morocco
In Morocco the guide-lines are more difficult to
use than the standard. To develop national
standards based on the World Health Organisation
standard presupposes the justification through
appropriate studies of the differences between
the standards and the guide-lines, and also the
evaluation of cost implicated when a level of a
parameter is set. In general this is not
possible in a developing country.

MR H. FISH
In many cases in developing countries, too much
sophistication in the processes of deciding
quality standards is unnecessary. Certainly I
have always worked on the basis that a World
Health Organisation guide-line should be adopted
as a standard where it was sensible and
practicable to do so. Where this is not
possible, guide-lines should be regarded as
targets for the future while interim judgements
are made on the best information reasonably
available.

MR P. G. WILLIAMS, National Institute for Water
Research, South Africa
Improvement in the bacteriological quality of
drinking water supplies has a dramatic effect on
reducing infant mortality once a certain socio-
economic level of the people has been reached.

What is the World Health Organisation doing to
define this socio-economic level? Is it defined
by personal hygiene, a minimum quantity of water
per person or some other factor?

Until this threshold has been passed
expenditure on improving water quality seems not
to be worthwhile.

MR A. VAN DAM, International Reference Centre
for Water Supply and Sanitation, The
Netherlands
The International Reference Centre has studied
the artificial recharge of groundwater (ref. 2);
cases have been found in over 20 less developed
countries and ten in less industrialized
countries. It is very important that experience
is collected, consolidated and disseminated, so
that all involved in water supply and sewerage
can profit from knowledge and experience
elsewhere. Conferences, journals and clearing
houses can all contribute to an improvement in
water supply and sewerage.

MR Y. N. ISKAROS, Alexandria Water General
Authority, Egypt
In my opinion the main aim in water resources
management is to face the leakage in the network
and also the unaccounted for water. This should
have priority during the rest of the world water
decade. If this target is reached there may be
no need to extend the decade.

For example in Egypt many projects for the
expansion and building of new treatment plants
have increased the capacity of production.
Feasibility studies have shown that unaccounted
for water has been estimated as 48% of
production; so if a leak detection programme
were started and the unaccounted for water
reduced, the capacity of production could be
doubled. Such a programme should run in
parallel with a gradual increase in the tariff
for industry and domestic use in slices of

PAPERS 1 AND 2: DISCUSSION

consumption.

The people in developing countries are used to having water free of charge from water stands in rural areas in a programme fighting against poverty, ignorance and illness where supplying water free of charge is a must. Education from an early age can also help to save such valuable water.

MR C. RADHAKRISHNAN, Ministry of Water Lands, Housing and Urban Development, Tanzania
In Tanzania a defluoridation research pilot scheme has been constructed using alum, lime and soda ash. In developed countries like Norway electric ion methods are being used. The World Health Organisation can communicate research activities carried out in various countries in the world using cheaper methods for easy implementation. Research activities being carried out in different countries can be improved without large investment.

DR T. VIRARAGHAVAN, University of Regina, Canada
On the subject of nitrates in drinking water, the US and pan-American primary drinking water standards, World Health Organisation guide-lines and guide-lines on Canadian drinking water quality all stipulate 10 mg/l of NO_3-N as the maximum permissible limit. The epidemiology of this value appears to be weak. Are there reliable epidemiological case studies relating to blue baby deaths due to the ingestion of water with nitrate levels of 10 mg/l NO_3-N or over?

Viruses in water are another problem. There is a need to develop simple virological assays for at least the gross estimation of viruses.

Trace organics in water is also an issue which must not be neglected. It is not sufficient to state simply that this is a problem only in the developed countries.

MR E. W. WARNER, Water and Sewage Authority, Trinidad and Tobago
The shift of the World Health Organisation from standards to guide-lines has put developing countries including my own at a disadvantage because there is an attenuation of the leverage with which to twist the arms of politicians to extract funds for water quality improvement and the maintenance of desirable standards.

MR H. FISH
The points made by Mr Van Dam about the exchange of knowledge on artificial recharge of aquifers and by Mr Radhakrishnan about the dissemination of research results are most relevant as indeed are the points made by Mr Iskaros about the control of leakage from water mains. In the UK much attention is being given to mains leakage.

On the subject of the World Health Organisation nitrate nitrogen guide-line for drinking water, as raised by Dr Viraraghavan, I understood that the 10 mg/l level was set for public health reasons additional to the blue baby risk. The guide-line is generally appropriate, but clearly there are many cases world-wide where conformity with the guide-line is totally impracticable at present. I can appreciate the point made by Mr Warner regarding the political effect of a shift from standards to guide-lines.

However, this gives no real case for insisting on standards where guide-lines are technically more suitable. Furthermore, engineers and scientists responsible for water management should always remember that, whereas rigid standards have administrative advantages in that they can be used by administrators, the proper interpretation and operation of guide-lines requires people with scientific and technological skills. There is a wide variety in standards applied to sewage effluents in different locations. While this might create difficulties for administrators, the resolution of these difficulties was one of the jobs for which the experienced and qualified water manager was paid to do.

DR K. M. OLAŃCZUK-NEYMAN, Gdansk Technical University, Poland
An increased number of iron-bearing bacteria in groundwater used with a lowered output has caused an increase in the amount of iron in the water. Correlation coefficients for concentration depend on the total iron in the water. The concentrations of ammonia nitrogen, chemical oxygen deficiency and of sulphates have been measured as 0.81, 0.72 and 0.62 respectively. The multiple correlation coefficient amounted to 0.95. Relations obtained from investigations have shown that the quantity of iron-bearing bacteria has a significant effect on the amount of iron in the groundwater.

MR J. E. J. GOAD, Binnie & Partners, UK
To emphasize the difficulties facing those responsible for water resource management when setting environmental quality objectives for waters receiving sewage effluent I would refer to three examples from my recent work experience.

First, in Jordan, sewage effluents from Amman's sewage treatment works constitute nearly 100% of the flows in the receiving watercourse in summer.

In Egypt, where the River Nile is the life-blood of the country, discharges of sewage effluents to the Nile are prohibited.

Last, in the UK, in south-east Dorset, sewage effluents from the area's treatment works are currently discharged to rivers, which have high environmental quality objectives which are generally attained.

MR J. A. SAUDER, ACROSS, Kenya
What are the socio-economic factors that make up the threshold below which investments in water supply do not bring about significant improvements in health?

DR A. M. CAIRNCROSS, London School of Hygiene and Tropical Medicine, UK
Before anyone takes the threshold-sanitation theory for the health benefits of water supply too literally, it must be pointed out that this model is essentially a theoretical construct, with no very sound epidemiological evidence to substantiate it.

It should also be remembered that water supply improvements can improve health in several ways, which depend on the local context; they may provide not only better quality water, but also better access to it. They may help to control not only diarrhoeal diseases, but also infections and skin and eye diseases, such as trachoma.

Trachoma is a major cause of blindness affecting many thousands of people at the very bottom of the socio-economic scale, who live miles from their nearest source of water. It has been shown (ref. 3) that such people do indeed suffer less from trachoma when their access to water is improved.

DR A. L. DOWNING, Binnie & Partners, UK
With regard to the fact that in attempting to control water quality some developing countries have set stringent water quality objectives and standards which in the short term are economically unrealistic, perhaps a phased approach such as was adopted for the Thames Estuary might often have been better. Compared with the Thames very high standards were set for Huangpu River, which flows through Shanghai in China. In that case the river is used for abstraction for potable supply downstream of the main waste water outfalls although the intention was to shift the intakes to upstream of the city. Does Mr Fish see this as a case in which a phased approach might have been more appropriate? Could he comment on policy in regard to standards in other parts of China?

The correlation shown by Dr Helmer between increasing expenditure on public water supply in Costa Rica and declining incidence of enteric disorder is impressive but can one be sure that the correlation necessarily implies a causal relationship? Had it been possible to divide the population into two similar groups, of which one received a better water supply and the other did not, then any difference in the incidence of disease could presumably have been attributed to the improved water supply or some factor directly associated with it. If this was not done is there not a possibility that some other non-water-related factor contributed to the apparent correlation?

I am interested in the conclusion in paragraph 14 of Paper 2 that an increase of one case of cancer in 100 000 people might be regarded as a socially acceptable risk. No doubt individuals will differ in their views about this but it is helpful to know the figure that formed the basis of World Health Organisation policy in the fixing of quality guide-lines for potable water supply. However, there remains the problem of having so often to infer the likely incidence of disease in humans in the long term from short-term tests in vitro (such as that of Ames) or on experimental animals. Has Dr Helmer any progress on that to report?

MR H. FISH
It is sensible to set high target environmental objectives, associated with a phased programme of step-by-step improvements over a stated time-scale. Mr Richardson has also pointed out the practical need for flexibility rather than rigidity in setting water quality objectives.

In the case of cleaning up the Huang-Pu River at Shanghai, a phased approach was appropriate and indeed the Shanghai authority has now adopted such an approach. With regard to other parts of China, the initial approach to the formulation of quality standards for river improvement had tended towards standards that are too high. However, this initial trend has now been overtaken by an awareness that it takes time, patience and a great deal of investment to secure real and lasting river clean-up.

DR R. HELMER
With regard to the health improvements due to water supply interventions demonstrated in Costa Rica and further qualified by the threshold saturation theory the amelioration of drinking water quality alone would not lead to immediate measurable health improvements but an array of concurrent measures is needed to achieve lasting effects. Similarly, it is difficult to attribute changes in morbidity/mortality to different areas of intervention (water quantity, quality, personal hygiene, food sanitation etc.). Although the epidemiological and statistical evidence is not confirmed to the point that direct casualty could be proved, there is ample indication that 20-40% of the reduction in gastro-intestinal infection rates could be credited to water supply and sanitation improvements.

The toxicological, epidemiological and chemical bases for safe concentration levels were established by the World Health Organisation (WHO) in the 'Guidelines for drinking water quality'. To find the correct limit for nitrates constitutes an extremely complex undertaking since general nutritional status, the extent of breast-feeding habits and the availability of alternative sources of water have to be taken into consideration. The setting of a national standard for nitrates is therefore a task that requires an exposure assessment for different geographical and geohydrological areas. In addition, the uncertainties of nitrosamine formation due to excessive nitrate levels in food and water have raised concern. In the case of potential carcinogens in drinking water, a point raised by Dr Downing, the setting of guide-line values has to rely on an extrapolation from animal tests, cancer risk models and other procedures which introduce considerable uncertainties into the quantification of acceptable concentration levels.

The changes in the WHO's approach from international standards to guide-lines for drinking water quality arose because the set guide-line values were considered to secure water that is suitable for drinking and all usual domestic purposes, and which would safeguard human health on the basis of lifelong consumption. There is ample margin, however, to set national standards at higher levels without endangering human health if the local geochemical or socio-economic conditions warrant it. Although this approach in many cases requires an additional step of exposure assessment at the national level, the WHO is ready to assist developing countries in this process if requested. Several regional and national workshops have already taken place to co-operate with health and water authorities in determining safe and affordable drinking water quality standards. It is envisaged that many national decade plans will take up the issue of drinking water quality over the next few years.

Several organizations of the donor community are also interested in supporting developing countries in their related efforts.

REFERENCES
1. CVJETANOVIC B. Health effects and impact of water supply and sanitation. World Health Statistics Quarterly, 1986, vol. 39, No. 1, 105-117.
2. INTERNATIONAL REFERENCE CENTRE FOR WATER SUPPLY AND SANITATION. Artificial recharge of groundwater. IRCWSS, The Hague.
3. CAIRNCROSS S. and CLIFF J. Water use and health in Mueda, Mozambique. Transactions of the Royal Society for Tropical Medicine and Hygiene, to be published.

E. H. HOFKES, J. T. VISSCHER, A. VAN DAM, International Reference Centre for Community Water Supply and Sanitation, The Hague, The Netherlands

TN1. Application of artificial groundwater recharge for rural water supply

Groundwater as a source of water supply has great advantages over surface water from streams, rivers, or lakes. Due to the long detention time underground, groundwater is generally bacteriologically safe. Water stored underground is available for abstraction even when drought conditions have depleted the surface sources.

2. Under suitable conditions it is often possible to augment the water yield capacity of groundwater aquifers by artificial recharge, i.e. controlled infiltration of water from surface sources into unsaturated pervious underground formations.

3. Considerable purification of the infiltrated water is obtained in the underground strata, as a result of various biological, bio-chemical and physical processes. Most suspended and colloidal impurities are removed from the water, and there is a virtually total die-off of pathogenic bacteria, viruses and other micro-organisms.

4. To ensure that the recharge water is bacteriologically safe when abstracted, the detention time of the infiltrated water underground must be sufficiently long - at least three weeks, and preferably one to two months (ref. 1). Thus the recovery means must be so located that a sufficient underground travel time of the water is assured, which depends on the transmissivity of the underground formation (Fig. 1).

5. It may seem inefficient to infiltrate water into the underground formation first, and then to abstract it for community water supply. However, the benefits of this operation in terms of improved water quality and assured quality are considerable. First, the process of artificial recharge effects a great improvement in the bacteriological quality of the water. Polluted surface water is converted into a groundwater resource that is safe for drinking and domestic use. Second, water stored underground in artificially recharged aquifers is largely protected from evaporation. Compared with water storage in open surface reservoirs, evaporation loss of water is reduced by up to 90%. Third, water taken from surface sources in the wet season and infiltrated for storage underground is available for abstraction in the dry season when rivers and streams have little water or fall dry. Availability of water throughout the year is particularly assured if deep underground strata are used for artificial recharge. Shallow groundwater resources are not so reliable and may dry up in the dry season.

WATER QUALITY EFFECTS OF ARTIFICIAL RECHARGE

6. Sometimes the water drawn from a river or stream for infiltration in the recharge area can be used directly, but often, particularly in the rainy season when the turbidity of river water is high, pre-treatment by settling or roughing filtration may be required to avoid rapid clogging of the infiltration area. Most of the settleable and suspended solids are then removed before the water is allowed to infiltrate the soil.

7. Various biological, bio-chemical and physical processes take place in the water during the artificial recharge operation (refs 1 and 2).

8. In most recharge means the water is in open contact with the atmosphere so that there is absorption of oxygen and release of carbon dioxide by aeration. Cascades over which the water falls and splashes can be built at the inlet of the recharge means to promote aeration. Some settling of oxidized organic matter and suspended solids takes place, forming a filter matting of deposited material on the bottom of the recharge means.

9. In the infiltration zone where the water enters the soil, impurities are filtered out, as particles larger than the soil pores are retained. Active colonies of predatory micro-organisms and bacteria develop and feed on the organic matter and other nutrients present in the infiltrating water. Suspended solids are removed by sedimentation and absorption on the soil particles. There is active oxidation of both organic and inorganic compounds. Various colloidal and dissolved impurities are retained by adherence on the surfaces of the soil particles. Bacteria and

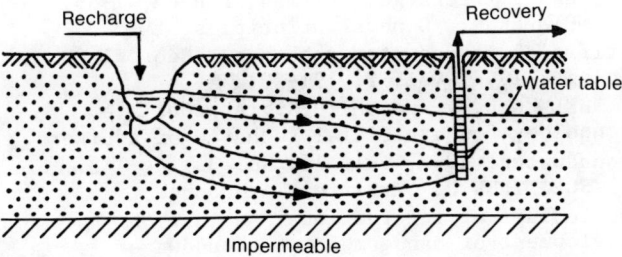

Fig. 1. Main features of artificial recharge scheme

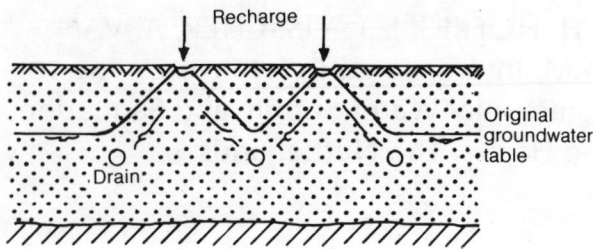

Fig. 2. Artificial recharge scheme using a shallow aquifer

other micro-organisms are also removed by absorption on the soil particles. The purification processes acting on the infiltration water are most active in the top few centimetres of the infiltration zone, and in the filter matting which forms on top of it.

10. After passing the infiltration zone, the water percolates downward in the underground formation to reach the groundwater table. Initially, there is oxygen available in the water and further oxidation of organic matters occurs. At greater depth, the oxygen content of the water may have been depleted and the environment then can become anaerobic. There is continued die-off of bacteria and other micro-organisms, because most of the suitable nutrients have been removed from the water and the conditions in the underground are adverse to microbial life.

11. Not all the chemical and physical processes working on the water during its underground flow effect improvement of the water quality. Some produce changes which make the water less suitable as a source of domestic supply. The water will leach out various constituents from the formation strata it traverses. For example, calcium and manganese carbonates can be dissolved by the water, especially if it is acidic. In underground formations containing peat or the remains of animal life, organic compounds are likely to be taken up by the water. Iron and manganese are leached out if they are present in the formation. In an anaerobic water environment, ammonia, sulphides and nitrite can be formed by reduction processes.

SMALL-SCALE APPLICATION OF ARTIFICIAL GROUNDWATER RECHARGE FOR RURAL WATER SUPPLY

12. The application of artificial groundwater recharge in small-scale schemes for rural water supply is of considerable interest, because these schemes can produce water that is bacteriologically safe and fit for domestic use, without requiring any special provisions for treatment of the water. The current trend to avoid wherever possible the need for treatment facilities in rural water supplies, of course, stems from the many problems experienced with water treatment for the drinking water supply of small rural communities.

13. There are various types of artificial recharge scheme that may be suitable for supplying small communities in rural areas of developing countries.

14. For artificial recharge of shallow aquifers, with the groundwater table relatively

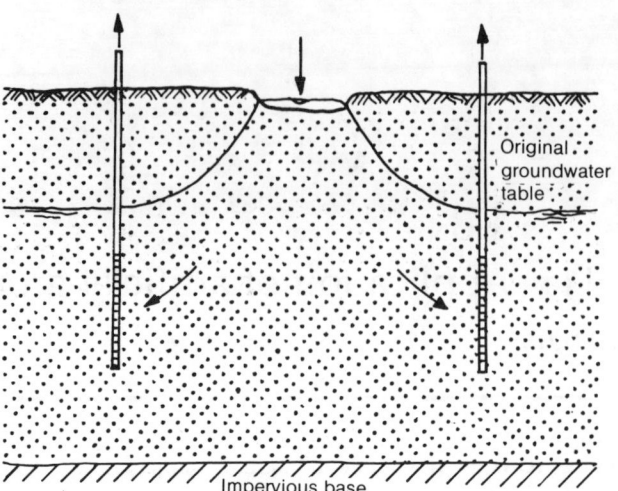

Fig. 3. Artificial recharge scheme using a deep aquifer

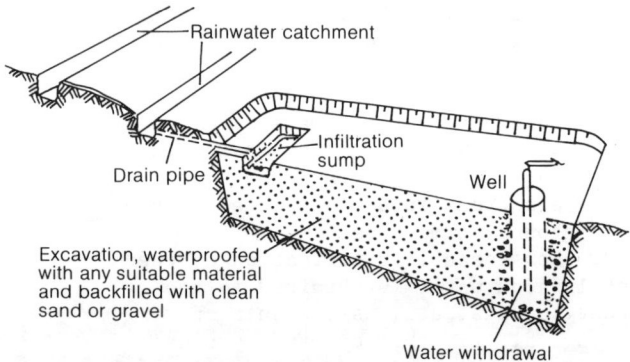

Fig. 4. Artificial recharge scheme using rainwater

close to the ground surface, the most suitable means of water infiltration is ditches (ref. 3). Collector drains can be used for recovery of the water (Fig. 2).

15. For deep aquifers recharge is usually effected by an infiltration pond or basin (ref. 3). A battery of wells surrounding the pond can be used to recover the water (Fig. 3).

16. Rainwater collected on a ground catchment can also be used as a source of water for artificial recharge schemes of small capacity (Fig. 4, ref. 4). Using a sand-filled excavation for recharge operations effects a considerable improvement of the water quality, and evaporation losses are greatly reduced when the water is stored underground rather than in an open-surface tank.

REFERENCES
1. HUISMAN L. and OLSTHOORN T.N. Artificial groundwater recharge. Pitman, London, 1985.
2. MEINZER R. General principles of artificial groundwater recharge. Economic Geology, 1946, vol. 2, 191-201.
3. HELWEG D.J. and SMITH G. Appropriate technology for artificial recharge of aquifers. Groundwater, 1978, vol. 16, No. 3, 144-148.
4. SCHOTTLER U. Artificial recharge. Proceedings of international seminar on development of management of groundwater resources, Roorkee, 5-20 November 1979. School of Hydrology, Roorkee, 1980.

P. O. LEHMSULUOTO, University of Helsinki, Finland

TN2. Contamination of potable water in rural areas: should it be systematically monitored as part of impact management of various interventions?

Microbial contamination of waters in rural areas may pose health risks (ref. 1). Health statistics of the Western Province of Kenya indicate that, although health surveillance, disease registration and reporting systems are incomplete, the major proportion of diseases is related to unsanitary conditions (ref. 2). Official disease incidence and prevalence statistics reflect only a minor part of actual episodes. The true account is much higher (refs 3 and 4).

2. Morbidity, subsequent disability and mortality cause great losses in productivity and national income as well as in expenditure of health services. Unsafe water plays an important role in these incidents. More than 70% of rural people in developing countries do not have access to safe water throughout the year (ref. 5), and an even higher percentage consume unsafe water, even though a safe water supply is provided (ref. 1).

3. There will be no increase in available freshwater resources. The amount available (totally and per capita) is decreasing for various reasons (e.g. population growth, increased withdrawal, deforestation and desertification), while the quality is simultaneously deteriorating (e.g. through contamination and pollution). Nevertheless, contamination has been widely considered unrealistic as an indicator of the true impact of programmes.

4. The major objective of various programmes is to upgrade rural living conditions with the ultimate goal of improved health. At present, most capital investments are allocated to the construction of technical facilities. Provision of sustained, safe water and its protection against contamination through integrated interventions has been grossly neglected. Promulgation of awareness of environmental sanitation and general hygiene together with primary health care connected to water supply and sanitation programmes, or vice versa, cannot be overemphasized in safeguarding the continuous beneficial use of facilities and in reducing susceptibility of water.

5. The degree of contamination (faecal coliforms) of water sources and household water has been evaluated in two locations: Elukongo and Moding. The stage from source to consumption at which possible contamination takes place was not studied. The degree of contamination of household water was used as an indicator of impacts. The zero guideline value for faecal coliform contamination as suggested by the World Health Organization (ref. 6) was considered too stringent for judgement (see also refs 5 and 7).

6. Four main types of water source examined were included in the report (ref. 1): 78 unprotected springs, 33 protected springs, 6 shallow wells and 2 boreholes. The numbers give an indication of the approximate distribution of various types of water supply scheme in the two areas. All together 216 households drawing water from these sources were also surveyed.

7. A majority (54%) of the unprotected springs had more than ten faecal coliforms per 100 ml of test water, which has been applied for as a relaxed guideline value for rural areas in the study. About 28% of the unprotected springs, and roughly 50% of the protected springs were uncontaminated. In Elukongo about 19% and in Moding about 43% of the protected springs exceeded the value of ten faecal coliforms per 100 ml. In Moding the spring protections were old, necessitating rehabilitation. Most of the shallow wells in Elukongo ran below the value of ten, while in Moding the shallow wells generally exceeded this value. In Moding the wells are old, unlined and without proper cover slabs. Both of the boreholes were uncontaminated.

8. About 84% of household water drawn from the unprotected springs, 68% from the protected springs, 89% from the shallow wells and 71% from the boreholes exceeded the value of ten faecal coliforms per 100 ml. Thus about 78% of all household water had more than ten, while about 47% had more than 100 faecal coliforms per 100 ml. If we consider a slightly relaxed guideline of one faecal coliform per 100 ml to be applied for, roughly 93% of household waters exceeded the limit.

9. There was seemingly no marked difference whether the water was drawn from an unprotected or a protected spring, from a shallow well or from a borehole. Water used in households was in most cases further contaminated after it had been drawn from a source, the most pronounced relative contamination being after it had been drawn from the boreholes. Could this be an implication of incomplete planning?

10. Due to short-term planning and rapid changes in programme formulation it was not possible to feed back the information to communities, and thus their response is not known. Neither are there regional studies on

water balance, especially on the effects of increased groundwater withdrawal, deforestation and population growth, nor on the effects on possible reforestation.

11. The results imply that the least advanced technology (e.g. spring protections) could as such, where applicable, be beneficial if only improved water supply is considered. However, the promulgation of awareness and sanitation measures will be of more importance the higher the technology used.

12. The impact of the present programmes and activities in the study area was reviewed. The very recent interventions do not allow speculation of the possible impact on health, but hopefully the survey may establish an adequate information base for possible future impact evaluation. It can, however, be stated that technical interventions did not have observable immediate impact on the contamination of household water. This information should give some momentum for future programme formulation. Complementary activities are necessarily needed for the provision of safe, uncontaminated water and for the improvement of hygienic habits and the proper utilization of provided water.

13. Such data on contamination may additionally be used for the integrated regional development planning of, for example, environmental health programmes, and water supply and sanitation programmes, and their phasing and implementation and the transmission of benefits to the proposed beneficiaries (ref. 1).

14. To cope with the major objective, various programmes should provide well-timed education and promotion to motivate people prior to the introduction of reasonable technical facilities with feasible reliability and cost, contrary to the present practice in which a high proportion of investments has been used for construction followed by trivial education, if any. The infrastructure for supporting the operation, maintenance and repairs of facilities and for the continuous motivation and development of hygienic behavioural habits has been lacking funds. The use of facilities and services in the sense of the programmes should be ensured. But above all the programmes should result in some real benefits for rural people by the proper use of facilities and services. Focus should be on the consumer (ref. 8).

15. Safe water can be assured and corrective action taken only through routine monitoring. There has been little encouragement for such monitoring (ref. 9), although it would be a small fraction of capital investment and maintennance costs and might improve the final cost-benefit ratio. Monitoring should represent the culmination of water supply and sanitation development programmes (ref. 5), i.e. it should critically evaluate progress towards the ultimate goal.

REFERENCES

1. LEHMUSLUOTO P.O. Survey on contamination of water sources and household waters as an integrated part of impact management. Case: Moding and Elukongo sublocations in Western Province of Kenya, East Africa. 1985, Draft, 1-69.
2. MINISTRY OF HEALTH, KENYA. Health Information Bulletin. 1981.
3. NORDBERG E.M. On the true disease pattern of Kibwezi division. AMREF, Discussion paper 1, 1981, 27.
4. NORDBERG E.M. The true disease pattern in East Africa, Part 1. E. Afr. Med. J., 1983, vol. 60, No. 7, 446-452.
5. LLOYD B. et al. Safe water in the third world. Society of Applied Microbiology, Symposium on microbial aspects of water management, 1984, 1-39.
6. WORLD HEALTH ORGANIZATION. Guidelines for drinking-water quality, vol. 1, Recommendations. 1983, 1-40.
7. FEACHEM R.G. et al. Sanitation and disease. Health aspects of excreta and wastewater management. World Bank Studies in Water Supply and Sanitation 3, 1-501.
8. HELLAND-HANSEN E. Introducing a dynamic planning concept for village water supplies. Proceedings of the Norwegian National Committee for Hydrology, Water Master Planning in Developing Countries, Study Case: Water Supply and Sanitation in Tanzania, 1983, 91-109.
9. WORLD BANK. Measurement of the health benefits of investments in water supply. Report of an Expert Panel, 1979, 1-10.

A. HAYES, MICE, Ove Arup & Partners, Birmingham, UK

TN3. Drilling in disaster areas

This Note discusses the effectiveness of water well drilling to provide emergency water supplies for refugee camps in disaster relief operations. It is aimed at the civil engineer who may have considerable experience of water supply, but not specifically of drilling. It is hoped that it will also be useful to those in the disaster management industry.

2. The Author draws on his experiences gained as a member of the Register of Engineers for Disaster Relief (REDR) while on secondments to the Save the Children Fund (SCF) in Somalia and Uganda, and Oxfam in Eastern Sudan.

3. Many recent disasters have occurred in the arid zones of Africa where drought, often exacerbated by military conflict, has led to widespread famine and displacement of people. The victims' first need is water, followed by food, shelter and medical attention. Camps spring up quickly for a number of reasons; the availability of water or ease of communication are rarely first considerations.

THE ROLE OF THE ENGINEER

4. The importance of the role of the civil engineer is little recognized by most of the aid and international organizations, although Oxfam, SCF and UNICEF are notable exceptions. Few engineers are employed by such organizations although their experience in water, sanitation, roads, building and infrastructure is ideal for disaster relief operations.

THE SEARCH FOR WATER

5. The search for water is often a case of making the best of what is available from both surface and underground sources.

6. The advice of a hydro-geologist is invaluable, although much information can be obtained by investigating existing water sources and wells, and by talking to local people. It is surprising what information is available from previous investigations, but gaining access to it is another matter.

7. In most areas there is water within 30-100 m of the ground surface. A well yielding only 1 lisec will provide 86 m^3 per day - enough to keep alive around 15 000 people - whereas a single handpump will support only about 500-2000 people.

8. However, in the chaotic and emotional conditions that prevail, rumour and hearsay become fact in a short time. Experts emerge whose credentials cannot be checked, and panic decisions may be taken which can adversely affect long-term operation.

CASE HISTORIES

9. The most obvious errors can be made. In Somalia, boreholes were drilled in land inundated annually by wadhi flood simply because it was more convenient to drill there. The rig was old and had no spares backup.

10. In the Sudan a drilling programme was started on an area of high land, far away from a known aquifer, and devoid of the vegetation associated with it. The Sudanese programme was based on the advice of a 3.4 expert. The Sudan rig was ill-equipped for water well drilling and stood idle for weeks waiting for equipment when it could have been used to renovate existing boreholes. Poor drilling techniques included the excessive use of bentonite, the use of an unsuitable well screen, failure to plug the end of the casing or to gravel pack and develop the wells.

11. A second rig was standing nearby waiting for clearance to embark on a well-publicized, but intended clandestine, mission into Eritrea. It was an ancient cable tool machine with new support equipment. The rig would have been adequate for the alluvial conditions at the Sudanese camp, but it was too slow and noisy to operate within reach of the Ethiopian air force.

12. Fortunately, a third rig operated by a Dutch development organization came to the rescue and renovated the existing wells. Oxfam air-freighted in a compressor for the first rig, a bailer was made locally, well casings were purchased in Khartoum, and suitable gravel pack was obtained after an intensive search of the countryside.

13. Thus a drilling programme which had achieved little in three months produced five successful wells in two weeks. Oxfam supplied Lister driven Mono pumps which were installed by a Sudanese contractor under REDR supervision. The wells produced over 700 m^3 per day of water and radically improved the lives of the 80 000 refugees on the camp.

14. The Ugandan situation was very different. The relief effort was to provide water to people in 'protected' camps in the war-torn Luwero area. It was a joint effort between the SCF, UNICEF and the Kampala Water Department.

15. There was an extensive system of delapidated boreholes in Luwero which was

Table 1. Water well drilling package

Drilling rig
- truck-mounted, four-wheel drive an advantage
- hydraulic top drive preferable
- winching capacity 5000 kg
- torque 4000 Nm
- rotary speed 0-150 rev/min
- complete with levelling jacks, drillpipe (breakout system and handling tools)
- fitted with mudpump of Duplex piston type, 140 gpm at 300 lb/in^2 including suction and delivery (10 lisecs at 20 bar)
- an augering capability an advantage

Compressor
- a trailer mounted (towed by the rig)
- 600 cfm at 150 lb/in^2 minimum (300 lisecs at 10 bar)
- including all hoses

Supply truck
- minimum 3 t capacity, four-wheel drive an advantage
- fitted with hydraulic lifting arm
- to carry drillpipe, casings, well screens etc

Testing pumping equipment
- trailer-mounted could be towed by field vehicle
- output slightly larger than typical discharge required
- with tripod and blocks to enable it to be used independently

Water bowser
- trailer-mounted towed by supply truck
- with onboard pump

Field vehicle
- Landrover or similar four-wheel drive

Ancillary drilling equipment
- drillpipe 3.5 in o.d. in 3 m or 6 m lengths including connectors, collars etc
- non-return valve for air drilling
- selection of rockroller bits and drag bits for alluvium
- down-the-hole hammer unit and selection of suitable bits
- line oils for air system for DTH and bit grinding tools
- fishing tools
- development tools for jetting, surging, bailing and lightweight pipe
- pipe wrenches and comprehensive tool kit including welding gear
- bentonite, polymers and additives or foam for air drilling
- fuel, oils, lubricants for all equipment
- 10 in conductor casing in 2 m lengths (6m)
- 8 in and 6 in temporary casing in 3 m lengths (42 m and 60 m)
- above with shoes and plug and casing clamps/slips and tools
- supply of well casing and screen appropriate to pumping equipment and ground conditions
- gravel pack material

renovated using an old truck-mounted maintenance rig to remove heavy old handpumps, clean out the boreholes, and then fit them with new handpumps. The programme was successful and valuable experience was gained in the logistics and organization of a programme over a large area. It was used to plan another Ugandan drilling operation so that the maximum use was made of the rig by co-ordinating the supply of materials, drilling and pump installation.

16. The driller is the key man and the success of a programme can depend on him. Ironically the driller in Somalia was proficient, but his intemperate behaviour offended local dignitaries, whereas in the Sudan the driller lacked water experience, but was a fine ambassador for his organization.

DRILLING EQUIPMENT

17. A drilling rig is a specialized item of equipment requiring a skilled crew, support vehicles, equipment and fuel.

18. The probability is that mixed ground conditions will be encountered. Reliability and versatility are essential requirements.

19. Rarely is a reconnaissance made by a drilling expert although the cost of his visit is insignificant when compared with the cost of the rig.

20. The type of pumping plant to be installed should also be considered. Handpumps require only small diameter casings, but motor driven pumps may require much larger casings to accommodate the pump.

WATER WELL DRILLING PACKAGE

21. A package of equipment is suggested in Table 1 for drilling wells of 4-12 in diameter up to 150 m deep in a variety of ground conditions. The suggestions are based on a truck-mounted rotary rig, but trailer-mounted, tractor-mounted and cable tool machines should not be discounted. The list is not exhaustive and specialist advice should be obtained before specifying.

CONCLUSION

22. Water is usually the most important aspect of a disaster relief operation; drilling programmes and the civil engineer have vital roles. Observance of the simplest advice offered here should ensure that drilling in disaster areas is not a drilling disaster.

ACKNOWLEDGEMENTS AND REFERENCES
1. HAYES A. Sudan - what a relief. The Arup Bulletin, 1985.
2. THOMAS et al. Refugee water supplies in Somalia and Sudan. WEDC 12th Conference, Calcutta, 1986.
3. SHAW and HAYES. Tackling water supply problems in Uganda with REDR. Waterlines, Oct. 1984.
4. The REDR field manual.
5. Malcolm Bonnell, Drill Sure Limited, Warwick: advice on drilling equipment.
6. Jim Howard and the Oxfam Water Team.

B. DENNESS, Bureau of Applied Sciences, Whitwell, Isle of Wight, UK

TN4. Water: supply and demand forecasting

The essence of successful enterprise is the accurate forecasting and matching of supply and demand. In the context of water resource management this means essentially forecasting the likely availability of water resources (supply) and population (demand). Deterministic methods of forecasting for both climatic change and population growth for the world as a whole are summarized. These models can be calibrated in terms of regional information.

CLIMATE CHANGE
2. The sea bed holds many of the clues leading to a description of past climates ranging from a few millenia to several millions of years ago. Beyond this, other geological information extends the study back to at least three billion years ago. More recently, isotopic studies of ice cores have led on to archaeological, historical and instrumental records of global temperature and related phenomena over periods of a few years only in the last century.
3. In separating natural from man-made (greenhouse effect) climate (ref. 1), a summary of this illustrated how a broad sweep of these time series data can be matched by one simple equation describing a sine series which summarizes the variation of global temperature over all timescales

$$G(t) = \sum_{n=N(T)}^{\infty} A(T) \cdot a^n \cdot \sin b^{1-n} \pi \frac{t}{T} \qquad (1)$$

which is zero-registered at time T_0 years. In equation (1) $G(t)$ is a time-based climatic index (e.g. global temperature), $A(T)$ is the amplitude of a reference periodicity T, $N(T)$ is the reference integer for periodicity T, a and b are absolute constants, here taken as 0.84 and 0.5 respectively, n is an integer (i.e. the reference number of a particular sine component) and t is time in years.
4. The graphical output from this equation can be shown (ref. 2) to correspond to the variation of global and regional economic progress since 1850 and 1280 respectively. It was proposed that the close correlations resulted from an agricultural link. It is known (ref. 3) that as global temperature, represented by the equation, rises so the major food-producing parts of the world in the mid-western states of the USA, north-west Europe and Kazahkstan in the USSR become drier. In the extreme, dryness reduces productivity. Therefore, as global temperature increases either global food productivity reduces and causes socio-economic strains or finance is diverted from elsewhere in the economy to sustain food production at a cost to other sectors.
5. Greater detail on the link between global temperature and economy has also been given (ref. 4) in the context of water resources forecasting, especially in the Sahel, thereby further substantiating the ability of the deterministic global climatic model represented by the equation to be calibrated in terms of regional water resources variation.

POPULATION GROWTH
6. Malthusianism qualitatively describes a growth system in which a population continuously grows slightly faster than its ability to feed itself. Broadly it is at the root of economic policy in many capitalist countries, as it was in the 18th century England of its origin.
7. To this can be added the notion that it is likely to take two people only half as long to have a bright idea as it takes one. Hence twice as many people take only half as long to generate the spark behind a technological revolution as it took the previously smaller population to conjure up the last one.
8. Combining these shows each successive technological revolution enabling man to manipulate his environment and produce food a little more easily and hence to breed more quickly (to keep expanding a little faster than food production). More specifically the expansion mechanism is measured by the expression

$$E_n = 320 \sum_{r=1}^{n} 0.5^{r-1} \qquad (2)$$

where E_n is the time expired since AD 1390 for the global population to double n times from its 1390 figure of 312 million.
9. Before about 1390 history does not provide a sufficiently accurate demographic record to extend equation (2) backwards with confidence although it is broadly in agreement with historical estimates. Since 1390 equation (2) matches population growth almost exactly, permitting global population forecasts to be

made for the future by adhering to the expansion mechanism unerringly pursued to date. For climatic analysis, the population growth model can be calibrated on a regional or national basis.

FORECASTING SUPPLY AND DEMAND

10. Both the climate (supply) and demand (population) models are time-based. Each can be calibrated regionally. Therefore, each is suitable for use in forecasting mode on a national basis. The balance of likely future supply and demand can thus be explored and provision made for anticipated divergences. Conventional forecasting fails on both counts.

11. There is no successful conventional method of climate forecasting. Current meteorological methodology based on general circulation models is suitable only for forecasting to several days ahead. Apart from the above method there is no known climate forecasting technique able to forecast many years ahead.

12. There is no successful conventional method of population forecasting. Current methodology based on the analysis of present growth in age/sex sectors is a blatant failure as is easily demonstrated by applying the method to historical data. Apart from the above method there is no known population forecasting technique, regional or global, able to forecast many years ahead.

REFERENCES

1. DENNESS B. The greenhouse affair. Marine Pollution Bulletin, 1984, vol. 15, No. 10, 355-362.
2. DENNESS B. The climate-energy-economy link. Energy Exploration and Exploitation, 1984, vol. 1, No. 3, 61-69.
3. WIGLEY T.M.L. et al. Scenario for a warm, high CO_2 world. Nature, 1980, vol. 283, 17-21.
4. DENNESS B. Water resource forecasting. Proc. 4th IAHR Conf. Water Resources Development and Management, II, Chiang Mai, Thailand, 1984, 1177-1194.

K. M. OLAŃCZUK-NEYMAN, PhD,
Technical University of Gdańsk, Poland

TN5. Occurrence of iron and thionic bacteria in groundwater of Quaternary formations

Groundwater intake localized in northern Poland contains mainly water of the Quaternary aquifer. An intensification of water consumption and, in particular, a considerable increase of the groundwater table depression has resulted in changes of the hydrogeological regime of water and significant variations in the contents of iron in the water. The results of the groundwater investigations related to the time during which the intake operated with lowered and variable output are analysed.

MATERIALS AND METHODS
2. Discussion of the results of chemical and bacteriological investigations of water samples taken from the well at the intake lasted for four years, and testings lasted for two years and included water in observation wells. Chemical analyses were aimed at determining pH, alkalinity, hardness, ammonia nitrogen, nitrate nitrogen, total iron, manganese, COD, sulphates, phosphates and oxygen. The iron bacteria of the well water were determined by means of membrane filters and the water in the observation wells was studied by biofilm technique. The preparations were stained in a contrast way by a modified method according to Rodina (ref. 1). The iron bacteria were counted under a microscope by the use of immersion (ref. 2). The thionic bacteria were detected by the application of Starkey's liquid medium with respect to Thiobacillus thioparus, and in the case of Th. thiooxidans by means of the medium of Waksman and Starkey (ref. 3).

GEOCHEMICAL CHARACTERISTICS OF THE ENVIRONMENT
3. The geochemical environment of the water intake region is to a great extent diversified. The superficial zone is made up of peaty formations, below which sand and gravel formations reach down to 40 m. In some profiles there are impermeable formations, like silt and mud. The iron content in the superficial formations amounts to 4-44 mg/g of soil; in the lower formations it is insignificant.

DISCUSSION OF RESULTS
4. A lowered, non-uniform consumption of water at the intake has caused a rise and variations in the water table and a reduction of the aeration zone thickness. The number of the iron bacteria in the water of the intake in use with a lowered output was maintained for the first two years at a constant level within the range of 1.5×10^4 to 3×10^4 cells/cm^3. This period was followed by a one hundredfold increase in the quantity of the iron bacteria in the water.
5. When the investigations were carried out the water was dominated by iron bacteria of genus Siderocapsa; the other genera were Sideromonas, Naumanniella, Leptothrix, Gallionella and Ochrobium. The thionic bacteria appeared mainly in the water of shallow wells (the depth was measured from the top of the screen) and seldom in the water of deeper wells. The contents of iron in the water investigated exceeded the permissible limit for drinking water.
6. Correlation coefficients have been determined for concentration dependences of total iron in water on the concentration of ammonia nitrogen, COD, and the concentration of sulphates; these amounted to 0.81, 0.72 and 0.62 respectively. The occurrence of some evident dependences of total iron concentration in water on the concentration of ammonia nitrogen and COD indicates the participation of micro-organisms in the mobilization processes of iron from organic compounds in this environment.
7. The disengagement of iron from ferro-organic compounds in the presence of iron bacteria takes place under, for them, appropriate conditions, whereas the ferrous ions are oxidized with the participation of extracellular polymeric substances produced by iron bacteria or oxidized chemically (ref. 4). The oxidized ions are liberated in the form of hydrated oxides on the surface, or capsules of the iron bacteria, or in the environment surrounding the bacteria. On account of some insignificant dimensions of the iron bacteria cells occurring in the water investigated (filamentous bacteria below 0.4%) and an intensive flow of water in the vicinity of the well, even after incrustation with iron oxides the bacteria still remain suspended in the water which is being used up. Hence also the contents of the total iron in the water may rise as a result of the growth and metabolic activity of heterotropic iron bacteria. This phenomenon was observed in the investigations. The lack of correlation, pointed out by many researchers, between the number of iron bacteria in water and the iron contained in it has also been shown.
8. Apart from the iron bacteria a very important role in forming the quality of water

World Water '86. Thomas Telford Limited, London, 1987

is played by the thionic bacteria. They were found mainly in shallow wells. The bacteriological investigation results relating to thionic bacteria have been proved by chemical results. It has been found that shallow groundwater is characterized by higher sulphate ions (above 100 mg/dm^3). In deeper water less colonized by thionic bacteria, the content of sulphates was lower. However, the results of investigations of water taken from the observation wells situated in the area of the intake within the depression zone have indicated that the number of iron bacteria in water decreases with the depth.

9. The phenomenon presented of the bacteria effect on the contents of iron in the groundwater requires further investigation.

REFERENCES

1. RODINA A. Mikrobiologiczne metody badania wód, p. 353. PWRL, Warszawa, 1968.
2. BEGER H. Leitfaden der Trink- und Brauchwasserbiologie, p. 94. Gustav Fischer Verlag, Stuttgart, 1966.
3. COLLINS V.G. Isolation, cultivation and maintenance of autotrophs. In Methods in microbiology, vol. 3B, p. 33, edited by J.R. Norris and O.W. Ribbons. Academic Press, London and New York, 1969.
4. CZEKALLA C. et al. Quantitative removal of iron and manganese by micro-organisms in rapid sand filters. Water Supply, 1985, vol. 3, Berlin B, 111-123.

E. F. RAMADAN, Drainage Research
Institute, Cairo, Egypt

TN6. Water and salt balance in the Fayoum, Egypt

The Fayoum Depression is situated in the western desert of Egypt, aboout 80 km south of Cairo. It is a huge geological hollow in the desert, covering an area of about 1800 km^2: its lower elevation is 44 m below sea level at Lake Qarun, which occupies the north-west section of the depression and covers an area of about 240 km^2.

2. The depression is embedded in the Eocene marls and limestones. The alluvial deposits overlying the bed rock consist of clays, sands and gravelly sands. The depression has an elevation of +27 m above mean sea level at the main water intakes, sloping down in the north-west direction to -44 m at the south shore of Lake Qarun.

3. The cultivated land area is 315 000 feddans (132 000 ha) and the harvested area is 556 000 feddans (234 000 ha), indicating a 177% cropping intensity.

4. The Fayoum is irrigated with Nile water entering the depression via two main canals. From these main canals the water is conveyed to the fields by means of secondary and tertiary canals (Fig. 1).

5. Drainage is by gravity. Most of the drainage water flows to the deepest point of the depression - Lake Qarun. Because it is a closed lake, evaporation is the main process for dispersing the salty drainage discharges into the lake basin.

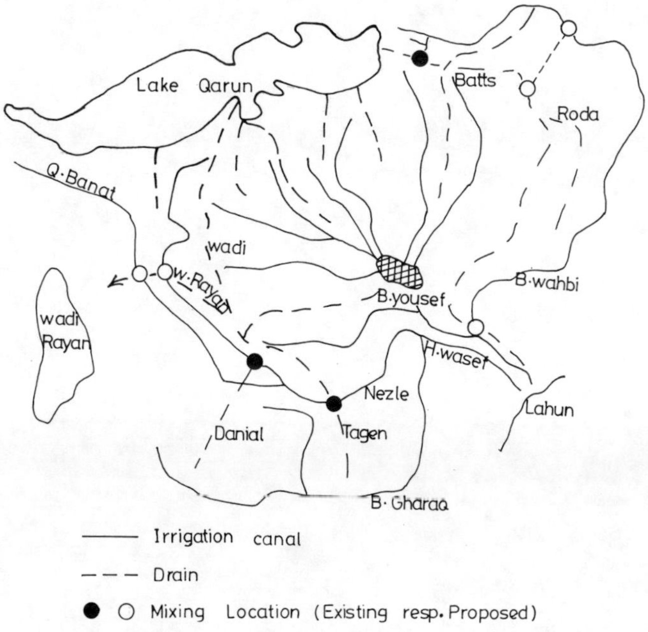

6. The amount of water has increased: Lake Qarun's water level rises year after year, destroying about 10 000 feddans (4200 ha) of the surrounding cultivated lands.

FAYOUM WATER AND SALT BALANCE MODEL PROJECT

7. The project started in April 1983. Its immediate objective is to study the water flow in the Fayoum Basin, the reuse of drainage water, the establishment of a monitoring network and the development of a water and salt model suitable for simulating various alternatives in the field of water management in the Fayoum Basin.

FSWB project investigations

8. The project is investigating

(a) total monthly inflow to and outflow from the agricultural lands in the Fayoum
(b) the total salt loads entering and leaving the area
(c) the area cropped per month and its potential evapotranspiration
(d) uniformity of water distribution in the main irrigation network
(e) the efficiency of water use in the Fayoum
(f) the re-use of drainage water in the Fayoum.

Water balance of Fayoum

9. An indicative water balance for the Fayoum was derived from the data collected. This balance is applied for the gross command area of the Fayoum of 148 000 ha. The maximum cropped area is 132 000 ha or 89% of the gross area. The equation of the water balance is

$$Ir = DR + ET_{crop} + \text{rest term}$$

The rest term is composed of evaporation losses, deep percolation and the change in volume of the soil moisture in storage.

FACTORS AFFECTING OPTIMAL WATER USE

Water distribution related to time

10. The irrigation intake per month or per decade is not properly matched with the actual water requirements of the crops. In summer deficits occur, and in winter there is a surplus of irrigation water.

Fig. 1. Fayoum Governorate: main irrigation and drainage network

Water distribution along main canals

11. The water distribution along the main canals is not proportional to the area served. Water duties are not uniform along the main canals. The result is that some areas close to the main intake have more water duty than the average, and the area at the tail has less water duty than the average.

12. The project proposed an integrated water management scheme, whose main components are rehabilitation of the existing water distribution structures, reconstruction of major irrigation works, strengthening of the operation of the main irrigation system and controlling of the water up to the tertiary level, re-use of drainage water, and drainage of water-logged soils.

Excess drainage flow

13. Annually about 30% of good quality drainage water (average salinity 1000 ppm) is evacuated to Lake Wadi Rayan and Lake Qarun where it evaporates without any productive function. It would be better to re-use this water in the downstream areas. The immediate effect would be to reduce the inflows to the main canals.

14. The project recommends that re-use of drainage water should be an interim solution to secure sufficient irrigation water (mixed drainage and irrigation) for the tail end area; it should not be diverted to new areas as the amount of drainage water for re-use is at a minimum in summer when the demand is at a maximum. In winter the situation is reversed: low demand and high supply.

Field application methods

15. Basin irrigation is used on most of the agricultural lands in the Fayoum. Furrow irrigation is also practised. The inaccurate levelling of the basins and the non-uniform gradients of furrows result in a variable application depth; low field application efficiency is the result.

SALT BALANCE OF THE FAYOUM

16. Salinity of irrigation and drainage water has been measured twice a month since 1980. The annual average salinity of the irrigation water is 290 ppm and the drainage water has an average salinity of 1000 ppm.

17. The average annual salt load of the irrigation water is 664 000 tons, and the average annual salt load of the drainage water is 680 000 tons, i.e. the salt balance is in equilibrium.

BIBLIOGRAPHY

MANHAROUS W.F. The hydrology of Lake Qarun. MSc thesis, Cairo, 1979.

Nippon Koei Ltd. Reconnaissance report on the Fayoum irrigation project. Nippon Koei, Tokyo, 1981.

RAMADAN E.F. The state of water management in the Fayoum. FWSB Project, Giza, 1985.

D. B. FIELD, BSc, MInstP, MIWES, G. F. MOSS, BSc, MICE, MIPHE, and S. H. WHIPP, BSc, MICE, WRc Engineering, UK

3 Distribution and collection systems

In recent years, emphasis in the UK has turned away from new construction towards operations and maintenance. The UK Water Industry has developed cost effective strategies for dealing with a range of problems including network simulation, leakage control, water mains rehabilitation, and sewerage rehabilitation, and is investigating others. Associated with each of these strategies is a variety of enabling tools which use the latest technology to aid implementation of the strategy. These strategies mesh together leading towards the overall cost effective operation and maintenance of water supply, distribution and sewer networks.

INTRODUCTION

1. Until a few years ago, most of the major works associated with water supply and sewerage were centred around new construction. These included development of new sources of supply, extensions and reinforcements within distribution systems to meet increasing demand and new consumers and the construction and reinforcement of sewer systems. Maintenance of existing systems tended to take a secondary rôle with problems being dealt with as they arose, usually on a local basis.

2. This situation has now changed in many countries with operations and maintenance becoming more important than new construction. This changing requirement was recognised in the UK during the 1970s. In order to ensure efficient operations and maintenance procedures using the latest technology, the UK Water Industry decided to establish a new unit within the Water Research Centre, its R & D arm. This unit, known as WRc Engineering, has a remit to investigate operations and maintenance problems associated with sewers and water mains. Working in close collaboration with the UK Water Industry, both directly and via the national Sewers and Water Mains Committee, procedures for network simulation, leakage control, water mains rehabilitation and sewerage rehabilitation have been successfully completed.

3. Areas currently being investigated include remote monitoring and control, management information systems, selection and specification of pipeline materials and construction. Some of the subjects listed above are major sectors of expenditure and can be broken down into numerous component parts. For example, sewer rehabilitation includes records upgrading, structural inspections and assessment, hydraulic analysis, model verification, economic analysis, renovation design and specification, reinforcement, and so on. Similarly, all of those listed above form part of a larger problem and that is the long term cost effective operation and maintenance of water distribution and sewer systems.

4. The important message is that in the UK we are aiming for, and achieving integrated solutions. Whilst the larger problems are worth tackling on their own, the solution of one aspect becomes an input to the next. Gradually, the solutions fit together like the pieces of a jigsaw. Some examples of the work undertaken to date are given below.

NETWORK SIMULATION

5. Network analysis - producing a model to mimic the behaviour of a water distribution network for designing network extensions - is not in itself new. WRc pioneered the techniques 15 years ago when electrical analogue models were used because the digital computers then available had insufficient capacity.

6. Network simulation today goes a lot further and is an essential tool for the efficient operation and management of water distribution systems. As well as simply modelling the behaviour of a water distribution system for one instance in time, network simulation can portray the behaviour of the complete system over 24 hours, a week, or other specified time. This opens the door to many applications such as planning network extensions, siting and sizing of storage reservoirs, pressure control and in optimisation of the operation of networks. In addition, network simulation provides an input to leakage control and mains rehabilitation strategies.

LEAKAGE CONTROL

7. There are five basic methods that can be employed for the control of leakage. Each method has the capability of reducing leakage to a correspondingly lower level, although each has associated with it a corresponding increased cost. A procedure (See Figs 1 and 2) for determining the most appropriate method in any given network has been developed and successfully employed in many systems in the UK and elsewhere. The procedure also enables the effort necessary for the economic operation of the chosen method to be determined.

World Water '86. Thomas Telford Limited, London, 1986

There are a number of methods of leakage control which can be employed:

o Passive control
o Regular sounding
o District metering
o Waste metering
o Combined metering.

Only one method is economic for a particular system.

Each of these methods will show further savings by pressure control.

The consequences of choosing the wrong method of control can be as expensive as not practising leakage control measures at all.

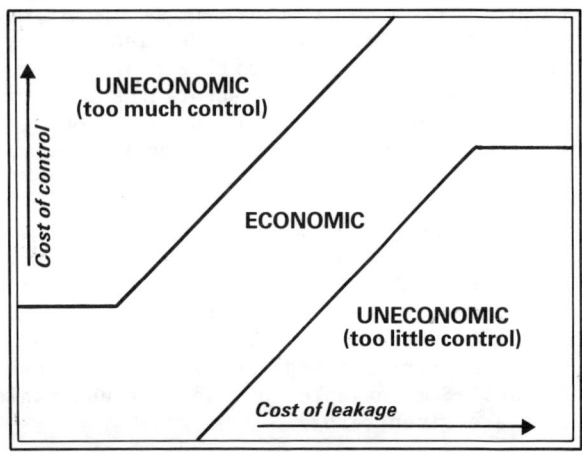

The economics of leakage control.

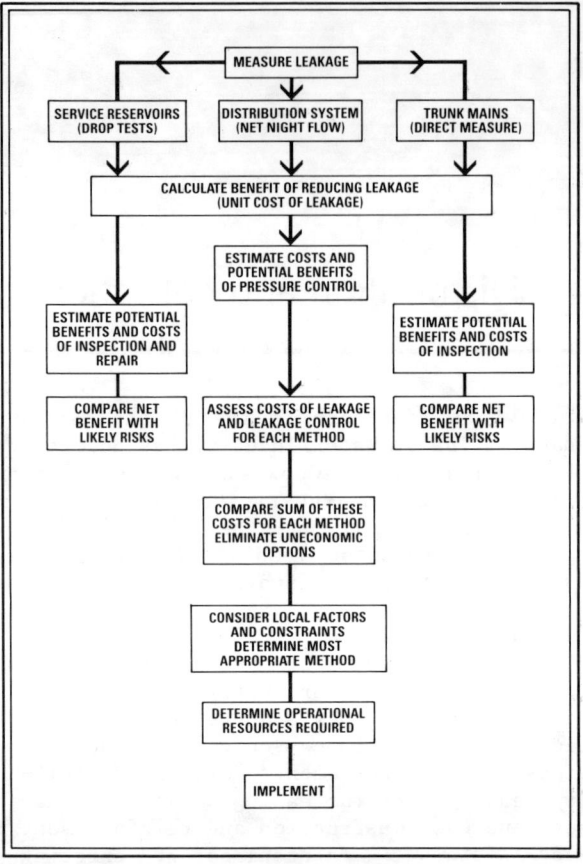

The procedure developed by the WRc.

Figs 1 and 2 - Determination of most appropriate leakage control method

8. A fundamental requirement for the employment of most methods of active control is network simulation. This enables districts to be selected, the siting and sizing of meters to be determined and the scope for leakage reduction by pressure control to be assessed.

9. Recent developments in the enabling tools associated with leakage control include:
o Production of a third generation leak noise correlator to accurately locate the position of leaks. The new device is smaller, lighter and more sensitive than its predecessors and use of solid state technology has made significant cost reduction possible.
o Flow measurement recorders - data loggers - capable of recording the output from a whole host of flow meters used for both district and waste metering. Waterproof and "truckproof", the equipment can withstand any conditions likely to be encountered in the field. The equipment records the data necessary for detecting changes in leakage levels and can also be used for step testing. In addition to recording flow, the loggers are also designed to record pressure, reservoir level, or any two of them simultaneously. It is, therefore, the same equipment that is used for the calibration of network models which is another part of the integrated solution.

10. These developments provide practical assistance for implementing effective leakage control policies and incidentally lead to a better general understanding of the way systems operate.

WATER MAINS REHABILITATION

11. In broad terms, the word "rehabilitation" is used to mean maximising the use of existing assets to cater for all estimated future demands. Deterioration of water mains causes very many problems, including leaks, bursts, poor flow and pressure, discoloured water, blockage of meters and customer complaints. A manual, "The Water Mains Rehabilitation Manual", which details the procedures (See Fig 3) and methods that are necessary for surveying the condition and performance of water mains, together with the policy and execution of cost effective rehabilitation methods, has been written and will be published in January 1986. It is beyond the scope of this paper to go into the detailed procedures contained in the Manual but these allow situations to be identified where rehabilitation can be justified on the grounds of improving service or on the basis of saving operating costs.

12. Specific pieces of equipment developed to enable the implementation of the Manual include

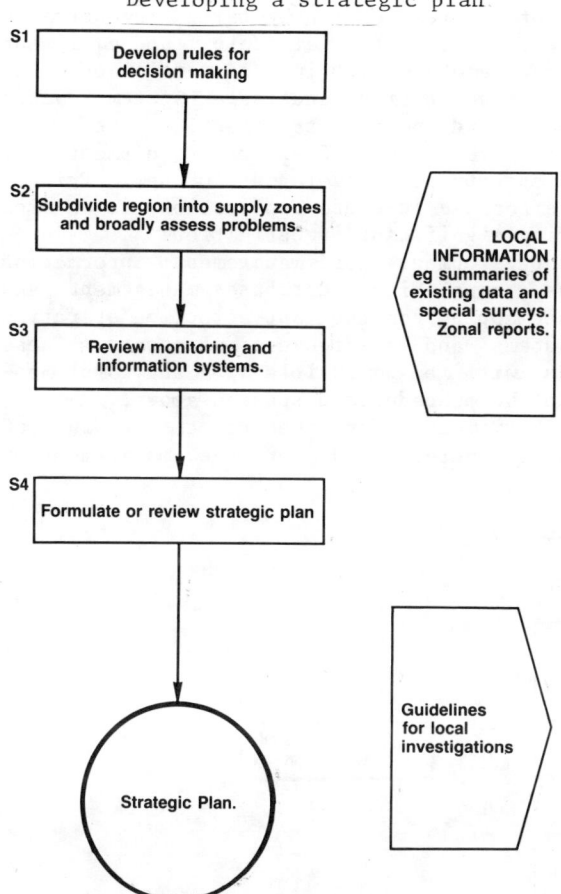

Fig 3 - Strategic planning for water mains rehabilitation

magnetic and ultrasonic devices for corrosion measurement and automated sampling equipment to quantify water discolouration problems. WRc has also been active in developing and promoting the use of fibre optic instruments for the internal inspection of water mains.

13. In addition, operational guidelines have been developed for the application of rehabilitation methods including chemical treatment, cleaning, sliplining and insitu relining systems. A method has also been developed for cracking and bursting an existing pipeline and replacing it with a pipeline of the same or larger diameter. This is one of several "no-dig" techniques currently being researched for pipeline replacement and is equally applicable to both water mains and sewers.

SEWERAGE REHABILITATION

14. Sewer system performance can be considered under three headings: structural (collapses), hydraulic (flooding) and environmental (impact on river quality). As with water main rehabilitation a cost effective approach relies on considering all three performance aspects simultaneously.

15. Sample surveys have built up a picture of current system performance and the backlog of required rehabilitation works. These have demonstrated the need for a radical change in approach if the limited funds likely to be available in the foreseeable future are to avoid a deterioration in levels of service.

16. The key factors at the centre of the recommended strategy include:
o Most of the expensive structural and hydraulic system failures are in the core of the network. It is in these areas that survey, analysis and pre-emptive rehabilitation should be concentrated. These are termed the critical sewers.
o Recent developments in hydraulic analysis of sewer networks and flow verification have greatly improved the possibilities of optimising hydraulic performance, minimising the need for replacing sewer lengths because of hydraulic overload.
o Further, established renovation methods can now be structurally designed and installed with confidence when used within their known limitations.

17. For the UK the most cost effective approach will be to retain as much as practicable of the existing system by optimising the hydraulic and environmental performance and maximising the opportunities for renovation. To do so it is necessary to consider the performance over complete drainage areas, and develop solutions which deal with both the structural and hydraulic problems concurrently. It is also preferable to maximise the opportunities for staged solutions so that early returns can be made on investments and upgrading of performance can be progressed on a broad front.

18. This approach (See Fig 4), together with detailed guidance on available methodology and equipment, has been set out in the Sewerage Rehabilitation Manual which was published in April 1984. The Manual procedures have been implemented by the UK Water Industry and the experience gained, together with the results of ongoing research during the last two years, has allowed useful revisions to be made in most sections of the procedures. These will be incorporated into a second edition to be published in the spring of 1986.

19. Work is continuing to make the investigations more cost effective. It has been clearly demonstrated that significant sums of money can be saved in rehabilitation works but the information gathering and interpretation efforts are considerable and can in many cases appear daunting to practising engineers. Attention is therefore now being shifted to using the latest information technology techniques to optimise the data collection and analysis methodologies.

20. There is also a major programme under way on providing a comprehensive manual on the design, construction and maintenance of long sea outfalls - again covering structural, hydraulic and environmental aspects - which will complete the jigsaw for urban drainage.

21. Enabling tools available to support the implementation of sewer rehabilitation strategy include:
o Portable equipment for measuring and recording flows in sewers;

o Software that enables calibrated models of sewer systems to be provided; and
o Devices for controlling or attenuating flow within systems.

RENOVATION METHODS

22. WRc works with contractors to investigate the problems associated with renovating sewers and to monitor/develop methods to overcome them. Such developments include spray-applied coatings, non-circular sliplining systems and sliplining in short lengths, pipe bursting (moling), water-jet cutters for removing protruding laterals and other obstructions from non-man entry sewers, techniques for grouting the annular gap between old sewers and new, and methods and equipment for remotely making lateral connections to drains.

CURRENT FUTURE WORK

23. Other areas currently being investigated in the UK include the methodology and equipment for the remote monitoring and control of water supply, distribution and sewer systems, water treatment and sewage treatment. The work includes evaluation of existing equipment, and where appropriate development of new. Primary sensors for use in these applications is an area requiring significant investigation.

24. Development of management information systems, including database management and digital mapping for the cost effective operation of systems and maintenance of assets. These systems will be compatible with and complementary to the procedures discussed above.

25. Equipment for the remote reading of customers' meters and for the management of

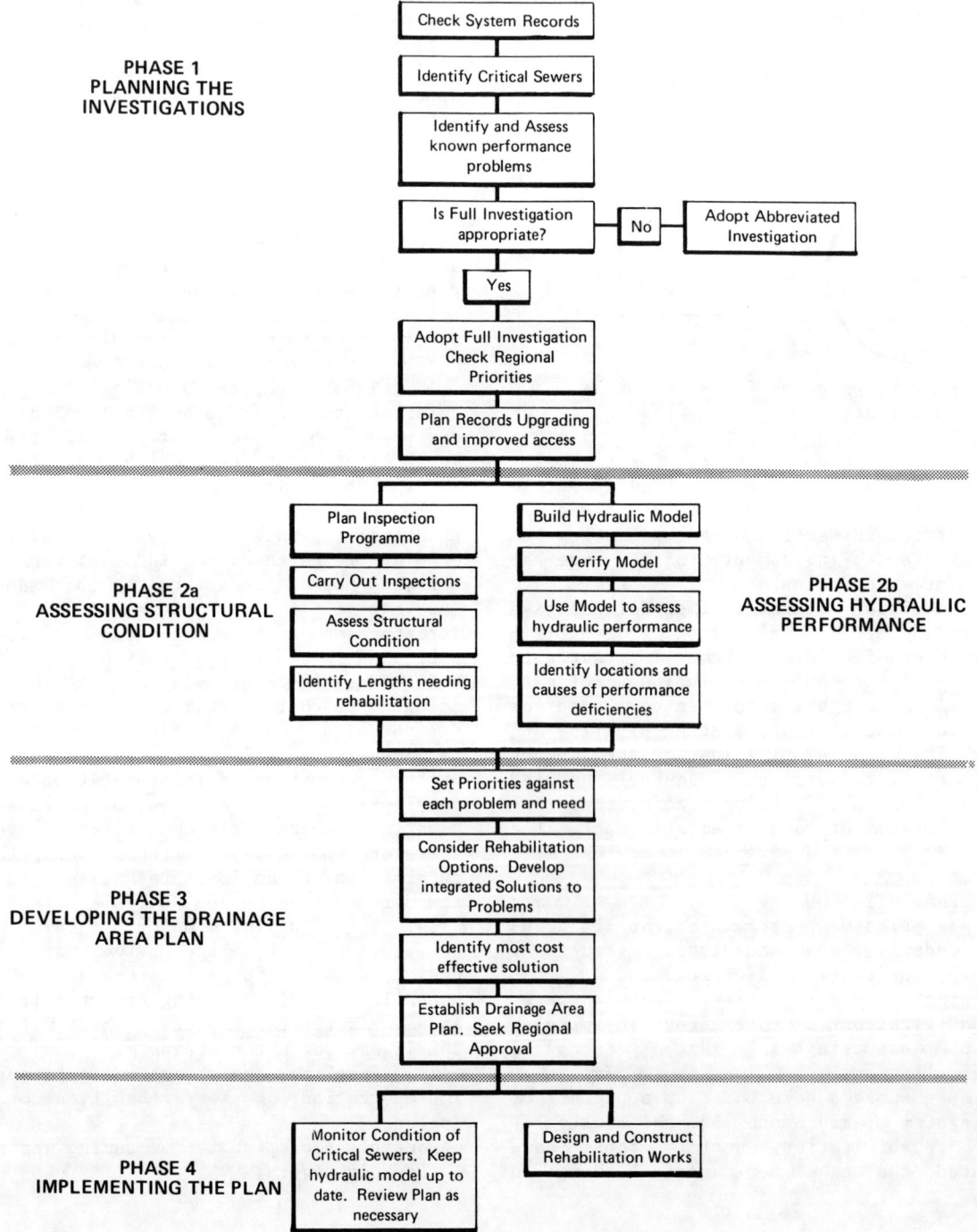

Fig 4 - Investigation flow chart for sewerage rehabilitation

demand, is now undergoing field trials. Systems using both telephone lines and mains cables are being evaluated.

26. Manuals for the selection, specification, purchase and construction of service pipes, distribution mains and trunk mains, giving guidance on what to use, and more importantly what not to use, in different circumstances, are also being produced.

APPLICATION OF RESULTS OUTSIDE OF THE UK

27. Whilst many of the results of research in the UK are likely to benefit other countries, technology transfer should not be attempted directly. What can be applied, and should be applied, is the approach for the determination of an optimum solution. By following the approach - the strategies developed by WRc and the Industry - relevant local factors which are different from the UK can automatically be taken into account. These local factors include management basis, temperature, climate, ground conditions, availability and cost of water, pipe materials, availability of manpower, etc. All of these can, and are, taken into account when developing a solution to a problem.

28. Feedback from involvement in overseas projects is of great benefit to WRc and helps to direct effort on the Research and Development programme. This is because overseas projects often introduce a new perspective which requires some lateral thinking to develop existing approaches and which can be used to strengthen the basic assumptions on which current recommendations are based. Work is currently under way in Bahrain, Oman, Brunei, Malaysia, Zaire, Indonesia, Turkey, Thailand and Abu Dhabi. By working in this manner, the results of the research of the UK Water Industry are made available worldwide.

THE WATER RESEARCH CENTRE

29. The Water Research Centre is a private company limited by guarantee. Its current turnover is around £17 million and its main customers are the UK Water Utilities. It also carries out a substantial programme in the environmental field under contract to the Department of the Environment. WRc employs around 550 staff working in laboratories at three locations - WRc Engineering (Swindon), WRc Environment (Medmenham), WRc Processes (Stevenage). It has no shareholders and does not distribute any profits. Indeed, the profits that arise through WRc commercial contracts are ploughed back into the R & D programme for the benefit of the UK, and subsequently worldwide water industries.

A. M. WRIGHT, PhD, Project Officer,
Water Supply and Urban Development
Department, World Bank, Washington, USA

4 Non-sewered sanitation in developing countries

During the first half of the Decade significant progress has been achieved in the promotion of policy shifts from a focus on conventional sewerage alone to a broader approach that also involves consideration of the use of non-sewered sanitation technologies where appropriate. However, several key issues remain to be resolved to pave the way for a wider use of non-sewered sanitation systems in large-scale replicable projects. These issues involve affordability, cost recovery, and institution capacity for their delivery.

INTRODUCTION

1. The second half of the 20th century has been characterized by gigantic technological strides. For the first time man has walked on the moon; the electronic chip industry has ushered mankind into an information revolution whose impact on our lives cannot yet be fully grasped; developments in genetics and medicine have brought test-tube babies, and there are real possibilities that man will influence the future course of his evolution. These developments have brought in their wake many amenities and comforts of life that have become commonplace in the advanced countries of the world; but in the developing countries many basic needs remain unmet. Among these basic needs is adequate sanitation. Poor sanitation and inadequate water supplies have created for the majority of people in the developing countries a vicious circle of disease, low productivity, and poverty. As a result half the work of a sick peasantry in such areas is believed to go to produce food to feed the intestinal worms that make them sick.

2. It was largely to address this problem that the International Drinking Water Supply and Sanitation Decade (the Decade, for short) was launched. Today, half-way through the Decade, it is gratifying to see the Institution of Civil Engineers convening this Conference to take stock of our achievements and to plan for the future.

3. During the first half of the Decade several sanitation systems were promoted by a number of international and bilateral organizations, including the Technology Advisory Group (TAG) at the World Bank. Among these technologies were non-sewered sanitation systems. In the course of this promotion, there have been successes and failures; lessons for the future have been learned, but some unresolved key issues still remain. The challenge of the second half of the Decade is to apply the ingenuity we used to get to the moon to pave the way for replicable, large-scale sanitation projects in the developing countries.

4. I would like today to briefly review the salient features of non-sewered sanitation systems in developing countries, and to highlight for your consideration some of the key issues that need to be addressed to provide a sound foundation for progress in the second half of the Decade.

DEFINITION

5. Sanitation systems can be divided into two broad categories, namely, sewered systems and non-sewered systems. Sewered systems consist of two types of facilities: private and public. The private facilities are those fixtures and structures that belong to the property owner; they are, as a rule, located within private premises and properties. Public facilities consist of sewers, conduits, and other appurtenances that are used to collect and transport wastes from individual properties to central points for treatment and disposal. Examples of sewered systems are conventional sewerage, small-bore sewer systems, shallow-trench sewers, and "simplified" sewerage systems.

6. In contrast, non-sewered sanitation systems consist only of private facilities except where they are designed as public conveniencies. Typically, they are small-scale, self-contained units for excreta and sullage disposal that are designed and built so that they do not interact with one another and are located entirely within the property or group of properties they serve. Since units are not interconnected, as is the case with sewerage systems, breakdowns and operational problems are restricted to units in which they originate.

7. Some, like the pour-flush latrine, Ventilated Improved Pit (VIP) latrines, aqua privies, and septic tanks, are designed for long-term, on-site treatment and final disposal. Others, like vault and cartage latrines and pan latrines, are intended for temporary on-site storage, followed by off-site disposal.

8. <u>Salient Features</u>. Non-sewered sanitation systems are characterized by technological simplicity and smallness of scale; their construction on a large scale involves simple repetitive building operations and a high degree

World Water '86. Thomas Telford Limited, London, 1986

of interactions with prospective users. As a group, they tend to cost less than sewerage systems, and they lend themselves to:
  (a) standardization of key operations and components;
  (b) standardization of training for key operations and processes;
  (c) use of low-level technicians;
  (d) decentralized management, delivery, and operation;
  (e) community participation and co-production with beneficiaries;
  (f) self-help approaches; and
  (g) on-the-job training for their construction.

9. <u>Areas of Comparative Advantage</u>. Like all technological systems, non-sewered sanitation systems have limits of applicability; and their choice should be dictated by local conditions. Based, however, on their salient features, they should have a comparative advantage over sewerage systems when the following conditions prevail:
  (a) willingness to pay and income levels are low;
  (b) property values are low;
  (c) water consumption is low;
  (d) space is available for on-site location of sanitation facilities;
  (e) soil properties are appropriate for the choice of technology; and
  (f) level of technical skill is low.

DECADE EXPERIENCE

10. <u>Options Promoted</u>. During the first half of the Decade, a number of non-conventional sewerage systems and non-sewered sanitation systems were promoted. Of the non-sewered sanitation systems, the principal types promoted were on-site systems, which included the pour-flush latrine (promoted mostly where water is used for ablution, notably in Asia), the VIP latrine (promoted where it is not customary to use water for ablution, mostly in Africa south of the Sahara), and the low-volume flush water closet system (designed for discharge into pits and septic tank systems). The low-volume flush toilet has so far been promoted mainly in Latin America.

11. These non-sewered sanitation systems were introduced initially in demonstration projects aimed at promoting policy shifts away from conventional sewerage systems alone. Increasingly, they are being used as part of a range of technologies for full-scale projects funded through direct national budgets or through external credits, loans, and grants.

12. <u>Strategies</u>. In rural areas non-sewered sanitation systems have been promoted with success mainly in compact settlements (as opposed to dispersed settlements) where communities are well organized to facilitate their effective participation through the contribution of labor and local materials. Success has been due, in part, to subsidies in the form of free supplies of slabs, cement, and fly screens; the arrangements have been such that no cash outlays have been required from the beneficiaries. Examples of project areas are Wangin'ombe in Tanzania (where more than 10,000 VIP latrines have been built through communal self-help in four years), in Zimbabwe (where more than 100,000 VIP latrines have been built in the past seven years using the same approach), and in Sri Lanka (where about 7,000 pour-flush latrines have been built through self-help programs in just under 18 months).

13. In urban areas non-sewered sanitation systems have been promoted as part of low-income housing projects, as part of city-wide sanitation projects in which sewerage has been used in some locations, and also as free-standing projects. In all cases, small-scale contractors have been used for the installation of the non-sewered sanitation systems. In India, for example, about 70,000 demonstration pour-flush latrines have been built in the past six years. Other projects have been undertaken in Indonesia, Botswana, Lesotho, and in various countries in Latin America, most notably in Brazil.

14. <u>Current Status</u>. Today, at the midpoint of the Decade, a lot is known about the pour-flush and the VIP latrines. It is true that some unanswered questions still remain about the risk of groundwater pollution, about the risk of filariasis where standing waters occur in VIP latrines, and about appropriate types of equipment for pit emptying. But, by and large, we have a working knowledge of basic principles and technologies for these sanitation options.

15. A lot also is known about the low-volume flush toilets which have a high potential for acceptance in urban and peri-urban areas because they can be readily converted from on-site status to a sewerage system. Some designs are currently being tested in the field and no serious problems are anticipated. What we do not know so well are the institutional, financial, and other non-technological aspects of delivery of large-scale projects.

16. <u>Impact</u>. The impact of non-sewered sanitation projects on policy shifts has been quite significant. At donor as well as recipient government levels there has been a growing recognition of the importance of sanitation and the role of non-sewered sanitation systems in meeting the sanitation needs of communities. It is now generally recognized that several non-sewered sanitation technologies provide the same health benefits as conventional sewerage at a fraction of the cost. Moreover, it is realized that in many urban and peri-urban areas of developing countries, where total coverage with conventional sewerage is impracticable for the present, it is feasible to use a zonal sanitation approach to solve sanitation problems. This approach entails the use of different types of sanitation technologies, including non-sewered sanitation systems, in different zones of a community, the choice being dictated by such factors as affordability, level of water supply, housing and population density, spacing between houses, soil characteristics, and available drainage. The zonal sanitation approach leads to the immediate application of a range of technologies in different zones of a city so that they fit into an overall long-term master plan.

17. In view of these developments, at the national as well as the international levels, non-sewered sanitation systems are gaining greater acceptance, thereby leading to

significant policy shifts from a narrow focus on conventional sewerage to a broader approach. A notable case at the national level can be found in India, where the Government has decided that conventional sewerage will not, for the time being, be provided in towns with populations below 100,000. Moreover, in a recent development, significant budgetary allocations were made for non-sewered sanitation systems for urban and rural areas, and a provision of US$230 million was made in the country's Seventh Five-Year Plan for the construction of about two million pour-flush latrines in rural areas. In Indonesia, one million units of pour-flush latrines have been planned to be built over the next five years, and in Brazil a project worth US100 million is under preparation for low-cost sanitation and water supply.

18. So far a comparable achievement in replicating successful demonstration projects on a large scale has not been made. Due in part to this, the gap that existed between water supply and sanitation coverage prior to the launching of the Decade still persists and has even widened in many places. If this gap is to be bridged, a number of key issues must be addressed.

KEY ISSUES

19. The engineering profession of today is very competent in the implementation of large-scale conventional sewerage systems. But it has limited experience in the delivery and operation of large-scale non-sewered sanitation systems. A major goal in meeting the challenge of the second half of the Decade is therefore to overcome this limitation. And this should be no challenge at all for a profession with such a monumental record of achievements in development. There are, clearly, several approaches that can be followed. Let me invite this conference to focus on a few key issues. These are affordability, acceptability, cost recovery, and institutional capacity. In considering these issues, I will urge you to keep two factors uppermost in your minds: the private nature of non-sewered sanitation systems and the changing demographic scene in the developing countries, where an irresistible trend in urbanization is shifting the majority of populations from the rural areas to urban and peri-urban areas.

20. <u>Affordability and Acceptability</u>. One of the constraints to large-scale replication of non-sewered sanitation systems is the twin problem of affordability and acceptability. Non-sewered sanitation systems are four to ten times less costly than conventional sewerage. Yet the chief complaint against them has been their relatively high cost compared to per capita income levels of target groups or relative to the prevailing inadequate alternatives that cost nothing in many places. It has been argued that non-sewered sanitation systems cost too much for application in some low-income housing projects, and that in general they are affordable only to small fractions of those who cannot afford conventional sewerage. Compounding the problem of affordability is the issue of acceptability. Non-sewered sanitation systems are sometimes regarded as second-rate alternatives to conventional sewerage. Hence, many who can afford them find them unacceptable. This problem is not limited to beneficiary groups alone, but it extends in some cases to government officials as well.

21. This Conference is invited to consider what can be done to overcome these problems of affordability and acceptability. We need to determine the extent to which the successful marketing methods in private enterprise can be applied to sell sanitation. We also need to take due note of the role subsidies have played in the cases where large-scale rural sanitation projects have been successful during the first half of the Decade.

22. Already efforts are being made to reduce technology costs through the promotion of inexpensive local materials, the use of lower standards of construction that do not compromise safety and health benefits, and the promotion of cost-saving delivery methods. But it is doubtful whether significant cost reductions can be achieved through these approaches alone. This Conference is invited to explore other approaches, such as the use of revolving funds for financing households. On the technological side, a possible option to be considered is the open-ended approach, which begins with an assessment of how much a community is willing to pay for improvements in its sanitation facilities and proceeds to provide a level of improvement the household can afford, having regard to physical conditions and local capacity for its delivery. Such a policy, however, is likely to succeed only where there is a well-defined policy on cost recovery.

23. <u>Cost Recovery and Sustainability</u>. If sanitation projects are to be replicable on a large scale, they should be at once affordable and sustainable. But, from the viewpoint of a national government, a prerequisite for sustainability is adequate cost recovery. We have very limited experience with cost recovery for non-sewered sanitation systems, and it is doubtful whether full cost recovery is an attainable goal for all at this time. A number of policy questions remain unresolved on cost recovery. For instance, full cost recovery is rarely demanded for conventional sewerage. But, since non-sewered sanitation systems are private goods, it is often argued that their cost must be fully recovered. This question needs to be examined not only for non-sewered sanitation systems but also for other sewered sanitation systems apart from conventional sewerage, because it raises equity questions in zonal sanitation projects. Many other questions need consideration, including the following:

(a) What, if any, are the public benefits from non-sewered sanitation?
(b) What costs should be recovered: capital, recurrent, or both?
(c) What mechanism should be used for cost recovery?
(d) What institutional facilities are required for effective cost recovery, given the private nature of non-sewered sanitation systems?
(e) How should one deal with defaulters?
(f) How should interest rates and repayment periods be determined for loans?

This Conference is invited to consider these and other questions of cost recovery. Another crucial issue is the strengthening of institutional capacity.

24. <u>Institutional Capacity</u>. Without capable institutions, cost-effective delivery of sustainable and replicable sanitation systems is impossible. One major institutional organizational problem that requires serious consideration is fragmentation of responsibilities. In some countries responsibility is split along technological lines, with one agency assuming responsibility for sewered systems and another for non-sewered systems. In other places there is a further split along certain functional lines, with one agency responsible for software aspects and others responsible for the hardware. These splits would not by themselves be detrimental to effective delivery if adequate coordination and clear definitions of lead agencies and accountabilities were available. But often coordination is weak, leading to gaps and costly duplication. Such problems need serious consideration at this Conference, particularly in the light of the inevitable trend towards zonal sanitation systems that I mentioned earlier. What type of reforms, if any, need to be promoted to improve interagency and intraagency relationships? Are institutional reforms the best approach? What about designing delivery systems to suit existing organizational structures? What experience is available or obtainable on these questions?

25. In addition to questions about institutional organization, other problems include lack of trained manpower to fill vacancies, the need for training of available staff for crucial operations, and the high turnover of trained staff. What recommendations can the Conference make on such institutional capacity questions?

26. A further institutional issue is the promotion of a proper mix between public and private enterprise that will be consistent with local practices. It has been suggested that non-sewered sanitation systems lend themselves to the use of small-scale private enterprise and decentralized management. This feature, coupled with the fact that their delivery entails implementation of repetitive operations, creates a large scope for creative delivery strategies, including the use of non-governmental organizations (NGOs). The standardised training methods used in training staff of modern fast-food franchises is one approach worth considering for possible application to the training of small-scale contractors for non-sewered sanitation systems. One could even go further to examine the extent to which franchising methods could be adapted to non-sewered sanitation system delivery.

27. Then there is the problem of inappropriate academic curricula. A number of initiatives are already underway to deal with these questions. In several developed countries special courses have been established at universities, especially for water supply and sanitation in developing countries. Such courses are even more necessary in the developing countries. This is where the International Training Network, executed by the World Bank and funded by the UNDP and a number of bilateral agencies, has begun to play a role. The Network is working with a number of institutions in developing countries that have agreed to serve as network training centers to organize research, workshops, seminars, and other training functions for decision makers, engineers, and others involved in the delivery of water supply and sanitation projects.

28. Another development to support institutions is the creation of two sector development teams (SDTs), one for Western Africa and the other for Eastern and Southern Africa. Managed by the World Bank, they are multi-disciplinary teams funded by the UNDP and a number of bilateral agencies that have been established to assist governments in developing sound sector policies and investment programs, in identifying and preparing projects, and in locating funds for their financing. In addition, SDTs will serve as a medium for harmonizing donor support in the sector, thereby enhancing donor coordination for institutional strengthening.

29. This Conference is invited to review the adequacy of these and other related efforts toward the resolution of the institutional problems so far encountered. In this connection, the issue of incentives for attracting and retaining trained staff also needs to be addressed. Then there is the definition of the role of civil engineering consultants during construction phase of projects. Should it remain the same as it is for conventional sewerage, or does it need to be redefined to include on-the-job training and guidance of small-scale local contractors who must necessarily be used on non-sewered sanitation projects? Here, indeed, is an opportunity for the civil engineering profession to play a leading role to harness the immense resources in man for his development. Consulting firms, in particular, have a responsibility to work with developing country governments to find solutions to these problems. The universities, local professional bodies, and decision makers must all join hands with the international and bilateral agencies to solve these problems once and for all.

CONCLUSION

30. It has been my objective to highlight for your consideration a few key issues that need to be resolved to ensure better coverage of communities with sanitation systems during the second half of the Decade. We have found that non-sewered sanitation systems are small-scale, self-contained sanitation technologies that are characterized by simplicity and lend themselves to decentralized management and the use of small-scale private entrepreneurs; and we have found that there are many situations in developing countries where they would be appropriate. However, experience with on-site disposal systems during the first half of the Decade has been mixed. Whereas notable success has been achieved in promoting policy shifts from narrow concentration on conventional sewerage, efforts in generating large-scale replicable projects are yet to bear fruit. This is not to say that the efforts have been in vain. It only means that the seeds so far sown

for large-scale projects are yet to spring out of the ground and bear abundant fruit.

31. But as it is with the planting of all seeds, there is the risk that the seeds for large-scale projects may never germinate unless we remove weeds and fertilize and water the soil.

32. This is in effect the task before us today. The Conference is being invited to address the pressing issues that have been raised as a way of contributing towards ensuring a brighter future for that half of the world that is yet to enjoy some of the fruits of technological advancement that has characterized the second half of the 20th century.

33. I have no doubt you will rise to the occasion and meet the challenge before you.

Discussion on Papers 3 and 4

MR G. J. MALAN, National Building Research Institute, South Africa
The Water Research Centre has monitored the flow in water mains to aid the development of network analysis procedures. Based on that experience what duration of the peak flow rate would the Authors of Paper 3 recommend for use by the designer? On what peak flow duration should the peak factor for design be based?

MR G. F. MOSS and MR. S. H. WHIPP
The selection of duration for peak flow rate relates both to the service which is likely to be received by consumers and also to the security of supply that is likely to follow. Generally a factor of 1.8 times the average flow taken over one year is recommended as the basis for the design of pipes. However, local considerations may require that, for risks to the security of supply or to allow for future growth in demand, a higher value may be used. Towards the ends of distribution systems generally the pipes will be hydraulically oversized and this will mean that the design factor becomes inapplicable.

MR T. A. STOKER, Parkman Consultants Ltd, UK
Through working with the Water Research Centre (WRC) on both water distribution and sewerage rehabilitation studies, I am well aware of, and very impressed by, their work. I feel it is important to WRC and to the water industry that WRC should have some involvement in as many projects as possible, both in the UK and overseas, so that continuing experience can be passed on and received on a broad front.

It is clearly not practicable for WRC staff to carry out all such projects themselves. Could the Authors of Paper 3 enlarge on their approach to collaboration with consulting engineers for work in the UK and overseas?

MR G. F. MOSS and MR S. H. WHIPP
The Water Research Centre (WRC) is committed to ensuring the transfer of technologies developed both in the UK and overseas. As Mr Stoker points out it is not practicable for the WRC staff to carry out all such projects alone. The approach that WRC has developed to enable it to implement and transfer technologies more widely has been to develop an operating consortium with leading public health consultants from the UK which is referred to as Advanced Water Technology (AWT). The WRC provides training to consultant staff in the principles and details of the technologies to enable a broad service to be offered. Generally, when AWT consultants are working on projects throughout the world, the approach adopted is that, in the initial phases of the contract, a reconnaissance is made of the conditions applying so that appropriate recommendations can be made to apply the technologies to those situations. These early phases will normally involve either WRC or consultants' trained staff. There will then be a phased implementation during which the technologies will be transferred. For example pilot studies may be undertaken to demonstrate on a small scale the techniques and methodologies and to effect training of local staff. Once the pilot studies are completed, the local staff have the ability to continue and to implement the technologies themselves. There would then only be a necessity to provide updating and training as the technologies are developed or as experience causes the approach to be modified.

MR L. M. SOLWAY, Cooper, Macdonald & Partners, UK
An example of the savings that it is said can be achieved as a result of leak detection surveys and the rehabilitation of mains is given in Table 1.

A benefit of £1 million per year on a capital investment of £393 000 is an astronomical return and one that should attract lots of bankers and institutional investors. Is this a typical UK or overseas example or an exceptional one? If it is exceptional, what do the Authors of Paper 3 think are typical savings achieved from leak detection and the rehabilitation of water mains? On what basis are the savings calculated?

Table 1

Cost of water leakage	£2 million per year
Cost of survey	£60 000
Cost of rehabilitation of mains	£333 000
Value of water saved	£1 million per year

World Water '86. Thomas Telford Limited, London, 1987

PAPERS 3 AND 4: DISCUSSION

MR G. F. MOSS and MR S. H. WHIPP
The saving shown by Mr Solway's example is impressive and demonstrates good payback on the investment provided in undertaking leak detection. The philosophy that the Water Research Centre (WRC) adopts in looking into leak detection is to establish what levels of leak detection and control are appropriate based on both the economic liability of undertaking leakage control and the physical causes that are giving rise to leakage. These can be brought together to determine an appropriate level of leakage control in the future. It is the view of enlightened engineers that leakage can be seen as a resource that can be tapped, where the economics of leakage control permit. In studies undertaken by the WRC in Europe, the Middle East and the Far East the results are similar to the example given by Mr Solway. The example is typical of the circumstances commonly experienced by the WRC. For one exercise in the UK the WRC showed that payback in 4 years was achievable. Overseas, particularly where the unit cost of water is high, a payback period of several months has been found to apply. In some instances the payback period is less than the time it would take to implement the methods of leakage control on the ground.

MR T. A. STOKER
Paper 4 poses more questions than it answers but it is essential that these questions are dealt with now if the second half of the water decade is to show more positive results in the field of low cost sanitation than the first half.

I agree that confidence in the capabilities of consulting firms is justified as far as intentions and endeavours are concerned, but in the case of low cost sanitation these are rarely sufficient to ensure final success. The message, now long promoted by the World Bank and others, that low cost sanitation is an essential element for overall strategies in developing countries has been understood and accepted by most design offices and technical organizations involved and has in turn been promoted by them in studies and reports, but still in too many cases implementation of recommendations has not yet followed.

For example, feasibility studies carried out in 1981 for water supply and sanitation improvements in Uganda included the zonal approach to sanitation outlined in Paper 4, the town centres already being sewered, the higher grade properties being served by septic tanks and the lower grade peri-urban areas by pit latrines. All systems required rehabilitation, and recommendations included the upgrading of pit latrines to the VIP design.

The construction of the major civil engineering elements of the project, by conventional design and tendering procedures largely in the hands of the consulting engineers, started in 1985, but as yet not even the first demonstration pit latrine has been established. Lengthy discussions have taken place involving ministries concerned with water planning, health and education and various local authority departments with similar responsibilities. These culminated in a national workshop of several days' duration to agree guide-lines on the subject. So far the intended recipients of these simple improvements are still unaware of their existence and hundreds of latrines must have filled and been replaced by unsatisfactory installations while the consultant has been powerless to help.

Paragraph 11 of Paper 4 confirms that increasingly 'non-sewered sanitation systems ... are being used as part of a range of technologies for full-scale projects'. However, in the light of the problems experienced this policy might be questioned and a separate approach to conventional and low-cost elements considered, at least following the establishment of the overall strategy.

It is pleasing to learn of success stories but the fact that in most cases these have resulted from subsidies, free supplies of components or self-help labour schemes appears further to confirm their differences from the conventional approach.

With regard to affordability, I feel that in many cases the simple solution requires even further simplification. The most unsatisfactory feature of most existing pit latrines, both hygienically and structurally, is the floor. If a strong and well-finished slab unit could be supplied cheaply for fitting to existing or new pits, this alone would achieve a substantial improvement. A vent-pipe and fly-screen would be a desirable addition but the design of the pit and the superstructure should be left to the locals who know best how to provide these at minimum cost.

On the question of cost recovery, the aim of the sanitation strategy is the improved health of the community at large and there seems to be no logic in adopting different approaches to on-site and off-site sewage disposal. If some element of cost has to be recovered, could this not be achieved through general water charges or similar systems? Separate cost recovery might be more expensive to set up and operate than the returns would justify.

I think it is essential to simplify and streamline institutional arrangements so that the implementation of low-cost sanitation, coupled with the necessary parallel health education, can be accelerated.

With regard to the germination of the seeds of non-sewered sanitation, I consider that the seedlings are now well established where they were originally sown, but that they must now be transplanted carefully but firmly into the places where it is hoped that their fruit will be borne.

MR B. GODFROY, Agence de l'Eau Artois-Picardie, France
In France conventional gravity sewers and central treatment systems only are eligible to public funding programmes. Individual on-site systems are of great interest to small communities and isolated homes in rural areas and the outskirts of larger cities. Two experiments using large public grants have taken place in the north of France to upgrade individual systems. Similar systems could be

used where appropriate in developing countries.

Methanization and fixed culture biological treatment are considered as the basic trend in industrial wastes control. At the same time, individual systems using septic tanks and soil dispersion have for a long time been a source of complaint although cost-benefit analysis has shown this to be the best solution to the disposal of waste water in small areas.

Generally individual on-site systems have been built by private people, where no public sewer system was available. No preliminary soil studies were performed, most contractors were unable to build such systems and no maintenance has been done. As a result most on-site systems have been the cause of home flooding, sanitary or pollution problems.

Local authorities therefore prefer conventional sewer and central treatment systems because public funding programmes, which are very important for rural areas in France, are devoted solely to conventional systems, and because most local engineers and consulting firms are experienced only in such systems. A regulation published in 1982 still recommends some out of date systems but a newer regulation recommends septic tanks and soil disposal. Traditionally, local authorities are interested only in public works: individual systems are considered as a private responsibility.

Most rural communities have therefore developed conventional sewers and central treatment systems, even if they are much more expensive than alternative systems, and they have formed waste water districts to be in charge of investment, operation and maintenance – nothing comparable exists for alternative systems.

In 1982 the Agence de l'Eau Artois-Picardie (the financial water agency for the north of France) suggested that two experiments should take place to test the feasibility for public bodies to be responsible for both upgrading individual systems and operation and maintenance. The districts chosen were the District Rural de l'Est – St Quentinois (DRESQ) in the Aisne department near St Quentin, 150 km north-east of Paris (1700 inhabitants in 530 residences), and the Sivom du Canton de Poix-de-Picardie (SIVOM) in the Somme department near Amiens, 150 km north-west of Paris (2000 inhabitants in 600 residences).

In both areas preliminary studies had shown that it was much cheaper to upgrade individual systems (only 5% of them met current standards) than it was to build a conventional system, and in both cases public officials were particularly interested in non-conventional systems which would cost less than conventional ones. The experiments comprised the upgrading of a given number of individual units (80 in DRESQ and 100 in SIVOM), the use of special public funds (75% of the cost was met by grants from the French government and the Agence de l'Eau, and 25% was a low interest loan), and operation and maintenance. They were carried out between 1983 and 1985 and are now being followed by comparable projects.

Preliminary studies have to be accurate and extensive. In one case, a high groundwater level did not allow the use of simple soil dispersion systems. This led to higher costs than were expected and was the cause of many problems during construction.

A public body can be responsible for this kind of project. As it is entirely new to the French public institutions, builders, engineers and public officials have to be given help. The administrative management of the experiments was particularly difficult in the DRESQ, which had been formed specially.

The extensive engineering efforts involved during both the planning and the building phases (e.g. investigating each home and the soil conditions) meant that the engineering costs were proportionally higher than those of conventional systems: 11–15% of the total costs for the two experiments compared with an average of 8% for conventional systems. (The engineering effort also included management assistance to the public body because of its lack of experience.)

At present no French regulations allow a public body to be legally responsible for the installation or for the operation and maintenance of individual waste water systems which remain private, so for the experiments contracts between volunteers and the DRESQ and the SIVOM had to be negotiated for this purpose. This phase is particularly important and much time and effort has to be devoted by officials and engineers in addressing public meetings, visiting people and so on.

Local builders rather than bigger public works firms were chosen for the experiments. The on-site technology was new to them and only a few were able to cope with the precise specifications of the contracts. Training in this new technology must be given in professional institutions.

The operation and maintenance of individual units includes a routine inspection and the removal and disposal of waste from all on-site systems. It costs 300 fr per year user which is comparable with the cost of conventional systems (1986 prices).

In such projects the leadership of public officials is a key to success, especially in rural areas, where everyone is looking at what his neighbour or the community mayor will do.

It is now proposed that the Agence de l'Eau should allow grants to individual units which will be an incentive to promote alternative systems: 35% subsidies will be available (which may also be assisted by the government) instead of 20% for conventional sewers. The following restrictions will be imposed.

(a) A preliminary alternative analysis will have to be made by the public institution in charge of the project. A long-term plan of implementation will have to be decided by local officials with an alternative for each zone according to the development area map.
(b) The public local institution (possibly a contractor) will have to be responsible for operation and maintenance and user charges will have to be established. (A new law is to be passed on this point.)
(c) Only rehabilitation and upgrading of existing systems will be eligible for grants, either through the public local

institutions or directly through home-owners. (In the latter case the public institution will be in charge only of preliminary studies and operation and maintenance.) Individual units in newly built houses will not be eligible for grants.

Many local communities are currently carrying out preliminary studies. It is expected that in the long term one-third of the rural population in France will not have a conventional sewer system available, so this move towards alternative sanitation technology is particularly relevant.

Individual water-dependent units or, more often, non-water-carriage excreta disposal technologies may be best in both rural and urban areas. This has been an important point in the context of the International Drinking Water Supply and Sanitation Decade.

It is interesting to note a convergence of recent experience in developed countries.

DR H. O. PHELPS, Water and Sewerage Authority, Trinidad and Tobago

In situations where an adequate water supply is available the question arises of whether or not the on-site disposal of waste water is more cost effective than the provision of small centralized systems for groups of houses or individual housing developments. This is especially so when the soil is impermeable or of restricted permeability, making it impracticable to design an on-site disposal field within the boundaries of the property. The discharge of siltage water to storm water drains is insanitary. Is it not a better policy to improve the designs and efficiencies of small treatment plants under these conditions? This has the added benefit that such small centralized systems may conveniently be linked together into a larger system as and when funds become available.

MR J. E. J. GOAD, Binnie & Partners, UK

Two points of general interest have arisen in Cairo, where relatively small areas of the city are not provided with water-borne sewerage but are high-density residential areas.

First, the government department responsible for sewerage, sewage treatment and disposal does not assume responsibility for non-water-borne systems. Proposed pilot schemes are therefore due to be implemented by the local authorities, the Governorates, who have not hitherto had any systems for this.

Night-soil collection for vaults and disposal is at present mainly in the hands of private contractors who charge for their services. Those who cannot pay do not have their vaults emptied, causing unpleasant conditions. Disposal is sometimes made to the nearest watercourse, again causing unpleasant conditions. Proposed improvements relate to more sanitary and simpler collection and conveyance systems, and effective and inoffensive treatment and disposal.

MR M. COFFEY, Manus Coffey Associates Ltd, Ireland

Privy designs to date have concentrated on systems which are appropriate for small volume productions in rural areas. In urban or peri-urban areas with a radius of perhaps 40 km where large numbers of privies may be required it can be much more economical to precast concrete tanks where the material control can be reduced by improved shapes and quality control to about a third of that needed for a conventional latrine.

Studies for Trinidad have shown that a precast concrete latrine tank complete with top slab and seat pedestal could be manufactured on a quality basis, delivered to site and deposited in a pre-dug hole for a prime cost of materials and labour of about US $26.

A simple transport system has been designed whereby a conventional truck can deliver 20 privies at a time to a delivery point and an extremely low cost transporter can deliver tanks to site over soft or rough terrain and on narrow access tracks.

MR N. J. DAWES, Binnie & Partners, UK

The non-sewered sanitation systems described in Paper 4 are eminently suitable for rural areas but they are more problematic in dense urban areas. I was particularly interested in the point made in paragraph 17 of Paper 4 that one country has adopted a policy of not having sewered sanitation in towns with populations of less than 100 000. Many would regard such towns as quite large, and they are likely to contain high-rise development as well as water-using industries for which non-sewered systems are clearly inappropriate.

There is a close relationship between the work of the sanitary engineer and that of the urban planner. In many cases the need for adequate sanitation — one of the many services essential for an acceptable life in an urban environment — has been neglected at the planning stage. Could the Author of Paper 4 elaborate on the relationship between the types of sanitation being advocated by the Technology Advisory Group and the wide drive to improve urban living standards in the poorer developing countries?

MR L. O. WILD, Laurie Wild Consultants Ltd, UK

The Authors of Paper 3 have made little mention of the day to day operational activities and the effect of these on the ultimate fate of the systems. Over recent years in the UK the staff on the ground have been reduced and this makes the need for the development of sophisticated equipment even more necessary. Do the Authors find that the water authorities recognize that although it is economic to reduce the number of diggers of holes there is a need to employ technicians to maintain new equipment and flow meters, pressure reduction valves and so on?

When I was involved with London's water supply I found that there was a slight leakage from every joint exposed. In each case it was only a small quantity but in fact there are over 13 000 km of pipes with a joint every 3 m and so it amounts to a lot of water. It means that there is a leakage level below which it is uneconomic and unacceptable to go.

Do the Authors have any views on the materials of pipes to be used in the future to save costs later? A notable manufacturer has advertised that plastic pipes are 17 times more liable to failure than are iron pipes. This means that there is another item to include in the economic assessments.

With regard to Paper 4, I do not think that enough is said about the people who are to be served. It is known that their conditions can be improved but is it thought that populations who for thousands of years have been disposing of their human waste without any technology can understand what is being done? How is it possible to convince them of the need for the construction and maintenance of water supplies that they need to pay for? Surely their water needs should be part of an overall improvement in their life style - it should be enough to irrigate their crops and to provide for commercial needs.

With regard to low technology disposal of waste, in many countries any water is a priceless product, and it therefore seems wrong that any liquid should be allowed to seep into the ground.

MR G. F. MOSS and MR S. H. WHIPP
We agree wholeheartedly in that, as more sophisticated equipment is used, utilities must recognize the need for training staff to higher levels to ensure the effective operation of the equipment. This trend is found to be welcomed by staff whose motivation in the job is often increased by the training. Mr Wild's experience in London's water supply is a good example of just what is meant by developing an economic level of leakage control. Mr Wild asks whether the Water Research Centre (WRC) has views on pipe materials to be used in the future. The WRC has embarked on a major research activity in collaboration with the industry and with manufacturers to review the pipes that are being used in the water industry and advice is being drawn together by a working group of the Sewers and Water Mains Committee of the Water Authorities Association. This work will enable engineers to specify appropriate materials for a given situation and to identify the shortcomings of certain materials in particular circumstances.

MR M. S. Z. KILANI, President, Water Authority, Jordan
There should be no differentiation between a person in the developing world and the developed world. The same service must be introduced to the developing world: piped safe water and sewerage networks and treatment plants. It requires the same radical solution to the health problems in the developing world as in Europe in the second half of the 19th century to eliminate epidemics for ever.

Unfortunately the international water decade has not succeeded as it was hoped because it must be linked with the financial aspects. Estimates of the total cost of the projects for water and sanitation all over the world are needed so that the decade's goal could be achieved by looking for financial sources to pay for these projects in phases (first, second, third etc.) with grants or with loans with easy terms.

The countries which badly need the water supply networks and sanitation projects are in general poor countries facing high debts, so they cannot afford to pay the instalments and the interest of the debt.

Therefore it is our duty to allocate the money from a fund from the United Nations or the World Bank, as a grant or an easy-term loan, so that it will be tempting for the countries to benefit from these grants and loans to implement the water and sewerage projects.

MR C. INGRAM, GKW Consult, Egypt
Concerning the use of flow meters for network simulation as a necessary prerequisite to leak detection GKW is currently carrying out a leak detection programme in a pilot area of Giza which is the western section of Greater Cairo lying between the River Nile and the pyramids (population of 4 million by the year 2000 and a network of 800 km of pipes.) The company intended to construct flow measurement chambers at 26 locations throughout the network and discussed at considerable length the type of meters to be used. Eventually a 'pitot' probe type of meter was selected which could be transported from measurement chamber to chamber. These meters have the advantage that

(a) installation is simplified by avoiding pipe cutting
(b) costs can be saved by buying a basic number of meters (in this case only six meters were used for the 26 measurement chambers) depending on the diameters to be monitored.

However, other disadvantages occur with the probe type of meter (blockages etc.).

What are the Water Research Centre's views and recommendations on the use of probe-type meters?

MR G. F. MOSS and MR S. H. WHIPP
It is the Water Research Centre's experience that the pitot-type meter has a high stall velocity and suffers certain problems in relation to the accuracy of its flow measurement. However, in the hands of an experienced operator it can be an effective instrument. The shortcomings of the pitot meters have encouraged the development of both turbine and electromagnetic measuring devices which can be used, as Mr Ingram suggests, at meter measurement chambers. The advantages of turbine and electromagnetic meters are that they have a lower stall velocity than the pitot meter and they provide a better resolution and accuracy through the range of flow velocities. In the future, effort is being put into developing alternative flow measurement devices. One key advantage of the electromagnetic device is that it has no moving parts. All such insertion measuring devices whether they be of the pitot, turbine or electromagnetic type only

PAPERS 3 AND 4: DISCUSSION

provide point measurements of velocity in the pipeline and there is still a requirement for good operator skill.

MR S. N. MAKHETHA, Urban Development Project, South Africa

The low-cost sanitation programmes need to be supported by appropriate training of engineers and scientists in the developing countries. Currently the training in developed countries is far from where the need is. Can the World Bank sponsor such training in the institutions of developing countries?

More software for low-cost sanitation and water has to be introduced and education of the population is important for the systems. It should be remembered that the selling point for sanitation is not the health aspects but mostly the comfort and convenience.

More technical research has to be made into improving the level of service of the low-cost options to make them more attractive to the users.

The problem of cost recovery should be viewed in perspective and by comparisons with the past trends in water-borne sewage systems.

MR S. KOUCHEK-ZADEH, Brian Colquhoun & Partners, Guinea

I support Mr Kilani about the need for finance from aid agencies to implement the water decade. However, whether a town should have sewage or non-sewerage schemes should be looked at individually rather than by having a policy of saying that towns with say a population of less than 100 000 should not have a conventional sewage system. Every town has its own unique problems and should be treated accordingly.

Dr Wright mentioned the lack of colateral for sanitation schemes, but the World Bank or any other lending agencies lend money to the governments and not to a particular department, e.g. the water department or the road department. When the money has been loaned the bankers make sure that they recover their money whether the scheme works or not.

DR G. GHOOPRASERT, Provincial Waterworks Authority, Thailand

In most developing countries, there is no sophisticated equipment. Therefore 'eye inspection' leak surveys are usually carried out. On the basis of the Author's experience, then, what is the percentage of leaks that can be detected by eye? If this percentage is high, is it worthwhile to purchase sophisticated and expensive equipment?

It may be difficult to implement a 'self-recovery' water supply or sanitation schemes in rural areas, since people are so used to free services. If their attitudes are not changed, then the programme may face failure.

A good example is the 'revolving fund' scheme that the Provincial Waterworks Authority has initiated for rural water supply in Thailand, whereby villagers can borrow money at low interest rates for improvements or rehabilitation of their waterworks. The response has not been very good, since it is a loan, albeit cheap, instead of a grant.

MR G. F. MOSS and MR S. H. WHIPP

The question of the levels of leakage control which are most appropriate and if leaks can be detected visually by routine inspection to discover the position of leaks is referred to as passive control, which is the least expensive method of leak detection. If there are many leaks that can be dealt with in this manner it would be a good starting point. However, experience has shown that this method of overcoming leakage would not necessarily be the optimum as the ability to reduce leakage in the long term would, in most situations, justify a more intense level of leakage control. The determination of an economically justified policy would be recommended but, in terms of dealing with leaks, those which are visually apparent would be repaired first. It is interesting that studies have recently been undertaken in the Provincial Waterworks Authority of Thailand on a pilot study basis and it would be interesting to learn of the experience that has been gained in terms of the approach and methodology to leakage control which has been found to be appropriate.

MR C. B. BUCKLEY, Welsh Water, UK

There has been no mention of water quality in Paper 3. It is important to recognize that leakage control must be accompanied by an overview which includes intensified monitoring to ascertain the effect of spasmodic episodes of poor water quality, which can result from the ingress of contaminated water, at times when the pressure falls as a result of high sudden demands - bursts etc. Such events can be bacteriological or chemical in nature.

Rehabilitation of mains using cement-mortar relining can also produce water of poor quality, especially in rural areas where long lengths of small diameter mains feeding areas of low demand cause excessive retention times. Extremely high pH values ensue which can lead to adverse consumer reaction, if not direct problems of skin irritation.

MR G. F. MOSS and MR S. H. WHIPP

It is the Water Research Centre's philosophy that leakage control forms a part of the total rehabilitation activities being undertaken on the water distribution network. These are aimed at either achieving improvements in quality of service to the customer or a reduction in costs to the undertaking. In this context leakage forms one part of the total rehabilitation activity. Water quality is an important aspect of customer service and Mr Buckley is correct in identifying this.

G. MOVIK, Norconsult International A.S., Oslo, Norway

TN7. Rural water supply in Rukwa region, Tanzania

The NORAD-assisted water supply and sanitation project in the Rukwa and Kigoma regions of Tanzania started with field work and data collections for the water master plans in 1979. The plans for both regions were completed and the final reports issued in 1982.

2. The implementation of water supply schemes started in early 1982. The community participation concept was adopted right from the start in Rukwa. In 30 remote villages a total population of approximately 45 000 people have so far received potable water from new water supply schemes.

3. Most villages use groundwater as the source, but a few surface schemes using gravity or hydraulic ram pumps have been constructed. No motorized schemes have been constructed due to the present shortages of fuel and spare parts. Water treatment is installed at surface schemes. A simple chlorine dozer is being tested as well as a combined horizontal roughing filter (HRF) and a slow sand filter (SSF).

EXPERIENCE
Drilling section

4. Average well depth is 60 m. Drilling started using an Atlas Copco mobile rig B-80, which was replaced in September 1985 by a new Danish KNEBEL HY76.

5. The most common procedure is to drill an 8.5 in borehole through the overburden, then to insert 7.5 in casing and to continue drilling in the bedrock with down-the-hole hammer and 6 in drill bits. Depending on the local conditions, mud, foam and so on are being used. The overburden is cased off, and the casing is also grouted in the bedrock when required. Thereafter the wells are cleaned and developed by airlifting before test pumping. As at 1 January 1986 some 280 wells had been drilled with an average success rate of 80%. The productive wells are fitted with handpumps and disinfected before use. Comprehensive well data are stored on a data base using personal computers. The drilling staff is organized in three teams, one drilling new wells, one doing development and test pumping, and the third rehabilitation and maintenance of old wells. One expatriate hydrogeologist was employed until late 1985 when the counterpart hydrogeologist took over full responsibilities. One expatriate driller is still employed.

Construction section

6. Construction teams are doing handpump installations, constructing storage tanks, gravity and piped water supply schemes as well as new offices and workshops. All unskilled labour is provided by local villages. The section has one expatriate engineer. On-the-job training has been emphasized and quality and workmanship have improved considerably.

Workshop

7. Until 1 January 1986 there was a separate NORAD workshop carrying out the maintenance and repair of 14 trucks, 2 tractors, 2 drilling rigs, 2 compressors, 29 Toyota Landcruisers and 13 motorcycles. The workshop was headed by one expatriate supervisor and staffed by eight locally employed mechanics. The workshop is now being fully integrated with the Regional Water Engineers' old workshop.

Handpumps

8. Five different types of handpump have been tested by the project: India Mark II, SWN 80 and 81 and Duba Tropic II and VII. (The Petro pump was introduced early in the implementation phase, but was quickly abandoned due to bad performance).

9. So far the Mark II is the preferred handpump. It has a simple design and is easy to understand. Training of local village scheme attendants to maintain the pump has been a relatively easy process. The Mark II is also much preferred by the local population.

10. The experience with SWN 80 and 81 is also positive. However, the PVC rising mains break frequently. The breakages normally appear close to the sockets. The pumps are easy to install and operate. Unfortunately, they are equipped with bearings which are complicated to produce or repair locally. Corrosion is no problem because they are made mainly of plastic materials.

11. Duba Tropic II and VII handpumps have been installed in two villages. They are sturdy and said to be reliable, but they are complicated to install and heavy to operate. The villagers have tried the Mark II handpumps for comparison and they have used every opportunity to complain about the Duba pumps and their heavy operation. Many villagers refused to use the pumps and went to the old local wells instead. After these local

campaigns the pumps have been relaced by the Mark II.

Hydraulic ram pumps

12. The Hydram pumps are driven by water. They need few spare parts and both pumps and spares are produced in Tanzania. This of course gives a very low operational cost compared with motorized pumps. The ram pumps are simple and easy to maintain by trained village scheme attendants. However, the pumps need high yielding and reliable water sources. The system works well in Rukwa for a water head of up to 65 m. One scheme is lifting water to 95 m, and there have been some breakages due to high stress in materials like bolts and drive pipes. Although theoretically the pumps can lift higher it is believed that one should look for other alternatives when exceeding a head of 80 m.

Water treatment

13. The chlorination plant is a simple batch dozer mounted on the storage tank. It has been in operation for 18 months without any problems. The only disadvantage relates to the availability and cost of chemicals.

14. The HRF/SSF plant has recently been put into operation. So far there is hardly any operational experience; it seems to function well. Initial construction costs are high, but operational costs are low and dependent only on local manpower.

COMMUNITY PARTICIPATION

15. The construction of water schemes has been carried out with the extensive use of community participation. Information campaigns, promotion meetings and discussions with villagers have been arranged during both the planning and construction phases and all unskilled labour has been provided by the villages.

16. Completed schemes will be handed over to the villages as their own property. They will be responsible for operation and maintenance with assistance from the Regional Water Engineer only when and if needed.

17. A comprehensive community participation approach is probably by far the most important factor in reaching the overall objectives of a rural water supply programme.

M. MULLER, MICE, National Institute for
Physical Planning, Maputo, Mozambique

TN8. Low-cost sanitation in Mozambique

Mozambique - one of Africa's poorest countries (GNP $160/capita in 1984) - has a relatively large urban population approaching two million, most of whom live in poor conditions on the periphery of the colonial cities.

2. Prior to independence in 1975, little attention was paid to sanitation in these areas. Post-independence, mass campaigns were launched to promote latrine construction. These resulted in extensive coverage. According to the 1980 census, 72% of the urban population had a latrine. This compares with an average of 54% reported for all urban Africa (ref. 1).

3. Despite the high coverage, there are still acute needs. Improved techniques are needed to ensure stability since many early latrines collapsed or were flooded. It is also necessary to ensure that latrines are hygienic and acceptable to the community and thus used by all. Coverage must be increased.

4. In the development of an appropriate technology, certain constraints were obvious

(a) traditional building material such as wood poles was in short supply
(b) imported material such as reinforcing steel was not readily available
(c) transport in peri-urban areas is often difficult
(d) costs had to be kept as low as possible.

5. The technical element of the solution is the conical unreinforced concrete slab. This is economic in its use of materials. Designed with a close-fitting lid for odour and insect control, smoothly finished to facilitate cleaning, its small aperture and well-designed footrests permit and encourage use by all members of a family.

6. The institutional element is production by small neighbourhood co-operatives in each community. With initial support in construction materials and tools and on-going support in management and logistics, these have proved able to produce and sell slabs self-sufficiently. Their community links facilitate the promotional and educational work carried out by health authorities and community organizations.

7. In some soils and under adverse groundwater conditions, lined pits are necessary. In these cases, the co-operatives construct the pit for an additional charge using blocks or stone.

8. The costs of a latrine slab and of a lined pit with slab are 800 MT and 4450 MT ($20 and $110) respectively. It is a measure of the success of earlier campaigns that, 20 000 slabs later, sales have never been a difficulty. Indeed, the immediate problem is that it is not possible to satisfy the demand. This leads to queues and causes management problems when co-operatives accept advance payments from their clients.

9. Detailed studies on utilization are notoriously difficult to carry out. Available information suggests that

(a) utilization is better with 'improved' than with 'traditional' latrines
(b) utilization is not complete, particularly with children
(c) maintenance is usually good although lost or broken covers are a significant problem in assuring a hygienic system.

10. One technical problem is deciding whether it is worth promoting the improved pit latrine in high-density communities without other urban improvements, such as plot definition and access ways. For communities where the population density is such that it is difficult to find space to build one latrine per family, experimentation is underway into types that offer a reasonable lifespan with multifamiliar use.

11. Lifespan is at the heart of the other technical problems. Improvements are necessary to reduce pit collapses, preferably without resorting to the use of fully lined pits. Further, what happens when the pit fills? The hot, humid conditions in Mozambique are probably as favourable as any to promote rapid digestion and ensure a reasonably long life but there will at some point be a need either to build new latrines or to empty existing ones. The former is an appropriate solution with movable slabs over unlined pits; the latter indicates more the 'sunken investment' of lined pits in high density communities.

12. Social aspects may begin to cause concern in the medium term. If the early latrine compaigns served as the first phase of mobilization, we are now in the second phase of expansion. A third phase can be foreseen - probably the most difficult - that of consolidation in which it will be necessary to boost coverage by improved latrines from the

World Water '86. Thomas Telford Limited, London, 1987

60-70% which seems to be reasonably easy to attain to a figure closer to complete coverage. Correct utilization and maintenance will continue to demand close attention.

13. Economic aspects will probably be crucial to this final phase and may also have an impact on the present one. The programme's success to date is derived in part from the particular situation in Mozambique whose closed market has boosted the effective demand for latrines in the absence of many other goods. Economic changes imposed by continued South African backed attacks on economic infrastructure and the 'free market' policies being demanded by donors such as the World Bank and USAID have already begun to erode this base. It will be ironic if the result is a 'healthy' economy in which the population cannot be afforded the cheap and efficacious means to improve their health offered by the improved latrine.

REFERENCE

1. United Nations General Assembly, Economic and Social Council. Progress in the attainment of the goals of the International Drinking Water Supply and Sanitation Decade. A/40/108 E/1985/49, 1985.

T. VIRARAGHAVAN, University of Regina, Saskatchewan, Canada

TN9. Research and demonstration needs related to on-site waste water treatment and management

Approximately 20-25% of the population of Canada and the USA use on-site systems for the treatment and disposal of household waste-water. These systems are in use on farms and in low density rural and suburban areas in many parts of the world. There are many relatively large institutional on-site waste-water treatment systems as well. Other small community waste-water systems, which are extensions of the on-site concept, are being built with public or similar management of these systems. A number of areas of the on-site waste-water technology require further investigation, research and demonstration for the successful design, operation and maintenance of these systems, and for minimizing their adverse effects on the environment. Some areas for further study are briefly addressed.

SEPTIC TANKS AND EFFLUENT TREATMENT
2. Septic tanks today are essentially the same in design and construction as they were 50 years ago. The removal of low suspended solids through septic tanks is usually attributed to possible re-entrainment of appreciable quantities of sedimented material in the effluent stream through hydraulic scouring or sludge gasification or both. The introduction of gas baffles would definitely improve the performance of the septic tank. However, the use of gas baffles is not widespread; there is a need to popularize them through demonstration units and recommendations in the codes of practice.
3. Separate treatment and disposal of black and grey waters is an area requiring more research and demonstration. In this connection, the concept proposed by the World Bank suggesting the use of a three-compartment septic tank with only toilet wastes entering the first compartment and grey water entering the third compartment requires further investigation and documentation; demonstration units will prove useful to popularize this concept.
4. There are conflicting findings about whether or not compartmentation of septic tanks is beneficial. There is a need to investigate this issue and to conduct further studies to resolve the question.
5. A number of areas for further research are related to anaerobic filter treatment of septic tank effluent. These include the following.

(a) Laboratory studies and studies at low temperatures (5^oC and 10^oC) to examine the performance of an anaerobic filter treating septic tank effluent.
(b) Long-term field studies to monitor the performance of a septic tank: an anaerobic filter system with respect to BOD, SS, nitrogen, phosphorus, indicator micro-organisms and virus removals.
(c) Anaerobic filter treatment of grey water.
(d) Field studies on further treatment of septic tanks: anaerobic filter system effluent using grass plots and other simple techniques.
(e) Studies to determine the relationship between effluent quality and a subsurface disposal system area.

6. Further research and demonstration are needed on denitrification of nitrified septic tank effluent through the use of alternate carbon sources such as grey water, kitchen sink and laundry waste-water and the use of anoxic systems such as anaerobic filtration.
7. The use of peat for treating the septic tank effluent is currently under investigation and some experimental field systems are in operation. Both laboratory and field studies show good promise; peat filtration of septic tank effluent may be advantageous under certain circumstances.

SOIL ABSORPTION OF ON-SITE WASTE-WATER
8. A number of areas related to soil absorption of septic tank effluent and other on-site household waste-waters require further study. These include

(a) longevity (survival)/failure studies of existing soil absorption systems, both individual and institutional in many areas, including techniques to rejuvenate them using hydrogen peroxide and other chemicals
(b) research into desorption of phosphorus already adsorbed into the soil, through fluctuating water-table/heavy rainfall
(c) research on viruses present in the effluent discharged to the soil, their lengths of time of survival in the soil and the water environment including travel patterns, elution of heavy rainfall and the impact of their

presence on public health; fundamental studies on adsorption of viruses on to soil particles, and competition for adsorption sites on soil particles with soluble organic matter including trave organics, phosphorus and ammonium ion
(d) identification of trace organics in on-site waste-water and their transport through the soil to groundwater
(e) septage disposal on land and its environmental impact.

ON-SITE WASTE-WATER MANAGEMENT
9. Poor operation and maintenance of the on-site systems is the major cause of failure of these systems. An organizational structure to undertake proper installation, operation and maintenance is essential for the success of such systems. India and many other developing countries have embarked on ambitious programmes for providing on-site systems for large urban and suburban populations. Serious pollution problems and outbreaks of disease can occur if necessary precautions are not taken. One main reason for the slow progress of the low-cost sanitation programme in India has been identified to be the non-availability of a proper institutional structure for the implementation and management of the programme. There is a need for a number of demonstration projects related to on-site systems in developing countries especially to demonstrate the institutional concepts and arrangements and to learn from local experience.

CONCLUSIONS
10. Research is needed either to adapt existing technologies and measure their effects or to find new, more efficient techniques suitable for local conditions. Many developing countries spend little on research/development related to low-cost sanitation. There is a need to change this attitude and attempt at indigenous research and development if large-scale programmes of on-site sanitation are to succeed in the developing countries.

Dr D. VILLESSOT, Intercos S.A., France

TN10. Aqualis technologies to solve drinking water distribution problems

Water supply of urban and rural areas in industrial or developing countries is still a problem and conventional technologies for water treatment and distribution are not strong enough to keep the problem under control. Network qualities, the nature of pipes, breaks, intermittent distribution services, network servicing and maintenance, and pollution of groundwater are among the more frequently occurring problems.

2. Studies have therefore been conducted to develop appropriate Aqualis equipment at the household, building and community levels. Using reliable least-cost techniques such as micro-filtration, GAC filtration and UV sterilization, new domestic, mobile water purification and water treatment units have been produced.

3. Special attention was given to UV sterilization to obtain the maximum efficiency on each unit.

(a) Low pressure mercury lamps operate in jacketed tubes.
(b) Automatic flow control devices to restrict flow to the maximum permitted for the particular unit are currently offered.
(c) Accurately calibrated UV intensity meters (sensors) activate a solenoid valve or warning device if the UV lamp fails.

4. Three major units are currently available

(a) Aqualis 3000: a mobile unit producing 3 m^3/h of potable water from surface or ground waters, including a generator set, pumps, staged filtration (down to 5 µm), UV disinfection and automatisms to monitor output according to UV lamp efficiency; options: chemicals feeding (alum, chlorine); treated water packing (3030)
(b) Aqualis 1500-6000: filtration and disinfection of water at the point of use including staged filtration, UV disinfection, warnings; monitored output: 1500-6000 l/h
(c) Aqualis 800: staged filtration (GAC + 5 µm) of stored water including cooling (roof tanks), UV disinfection, UV sensors, warnings and water storage under temperature control.

5. These units must be selected and installed according to specific environmental parameters which have to be carefully studied and identified before implementation.

DR F. F. PADERNAL, Ministry of Public
Works and Highways, Manila, Philippines

5 Trends and issues in rural water supply: a case study

Developing countries like the Philippines undergo a rural water supply development pattern involving all or a combination of the following activities: master planning, programming or budgetting, preliminary and detailed engineering, formation and registration of water-users associations, construction and repair/rehabilitation of water supply facilities, water quality analysis and surveillance, assistance in operation and maintenance, implementation of training programs, equipping project implementors, research and development, engagement of consultants and periodic review. Donor countries and lending institutions have increased considerably their financial assistance to the Program, hence, project implementation is being accelerated and achieved quantitatively and qualitatively. In the process of constructing, operating and maintaining the water systems, there are inherent and man-made problems encountered. However, technical, institutional, managerial and political approaches that can resolve these problems and constraints are also available.

BACKGROUND

1. During the 18th-19th centuries, there was the Carriedo Waterworks that served the Manila area. Provincial towns had to rely on some dug/drilled wells and developed springs. With the coming of the Americans at the turn of the 20th century, improvements were instituted. Carriedo Waterworks became the Metropolitan Water District in 1919 and undertook the construction of the Balara Filtration Plant, its biggest feat. This plant is still in operation until now. Bigger towns and cities in the provinces boasted of new water systems, while less populated areas were installed with hand pumps.

2. But with the World War II, many water supply systems facilities were razed and badly deteriorated for lack of maintenance. At the same time, the population increased that water supply could not cope up with the demand.

3. In 1955, the National Water and Sewerage Authority was created to centralize and consolidate under its control direction and general supervision of all waterworks and sewerage systems in the country. A massive but too centralized program of improvement and rehabilitation of waterworks began.

4. By 1964, an interim improvement program was carried out within the Manila area. In the countrysides, water systems were re-built and put into operation. For the rural areas, needed wells for the village people were constructed.

5. By the end of the 1960s, NAWASA was just spreading itself too thinly. It just couldn't handle the growing metropolis and the provinces at the same time. It is important to note here that the pace of water supply development became more directed towards the metropolis and provincial urban centers to the detriment of the rural areas. Investments in rural water supply was minimal and therefore rural water supply development was slow. Rural sanitation was neglected.

6. By the 70's redirection of water supply goals was necessary so that the provision of potable water to the entire country, and therefore to every Filipino household became a prime policy of the state.

7. In response to the need for complete water supply coverage in the Philippines, the water supply sector was divided into three responsibility areas: firstly, dealing with the Metro Manila and its contiguous areas, is the Metropolitan Waterworks and Sewerage System (MWSS); secondly, dealing with the larger cities and municipalities with populations of 20,000 or more, is the Local Water Utilities Administration (LWUA); and thirdly, dealing with the rural population and the smaller provincial urban communities, is the Rural Waterworks Development Corporation (RWDC).

8. Other agencies that have set responsibilities in the water sector are the Ministry of Public Works and Highways (MPWH), the Ministry of Local Governments (MLG), the Ministry of Health (MOH), and the National Water Resources Council (RWDC).

9. The MPWH is the central coordination office for all water supply plans, programs and policies. Through its Project Management Office for Rural Water Supply (PMO-RWS), it is the principal implementing arm for engineering and construction of the rural water supply projects. As such, it plays a key role in the development of water supply and sanitation projects for the rural areas.

10. The MLG assists thinly populated rural areas without access to potable water and sanitation facilities through its Barangay Water Program which is supported by USAID.

11. The MOH is responsible for promoting safe water supplies, concurring the sites of water sources and exercising surveillance of water quality.

12. The NWRC is the planning agency for water resources development and management.

13. All these agencies coordinate and cooperate with one another with the end in view of providing water that is safe, accessible, adequate and affordable to the broad masses of the Filipino people.

TRENDS

Accelerated Pace in Rural Water Supply Development

14. By the end of the '70s, the water supply situation had improved considerably with the agencies concentrating on their areas of responsibilities. Metro Manila and other key towns in the country began to experience a much improved delivery of water supplies to its constituents.

15. However, despite the abundant natural water resources available in the country, much of the rural areas still depended on rainwater, open water holes and other sources of water often of doubtful yield and quality. As a consequence, these small communities' economic progress, which depended on adequate and reliable water supplies, was stunted. More improtantly, however, was the effect on the lack of potable water had on the health and well-being of the rural folks.

16. Aware of this, the national government took responsive

measures to improve the rural situation. It infused the much needed foreign and national funding, expertise and technologies to bridge the huge water supply gap between the urban and rural areas, specifically the unserved and underserved populations.

17. About the same time the Philippine government embarked on an accelerated and earnest pace of rural water supply development, the United Nations declared 1981-1990 as the International Water Supply and Sanitation Decade and vowed to make safe water available to all by 1990.

Planning for Rural Water Supply and Sanitation

18. Obviously, the disparity between the development thrust and efforts in the urban and rural areas was too much to be ignored any much longer. To correct this disparity was also a move to correct the single water supply development thrust by combining sanitation and public health enhancement as a second and equal thrust, particularly in the rural program.

19. The national government, through the coordination of the MPWH formed an Inter-Agency Committee representing all entities involved in water supply. This Committee formulated the country's "Integrated Water Supply Program, 1980-2000" which set out the general policies for the water supply sector and indicated short, intermediate and long-range targets. This Program therefore is a coherent and comprehensive guide for all entities on the national government's water supply development plan.

20. In 1982, within the context of the over-all Program, a more detailed plan was developed. Called the "Rural Water Supply and Sanitation Master Plan," this plan provided specific policies, targets and action programs for the provision of water supply and excreta disposal facilities to service the rural communities where the need for these basic services has been identified to be urgent. The plan offered a better balance of opportunities between urban and rural through the improvement of the key elements of social infrastructure that support the country's agri-based economy. Lastly, the Plan also provided the framework for coordinated development activities by all entities concerned within the rural sector.

21. By 1985, because of changes in development conditions in the country, the MPWH activated the inter-agency committee on water supply to assess the requirements and policies of the sector to make them more responsive and realistic to the economic and social changes taking place The result is an updated Integrated Water Supply Program, 1986-2000, that presents a new program of implementation, a corresponding financing plan, as well as a number of recommendations on how best to carry out these programs institutionally, financially and technically.

Project Development Process

22. Prior to the 80s, there were only two major activities being undertaken in the rural water supply sector, namely: annual programming of targets and investment requirements, and construction, per se. Institutional; and operation and maintenance, among others, were omitted.

23. The trend of the 80s is the delivery of this basic social infrastructure in a package of complementing activities vis-a-vis its integration with the sanitation sector.

24. A development pattern was established on a various level, viz: master planning, programming or budgeting, preliminary and detailed engineering, formation and registration of water-users associations, construction and repair/rehabilitation of water supply facilities, water quality analysis and surveillance, assistance in operation and maintenance, implementation of training programs, equipping project implementors, research and development, engagement of consultants, and periodic review. In addition, continuous development of new feasible or "bankable" projects is being pursued.

25. On a project level, the development process can be aptly described as follows. The process starts in two ways, to wit: (a) request is initiated by the proposed end-users; and (b) as part of the routinary functions of the agencies involved in water supply, potential projects are identified. Requests may be in a form of letter, telegram and the like and sometimes verbal requests are considered.

26. In any case, initial dialogues with local officials (i.e. Barangay councilmen, Mayor, Municipal Council) and some members of the community are conducted. The purpose of this activity is to determine preliminary acceptance of the project by the community in general as well as to assess the physical water supply situation of the area.

27. At this point, the proposed service area is determined. A preliminary calculation of the extent of the service area is made. Likewise, the target year of implementation is set.

28. The existing population is determined next. This is projected for five to ten years corresponding to the life span of the system. Then service population of the system is taken. It should be said that the service population does not always equal the actual community population as there are cases where some households ceased to be served by public systems for various reasons like the development and use of their own private water systems.

29. Knowing the service population and the proposed level of service (Level I or Level II), together with the preliminary cost estimates, the likely water consumption or demand is then determined.

30. Analytical and qualitative approaches are then made to select the type of water source to be used for the proposed system. Afterwhich, the yield or capacity of the source is ascertained. If the water demand is less than the capacity of the source, water samples are obtained and subjected to various quality tests. If the water demand is greater than the capacity of source, however, the proposed service area or service population is reviewed and adjusted, or additional sources are located.

31. If the water sample subjected to chemical and physical tests is satisfactory, the water supply capacity is resolved. But if the water quality is found to be unsatisfactory, another suitable source is selected and if there is none, then treatment works are considered.

32. After determining the water capacity, the design of water facilities is made. Subsequently, an estimate of the project cost follows.

33. Once the technical inputs are available, the institutional phase of the development process begins. The proposed project is presented to the community and a series of dialogues are made between the government and the proposed Rural Waterworks and Sanitation Association (RWSA), the project beneficiaries.

34. When an agreement is reached the proposed RWSA submits registration documents and eventually, the RWSA is duly formed.

35. Upon approval of the RWSA, the construction of the water supply system starts. Meanwhile, the RWSA members are trained on how to operate and maintain such system. Following the completion of the construction of the project, this is turned over to the RWSA for operation and maintenance.

36. The project is then subjected to a review to ascertain if any deviations from the project plans have occured. If this is found positive, necessary corrective measures are instituted to achieve the desired ends of the plans.

R & D in Rural Water Supply and Sanitation

37. The application of research and development is a new approach to rural water supply and sanitation development, this activity commenced only in 1984. Through its application, it is expected that resources, that is manpower, materials, equipment, money and technology will be optimally used at the least possible cost, and with the desired quality and result. R & D ultimately makes water facilities cheaper to build, operate and maintain, thus making water more easily accessible and affordable to every Filipino in the countryside.

38. Research is, therefore, considered a very important component in project development and works toward the following objectives: (a) to improve the efficiency and effectiveness of water supply facility and applied construction technologies; (b) to evaluate the viability for

adoption of new materials, equipment and technologies; (c) to determine the most effective technique in using new materials and equipment to attain the least cost possible methods in the implementation of the rural water supply projects.

39. The study areas being undertaken are as follows: a) Laboratory and field testing of various types of handpumps and screens; b) Study on salt water intrusion and ferrousity of water; c) Studies on the use of ferro-cement reservoirs and the applicability of rainwater collectors for simple rural water supply systems; d) Study on the installation of PVC and fiberglass casings and screens; e) Study on the choice and utilization of appropriate drilling equipment and methods for wells at varying hydrogeologic conditions; f) Study on well development methods and equipment; and g) Periodic evaluation of design standards.

40. R & D is looked upon as a continuous undertaking and will in the future concentrate on completing the study areas mentioned and determine replicable projects. The benefits derived from R & D are in the improvement of local technology; development of standard specifications, products and processes; the development of more appropriate water facilities; the development of facilities that would be affordable and easily maintained; and lesser cost.

Emergence of Training as an Important Activity

41. The last decade saw training emerge as one of the more important activities engaged in by the public water sector, especially for the rural program of water supply and sanitation development with the formulation of the master plan in 1982.

42. The government's policy of self-reliance has made basic social services such as water supply a joint responsibility of both the government and the end users or beneficiaries of the service. As new caretakers of the water facilities, especially in the rural water supply program, the beneficiaries have to be trained in water system's operation (i.e., technical and accounting) and maintenance. Also, now that the consumers pay for water services, the public water utilities are constantly pressured to provide better, more reliable and adequate water service, together with the improved facilities. Hence, the need for training the personnel of such utilities in every aspect of water supply management and operation.

43. Manpower development in the sector are, therefore, geared to the following: a) Training of national government personnel that provide assistance; b) Training of the personnel of the local water utilities that receive such assistance; and c) Training of the end users or beneficiaries of the new water system facilities.

Use of Microcomputers in Rural Water Supply

44. Since 1984, the concerned agencies have been utilizing microcomputers to expedite and simplify their operations.

45. Such microcomputers are being used in the design of low-cost rural water supply systems, in inventory control, in budgetting and project accounting, in water quality analysis and in data management. Other applications are being developed.

Increased Financing Aids and Technical Assistance

46. Recent years saw favorable trend in the increase of financing aid and technical assistance to the country's rural water supply development program from the major financial institutions of the world.

47. To give one an idea consider **Table 1**.

ISSUES

48. In the course of implementing the rural water supply program, several issues have cropped up that can be broadly identified into technical, institutional and financial issues.

Technical

49. The current Rural Water Supply and Sanitation Master Plan defines three levels of water service to rural communities as seen in Table 2.

50. The results of surveys conducted show that the present service level in most rural communities is considerably higher than that which can be provided under the present level I criteria. For some level II systems, design criteria combined with higher service levels result in construction costs that can be afforded only by limited communities. Furthermore, the per capita water use rate being applied to this level II systems becomes too high for the relatively low income rural target population. The existing service level standards are therefore, only appropriate in areas where there are more than 50 households per service point within a 250 m radius, or in exceptionally affluent villages. There is a pressing

Table 1

Name of Project	Cost (In P1000)	Financed By
1. First Rural Water Supply & Sanitation Project	P484,700	IBRD
2. Rural Water Supply II Project	110,566	OECF (8th yen)
3. Rural Water Supply I Project	245,386	OECF (7th yen)
4. MPWH-NIA Muleta Nupali Irrigation Project, W.S. Component	11,210	ADB
5. Third Davao del Norte Irrigation Project, W.S. Component	24,317	ADB
6. ALLAH River Irrigation Project (MPWH-NIA) (Supplemental Agreement) Water Supply Component Rural Water Supply III Project	6,477.31	ADB
7. Rural Water Supply III Project	288,980	OECF (13th yen)
8. Proposed Asian Development Bank (ADB) Assisted Island Provinces Water Supply Sector Project	646,430.33	ADB

Table 2

PRESENT WATER SERVICE LEVEL CRITERIA

Parameter	Unit of Measure	Service Level I	II	III
Per capita supply	l/d	30	60-75	100
Household per service point	Number	15-50	4-6	—
Maximum carrying distance	Meter	250	250	—
Type of service	—	handpumps, standpipe developed communal spring faucets		house connections/ individual taps

need for intermediate service levels which should represent an affordable improvement over the existing set-up.

51. A second technical problem is the high cost of acquiring new well drilling equipment. Although the government is presently augmenting and upgrading its equipment, the private sector has lagged behind. Also, since the thrust of the rural water program is toward smaller communities and islands, there is the need for more suitable and easily transportable equipment for use in these far flung areas.

52. Another problem is that many of the wells constructed by either the local or national governments through force account or by contract were developed unappropriately that these become unoperational six months to a year after its construction. Solutions may lie in the improvement of the present methods of bailing and the use of air compressors; and in the acquisition of better equipment.

53. A fourth issue is the result of excessive denudation of forests. With the alarming rate that such is being done, a marked difference was observed in the lowering of the static water levels in several areas even when considering the normal water level fluctuation that goes with the seasons.

54. Lastly, there is the water quality issue. Through studies, the deterioration of water quality is not only due to environmental pollution or contamination, but also to groundwater salinity and ferrousity. Most wells located in coastal towns were found to yield saline water with chloride concentration. varying on the distance of the well from the seashore. Preventing such salt water intrusion is by avoiding over pumping of water by major users like industries, as well as in constructing wells reasonably away from the seashore or the contaminated well.

Institutional

55. The entire water supply program of the country is being handled by several agencies of government. The rural water development portion is also divided among these agencies, even if some of these agencies' main thrust is the urban areas. There is multiplicity then of government agencies involved in the rural water supply development program. This duplication of functions, overlaps in responsibilities are not only a waste of resources but also make physical and policy coordination difficult to the detriment of the program. The rural program should therefore, be solely given to a specific agency.

56. A second institutional issue is the fact that coordination in project planning and implementation are not quite observed in the provincial level down to the municipal level. This has slowed down some aspects in project planning, implementation and even operation and maintenance and in the formation of end-user associations.

57. This leads us to the third issue involving water user and maintenance arrangements. There are about 10,000 projects being implemented every year. There must be a corresponding number of RWSA that must be formed and organized who will own, operate and maintain the newly constructed water facility. The backlog in RWSA formation has led to improper operation and maintenance of several systems since many projects are being constructed without a duly organized RWSA.

58. In cases where there is an RWSA to operate and maintain the system, there are also other difficulties being faced like: collection of water fees, high cost of O & M especially for pump systems, management of RWSA and expertise in system operation.

59. As for data management, more realistic national, regional and provincial figures are needed on the following: existing number of water systems, by service level, and by agency; physical status of projects; water quality water supply "gap"; hydrogeologic conditions; and inventory of equipment. And because of so much data available there is a need also for centralized data management so such data may be systematically collated, stored and made available to possible users of the information.

60. In selecting a project for water supply development there is a criteria to follow. Despite the presence of such a criteria, these are disregarded when political considerations are stronger. On the other hand, selected project areas are abandoned in a few areas because of civil disturbances. In either case, project timetables are re-set and, thus, adds a snag on the total program.

Financial

61. Many of the water systems being constructed are beyond the rural folks to support. Although such systems belong only to the level II category, the cost of such system has to be recovered thru amortization in which case it is 90 percent of the construction cost because the rest is generally financed by the government. Add here O & M costs, cost of spare parts, materials and chemicals, electric bill and personnel expenses — and the water bill becomes unaffordable by rural standards. A reasonable package of schemes is being worked out by the government.

62. Next is project funding as a financial issue. Since the rural water program was neglected until recently, there is much catching up to do. Yet, whatever funds that are available are spread out, sometimes too thinly. In accelerating the program, more project funding is badly needed locally or from foreign sources. Although foreign sources may generally be sufficient, the local counterpart funds' release becomes dependent on the prevailing economic situation in the country.

63. A third financial issue is the disparity in financing arrangement. Governmental agencies involved in water supply development have different schemes in recovering their financial investments. For example, for water facilities development, the Local Water Utilities Administration charges borrower water districts a graduated interest rate of ten-twelve-fourteen percent; the Rural Waterworks Development Corporation charges much lower rates to borrower water associations; and local government builds these facilities almost with outright grants. This disparity in financing arrangement leads to a disparity in water rates charged the consumers. But the most apparent effect is in the consumers' attitude where most, if not all, would like to get away from the responsibility of paying for water as a commodity.

CONCLUSION

64. The present trends in rural water supply development in the country show how much has already been done and how much can still be done if the issues presented can be resolved.

65. Still, it is believed that rural water supply development should be given the premium attention it deserves for rural communities are the take-off points for economic development, the Philippines being an agri-based country. So much investment has to be poured into this sector, but with systematized and organized government support as well as support from the United Nations system and other cooperating agencies, the future looks bright for the rural water supply sector in the Philippines.

A. L. DOWNING, DSc, MA, BSc, FIChemE,
G. R. GROVES, BSc, MSc, PhD, Binnie &
Partners, and C. J. APPLEYARD, BSc, formerly
Binnie & Partners, now Bostock Hill &
Rigby, UK

6 Treatment and disposal of industrial waste waters

Studies of appropriate means for treatment and disposal of waste waters from a variety of industries and the means recommended or adopted are briefly described. Local factors had a considerable influence on the chosen methods. These ranged from treatment of mixtures of industrial waste waters, after little pretreatment, in combination with domestic sewage using long established conventional methods to segregation of individual waste streams for treatment at source by advanced methods including some of novel design.

INTRODUCTION

1. The effective design of facilities for control of pollution from industrial waste waters usually demands selection of the most advantageous option from a range of possibilities, the extent of which depends considerably on local circumstances.

2. Often there are at least two possible receptors for the waste water for which the quality standards to be met differ and usually consideration must properly be given to the scope for reducing costs by a variety of measures ranging from at one extreme segregation of more concentrated waste streams for individual treatment at source, especially for recovery of valuable by-products, to at the other advanced "end-of-pipe" treatment designed recover a large proportion of the water flow for re-use. The adoption of 'in-plant' methods to achieve conservation of water, minimisation of waste and conversion of useless waste to by-product of value often can be justified on economic grounds alone. Thus the identification of appropriate opportunities is a necessary first step in the development of an optimum wastewater strategy for an individual industry or group of industries. Once this step is complete other important considerations can then be reviewed to determine the most effective design of facilities for overall control of pollution.

3. This paper briefly reviews some studies of diverse situations, which induced widely different approaches and solutions to the design problem.

COMMUNAL TREATMENT OF MIXED INDUSTRIAL AND DOMESTIC WASTEWATERS

1. This case concerns a large industrialised city in eastern Asia in which the flow of industrial wastewater greatly exceeds that of domestic sewage. The pollution control authority was fully conscious of the advantages often to be gained by treating strong wastes at source but because not much pretreatment had been previously demanded of industry and there was a strong desire to reduce quickly the severe pollution of a local river to which waste waters were being discharged, primary interest centred on the possibility of intercepting all wastewaters and treating them in a large communal "end-of-pipe" facility.

2. The major industries included pulp mills for the production of paper and artificial cellulose fibre from wood and other raw materials and owing to the discharge of pulp liquors the mixed waste waters were black in colour and foamed readily on agitation. The pollution control authority required an assessment of the facilities required to reduce BOD and COD by about 80% and particularly to remove colour and foaming potential. Because there seemed a reasonable chance that biological treatment would reduce colour and foaming potential as well removing BOD and COD and that such further treatment as might be necessary to complete reduction of these last two characteristics to satisfactory levels would benefit from the general reduction in organic content, attention was centred initially on assessment of the amenability of the waste waters to treatment by the activated-sludge process.

3. A problem was that because of the remoteness of the site it was not feasible to set up local test facilities. Instead representative samples were taken of the individual waste waters and these were flown as soon as possible to our London laboratories and mixed in the proportions in which they were expected to arise in practice. Because of the necessity to minimise expense the volume of mixture available for tests was limited to just under 20 l. To work with such a small volume demanded miniaturisation of the experimental reactor. The type chosen initially was one first used about 20 years previously in studies by one of the authors (ref. 1) in which the aeration unit (1-l capacity) was composed of porous polythene. The porous polythene permitted the treated effluent to flow through the aeration unit into a collecting vessel but

World Water '86. Thomas Telford Limited, London, 1986

retained the sludge particles, thus dispensing with need for a secondary clarifier and sludge return system. However, maintenance of the low flows required at the chosen rate proved troublesome and eventually the continuous flow plants were replaced by even smaller fill-and-draw units. Though not enough was done to determine optimum design features the studies indicated that the mixed waste waters could be treated to reduce BOD to not more than 60 mg/l, the standard contemplated, at sludge loadings up to about 0.2 g BOD per g sludge per day. Foaming potential was judged principally by a simple test involving measurement of the height of foam after aeration under standard conditions. Various instrumental methods involving measurement of light absorption at one or more wavelength were used to examine colour but eventually were abandoned in favour of a simpler visual inspection and comparison with a well-purified effluent from biological treatment of a predominantly domestic sewage. As assessed by these tests foaming potential was reduced to some extent by activated-sludge treatment. Colour tended if anything to be increased.

4. Various ways of removing colour both from the raw waste water and treated effluent were examined including oxidation with the individual reagents ozone, hydrogen peroxide and sodium hypochlorite; adsorption on peat, granular and powdered activated-carbon, magnetite and activated alumina; and chemical coagulation using iron or aluminium salts aided in some cases by addition of polyelectrolytes.

5. Of these methods chemical coagulation was by far the most effective both with raw and biologically treated waste water. Such treatment was also effective in reducing foaming potential and as would be expected removed the majority of suspended matter and with it much of the BOD and COD of the waste waters. Dosages required to reduce colour close to that of a well purified domestic sewage effluent were, however, quite high at around 30 mg/l as Al or Fe and even with the biologically treated effluent the volumes of sludge produced were large. Some typical results in which colour in this instance was measured in degrees Hazen are given in Table 1.

6. It was concluded that if clarification of black liquors at source continued to be regarded as impracticable, then the best method of treating the combined waste waters would be by treating them first by the activated sludge process, at loadings which might well be up to 0.5 g/g day, and then by coagulation of the effluent with 30 mg/l alum (as Al). However, because of the high dosages of coagulant required and the large volumes of sludge produced a study should be made of means of eliminating black liquors at source since there were good prospects of identifying methods which would be highly cost effective.

Table 1. Effect of treatment by the activated sludge process (A) followed by chemical coagulation and sedimentation, (B) on a mixed industrial and domestic waste water

Determinand*	Before	After A	After A & B
pH value	6.2	6.8	5.25
Suspended solids	164	119	
Colour (°Hazen)	300	600	65
BOD	340	28	
COD	1210	600	120
Sulphide as S	0.2	LT0.1	
Sulphate as SO_4	1375	1375	
Phenols	0.76	0.15	
Lead as Pb	0.15	0.32	LT0.02
Zinc as Zn	13.1	6.4	3.9
Chromium	0.04	0.03	0.05

* mg/l unless otherwise stated
LT: less than

TREATMENT OF WASTE WATERS FROM OIL REFINERIES PETROCHEMICAL COMPLEXES AND ASSOCIATED INDUSTRIES

1. This case concerns the building of a large industrial complex in the Middle East and the design of facilities for the treatment of wastewaters that would arise from the complex. The problem was unusual in that the characteristics of the wastewater had to be projected from first principles taking into account the major industrial processes planned for the new city (the primary industries), the most likely secondary industries which would "spawn" from these primary industries and the large number of support industries that would be needed prior to, during and following the building and operation of the complex. The primary industries planned for the new complex were mainly concerned with oil refining and petrochemicals and fertiliser manufacture although others were planned to make use of natural gas including an iron and steel manufacturing plant (with rolling mill) and an aluminium smelter. In total 13 primary industries were projected together with over 30 secondary industries and up to 100 support industries. Although the complex was to be located adjacent to the sea (thus permitting use of sea water for cooling purposes) all process wastewaters had to be treated at a central facility to a standard that would allow their unrestricted re-use for landscape irrigation. Major issues considered and resolved in the development of an optimum strategy included:
- most probable average and maximum wastewater flow rates arising principally from primary and secondary industries;
- most probable range of contaminants present in the individual waste water streams and most probable average and maximum concentrations of each significant contaminant in the mixed waste waters;

- the extent of treatment required to comply with the unrestricted re-use requirement;
- whether industrial wastewaters should be treated in admixture with sanitary waste waters arising from the new township (population greater than 250,000) to be built at the same time as the industrial complex;
- the extent of pretreatment to be carried out at individual industrial premises.

2. Waste water flows and contamination levels were projected on the basis of anticipated pollution loads per unit of production having regard to modern processing technologies (high standards of process design including in-plant control procedures were assumed to minimise, as far as practicable, water usage and material wastage). This procedure involved over 50 individual pollution load calculations. Pretreatment requirements were established therefore on this premise and also with due regard to protecting the materials of construction of the collection system and the proposed centralised treatment facility.

3. It was concluded that the total quantity of industrial waste water would amount to about 60,000 m^3/d and contain, amongst other components, up to 100 mg/l of oil contamination and 250 mg/l of biochemical oxygen demand. In view of the oily nature of the waste water, a treatment strategy was chosen based on initial physical removal and recovery of oil compounds using gravity separation and sand filtration, followed by biological treatment. Also it was decided to segregate all sanitary waste waters for separate treatment to ensure optimum conditions were maintained at all times for oil removal prior to biological treatment in a combined biofiltration - activated-sludge system. The treatment plant to be provided would also include tertiary filtration in dual media filters and ozone disinfection to ensure a final effluent quality suitable for re-use for landscape irrigation. Ozone was also to be used for COD control.

4. A major complication in the development of an optimum strategy was a requirement to provide waste water collection and treatment facilities within a 2 year design and construct time span. This necessitated an initial design of temporary facilities to accommodate approximately 30 percent of the ultimate design flow. To provide this capability a scheme was conceived incorporating the emergency storage component of the final scheme (as aerated lagoons for biochemical treatment) and skid mounted package dual media filters which would subsequently be incorporated into the final scheme.

DISPOSAL OF WASTE WATERS FROM AN INDUSTRIAL AREA

1. In this case various strategies were developed for the disposal of wastewaters from an industrial complex in Australasia. The industries included oil refining, metal refining, chemicals and fertilisers manufacture, iron and steel production, abattoirs, hides and skins processing plants and various other trades associated with the meats, hides and skins industry. The principal options included:

(a) Collection of all waste waters and discharge via pipeline to ocean waters in excess of 5 km offshore.
(b) Individual pre-treatment followed by collection for combined discharge to ocean waters about 1 km offshore.
(c) As (a) above but in admixture with primary settled sewage and discharging about 4.5 km offshore.
(d) Individual pre-treatment to a higher standard followed by discharge to sea via short individual pipelines (about 200 m or less).

2. For industries in the northern part of the area the optimum strategy, on a technical and economic basis, involved protein recovery at individual abattoirs, physical and chemical treatment only at other establishments followed by combined collection and discharge to sea via a short outfall.

3. For industries in the southern part of the area, which included the main primary industries, the optimum strategy involved conjunctive use of a proposed sewage effluent outfall. A scenario based on individual pre-treatment and continued discharge to sea via individual pipelines was chosen however for practical reasons.

TREATMENT OF OIL REFINERY EFFLUENT BY DAF, MICROFILTRATION AND REVERSE OSMOSIS

1. At another oil refinery waste waters were treated in a 50 m^3/d pilot unit consisting of dissolved air flotation (DAF) with lime dosing, tubular microfiltration and spiral reverse osmosis membranes. The pilot unit was operated on-site for three months to provide operating and design data for the full-scale plant. The effluent contained typically 30 mg/l oil, 100 mg/l suspended solids (SS), 100-150 mg/l total organic carbon (TOC), and about 2 g/l TDS. Flocculation was carried out using lime to pH 10 with the addition of 5 mg/l of polyelectrolyte. The DAF unit operated with 25% recycle of the underflow and at a saturator pressure of 5.5 bar. The microfiltration unit was a multitubular array of 12.7 m^2 filtration area (see last section). The reverse osmosis unit consisted of a 2 : 1 taper with four 4-inch diameter spiral elements (FilmTec BW 4040) in each pressure vessel.

2. A summary of the plant performance is given in Table 2. The pretreatment stages of DAF and microfiltration removed the oil and suspended solids content of the effluent and gave a good quality feed to the reverse osmosis section. The reverse osmosis feed had a silt density index (SDI) consistently below 5. The microfiltration fluxes were above 300 $l/m^2 h$ for 99% water recovery at an inlet pressure of 1.2 to 2 bar.

3. The reverse osmosis plant operated at 65-85% water recovery and produced high quality product water. Specific membrane fluxes (25°C) were consistently in the range 0.7-0.9 $l/m^2 h$

Table 2. Treatment of oil refinery effluent

Determinand*	Effluent	DAF product	Micro-filter product	Reverse Osmosis product
pH value	7.8	10.0	6.5	6.5
Conductivity mS/cm	2.72	2.80	3.00	0.16
TDS	2.10	2.00	2.66	0.12
Na	650	650	650	15
Cl	1065	1065	1560	28
TOC	116	80	80	15
Oil	35	ND	–	–

* mg/l unless otherwise specified

bar. The membranes were cleaned at 3-4 weekly intervals to remove organic foulants.

TREATMENT OF METAL REFINERY WASTE WATERS

1. The effluent from a nickel refinery had been lagooned for a number of years but leakages caused contamination of the surrounding area. A reverse osmosis treatment plant was installed but fouling of the membranes mainly due to organic materials (humic acid and bacteria) caused problems.

2. Pretreatment of the effluent using flocculation with 20 mg/l Fe^{3+} as ferric sulphate in the pH range 4.5 to 6.5 gave good organics removal. However, pilot flocculation and settling trials gave poor results with floc carryover at economic settling rates.

3. Pilot trials were then made using cross-flow microfiltration (see last section). The feed was coagulated with ferric sulphate before filtration (Fig. 1). The pilot unit was operated in the feed and bleed mode at selected water recoveries. Typical operation was carried out at inlet pressures of 2-3 bar and a minimum tube velocity of 1 m/s.

4. A summary of the results of the pilot operation are given in Table 3. The colour absorbance (AO) was used as a measure of the organic content of the effluent.

5. The microfiltrate quality was excellent at 15-25 (x 10^{-2}) absorbance units, less than 0.05 mg/l iron and negligible suspended solids.

Table 3. Microfiltration performance on nickel refinery effluent

Water recovery %	Micro-filter flux $1/m^2 h$	Feed Fe mg/l	Effluent colour 10^{-2} AU	Product Fe mg/l	colour 10^{-2} AU
25-35	100-120	30	160-185	0.02	10-25
50	90	45	135-140	0.02	15-25
90	70-80	140-175	–	0.02	15-25
95	40-50	360-510	160-170	0.02	10-30
99	40	1500	170	0.03	20

The quality was consistent over the water recovery range to 99%. Trials carried out using the microfiltrate as feed to test reverse osmosis membranes indicated very low fouling rates over two months of operation.

6. High conductivity wastewaters at another refinery were treated to recover water for reuse and produce a concentrate for recycling back to process. The wastewaters were from various sources and consisted mainly of ammonium sulphate solutions of concentration range 1-8 g/l. High rejections of ammonium were obtained by operating the reverse osmosis membranes at a pH of below 7.

7. A pilot unit was operated on-site for four months to collect desing data for the full-scale plant. The pilot-plant layout is given in Figure 2. The reverse osmosis plant consisted of twelve 4-inch diameter spiral reverse osmosis elements (FilmTec type BW 4040) arranged as a 2 : 1 taper with four elements per pressure vessel.

8. The microfilter pretreatment stage gave excellent performance producing a consistently high quality water for spiral reverse osmosis membranes at SDI values of below 4 from wastewaters of SDI above 6. Microfiltration fluxes were above 400 $1/m^2 h$ at 98% water recovery and an inlet pressure of 1.2 bar.

9. The reverse osmosis unit operated consistently and without evidence of fouling. Production rates were very good and the unit coped well with the wide range of TDS concentrations in the feed. Specific membrane

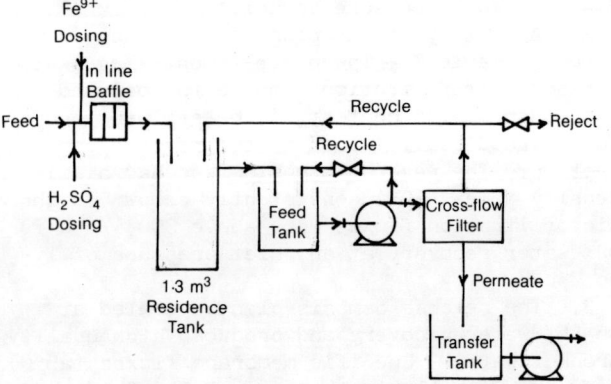

Fig. 1. Pilot scheme for nickel refinery wastes

Table 4. Reverse osmosis of metal refinery effluent

Determinand		Feed	Concentrate	Product
pH value		6.7	6.8	6.4
Conductivity	mS/cm	3.63	12.9	0.04
TDS	g/l	2.26	9.6	0.035
NH_3	g/l	0.57	2.36	0.006
SO_4	g/l	1.95	8.10	0.022

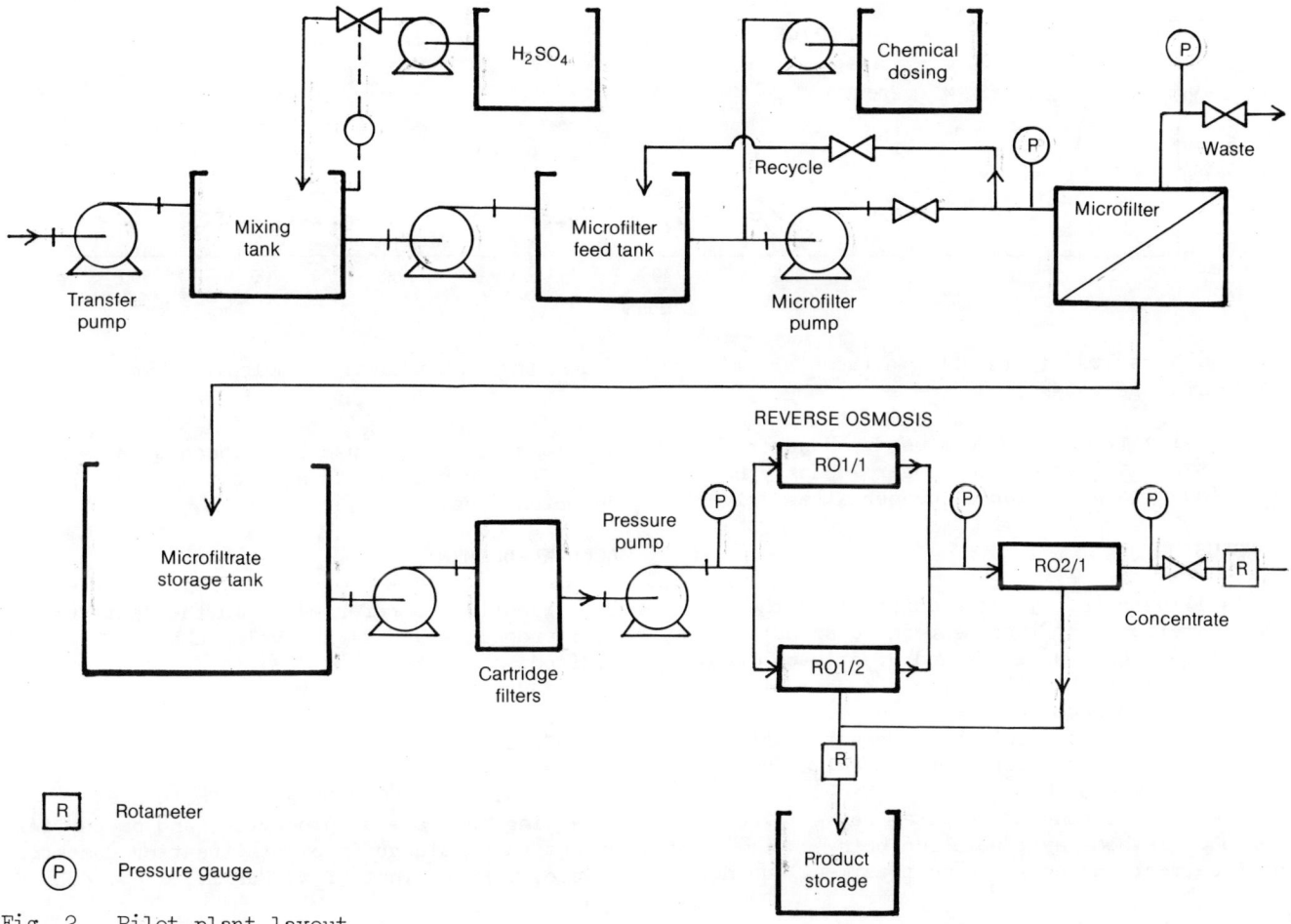

Fig. 2. Pilot plant layout

fluxes (25°C) were 0.8 to 0.9 l/m² h bar. A typical performance result (daily composite sample) is given in Table 4.

BIOLOGICAL TREATMENT OF AGRICULTURAL DRAINAGE WATER

1. In this case the waste water was of a rather unusual kind, being the drainage water from a fruit growing area, which contained a variety of pollutants, of which the most damaging in the local context was selenium, mainly in the form of the hexavalent ion selenate. Deformities in the young of birds inhabiting a reservoir into which the drainage water flowed had been attributed to the presence of this constituent. Other constituents included high concentrations of sulphate, nitrate, and borate and trace concentrations of various heavy metals. Though it seemed conceivable that a study of fruit farming practice might reveal ways in which the leaching of selenium, a natural constituent of the local soil, might be reduced there appeared to be no prospects of success within the short period within which action had to be taken.

2. Accordingly a laboratory investigation was undertaken into reducing selenate to the permitted limit of 10 µg/l. A variety of physico-chemical methods was tried without success including chemical coagulation, adsorption on activated carbon, alumina and iron filings, coprecipitation with barium sulphate and ion exchange. The first substantial progress was made using biochemical methods. It was found that in the absence of dissolved oxygen selenate could be converted by facultative micro organisms into organo selenium complexes and by bringing these into contact with microbial populations maintained in a suitable reactor the selenium could then be assimilated into the biomass. Any biomass particles containing selenium escaping from the reactor could be removed by filtration and for this purpose a special form of cross-flow microfilter could be used. The microfilter was of a novel design in which the internal surface of a porous tubular support is coated with a layer of suitable filtering medium. A variety of substances can be used to form the medium including traditional water treatment coagulants (eg iron or aluminium salts). Liquid to be filtered is passed across the surface of the medium under pressure and at high velocity, the shearing effect of which reduces fouling.

3. The performance of a two stage laboratory biological reactor based on this approach is shown in Figure 3. The final dissolved selenium content of the effluent from the second stage was not as low as desired but it was then shown that by addition of a third stage, the selenium content could be reduced to or only just above the target limit of 10 µg/l.

4. Subsequently the viability of this method was confirmed in a larger (100 m³/day) experimental plant on site. The reactors had to be "fuelled" with degradable organic matter

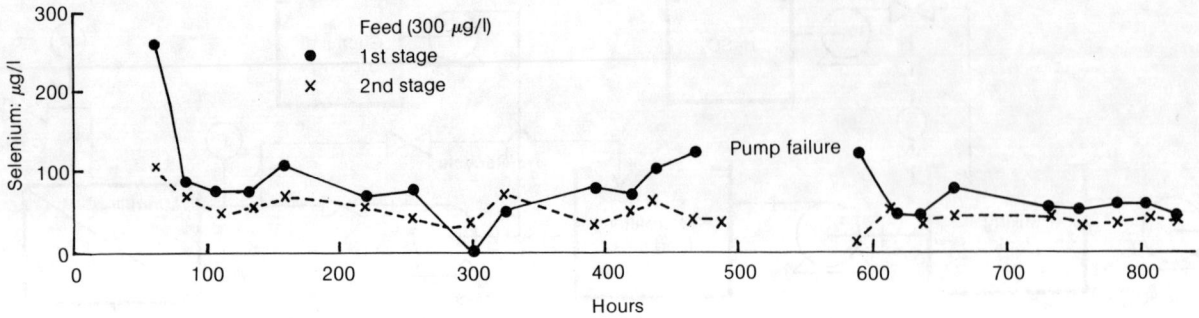

Fig. 3. Results of operating a two stage biochemical reactor for removal of Selenium from farm-drainage water

and originally methanol was used. However, waste streams from a local sugar industry have been found to be suitable cheaper alternatives.

CONCLUSION

As is illustrated by the studies cited the chosen methods for treatment and disposal of industrial waste water depend greatly on local circumstances. Usually there are economic advantages in reducing the extent of "end-of-pipe" treatment, by segregation of more concentrated wastes for pretreatment at source, especially where this can be combined with recovery of biproducts or water reuse. The range of methods available for both pretreatment and end-of-pipe treatment and new developments in the use of membranes and microfilters are providing exciting opportunities.

ACKNOWLEDGEMENT

The microfilters referred to in the text were of a proprietary design developed by EPOC Limited.

REFERENCE

HOPWOOD A.P. and DOWNING A.L. Factors affecting the rate of production and properties of activated-sludge in plants treating domestic sewage, J. Proc. Inst. Sew. Purif., 1965, 435.

Discussion on Papers 5 and 6

PROFESSOR J. PICKFORD, Loughborough University of Technology, UK
During a visit to the Philippines several years ago I was very impressed by LWUA's training programmes, especially those concerned with members of local committees (committees of lay people). Can Dr Padernal say something about these programmes?

MR E. W. WARNER, Water and Sewerage Authority, Trinidad and Tobago
Dr Padernal indicated that they were constructing about 12 000 rural water supply systems per year. In Trinidad and Tobago we have been constructing such systems although not on that scale. However, on account of our efforts to increase agricultural activity the problem of pesticides entering the water has arisen. Consequently we sometimes have to install pressure filters with activated carbon to adsorb these pesticides, thus increasing the capital cost of the final system. We are therefore pushed into constructing installations involving much higher capital expenditure so that the initial intention is almost defeated, that is, a low-cost rural water supply system. What is the experience in the Philippines with pesticide contamination of small rural water supplies?

How is surveillance of the water quality of the numerous water supplies taken care of?

MR A. J. WOODFORD, Sir Alexander Gibb & Partners, UK
In his Paper Dr Padernal deals with trends and issues in rural water supply, but he does not indicate how the downstream sanitation problems are dealt with. What is the approach to such problems arising from the 10 000 water schemes that are completed annually?

MR N. V. P. SARATHI, Central Water Authority, Mauritius
I fully share the views of the Author on the expectations of some funding agencies relating to the rate of return on the investment and cost recovery.

Is the rural consumer made to pay for the true cost of providing the water?

DR L. R. J. VAN VUUREN, National Institute for Water Research, South Africa
I have studied Dr Downing's Paper with special interest because of my own research experience in the field of water reuse. Since the beginning of the water decade, however, I have directed my research into rural areas.

In his paper Dr Downing refers to several high technology processes which confirm the greater complexity of industrial effluents in comparison with domestic effluents. The philosophy of treating industrial effluents at source is generally the best approach and this treatment should exploit or utilize biological processes to the full before embarking on membrane processes. Membrane processes have the great advantage of product recovery but pollutants, dissolved salts and toxicants are concentrated which pose secondary problems.

South African experience has shown that the integration of physical, chemical and biological methods has distinct advantages for treating industrial sewage. Biogradability is enhanced by chemical pretreatment.

Finally I believe that dissolved air flotation technology has great merit and should find wider use for treatment of industrial effluents.

DR A. L. DOWNING
I agree that the most cost-effective method of treating many industrial waste waters does not involve the use of membrane processes.

Usually the strongest cases for their use are in situations where they can be applied to relatively concentrated waste streams to recover a useful by-product, or to the final treatment of already well-purified waste waters to recover high quality water. In either case it is vital to consider carefully the various options for integrating them into the overall scheme of treatment, to arrive at an optimum process design.

Such consideration must certainly take account of the disposal of those concentrates that cannot be reused. Such disposal can be a problem although there are usually several possibilities. For mainly inorganic concentrates they can include dumping on properly managed authorized industrial tips, dumping in deep ocean waters, storage and trickling into sewers offering high dilution at a rate which does not cause infringement of effluent standards and the use of so-called chem-fix processes, in which they are reacted

PAPERS 5 AND 6: DISCUSSION

with reagents to produce granular solids (from which impurities are not readily leached) which are then applied to land. For non-recoverable organic concentrates incineration is a further possibility.

I also agree that dissolved air flotation is a very useful process for treating waste waters containing suspended impurities. In my experience its use is already growing and it will continue to expand for some time to come.

DR S. GASHI, University of Kosova, Yugoslavia
We have used precipitation and adsorption (coal fly ash, clays etc.) and both methods give good results in the recovery of heavy metals (lead, cadmium, zinc and copper) from acidic waste waters.

What do you do with the raw materials after treating such waters?

What do you think about the application of reverse osmosis on an industrial scale for the problems with which you were dealing?

DR A. L. DOWNING
There is a problem when adsorbents of the type mentioned that cannot be readily regenerated are being used. There may be no alternative but to use them as landfill. The associated costs can often outweigh the initial advantage of using relatively cheap materials. Also some such adsorbents are wastes from other industries and although they may remove certain pollutants they may also release others, e.g. chloride in the case of colliery spoil and arsenic in the case of some fly ashes. If it is possible to regenerate, as in the case of ion exchange, then there may be the same sort of problem as that of disposal of concentrates from membrane processes but sometimes valuable materials can be recovered from concentrates, the revenue from which may offset the costs of the disposal of the final residues.

It would rarely be economical to apply reverse osmosis to large flows of dilute metal-bearing wastes, but the process can be of considerable value when used to treat smaller flows of more concentrated effluents.

For the more dilute wastes, often rinse waters, chemical precipitation with lime after reduction of chromates (if present) followed by sedimentation and if appropriate sand filtration is often an economic approach. Electrolytic techniques are also popular; a new type of cell using a fluidized bed of glass beads between the electrodes to reduce concentration gradients is particularly effective.

DR H.-J. J. KOELLER, PREUSSAG AG Kohle, West Germany
Dr Downing, you have discussed tertiary sand filters incorporated into the line of waste water plants. Afterwards you called them dual-media filters.

Are all tertiary filters dual-media filters? What are the upper materials of such filters? What is the benefit of this arrangement and is it economical?

DR A. L. DOWNING
The dual-media filters were filled with anthracite and sand. They are in common use for water treatment but are less commonly applied to effluents.

Effluent treatment is often effected by the use of single-medium (sand) filters, frequently in an upflow mode when high throughputs are required and it is necessary merely to 'polish' the effluent to reduce suspended solids to meet limits of typically 10 mg/l.

In dual-media filters the anthracite is coarser and lighter than the sand and thus lies at the top of the bed. This allows suspended impurities to penetrate deeper into the bed and restricts clogging of the upper layers, thus making better use of the available filtering capacity.

This type of filter was chosen for the plant described because the effluent was to be reused for irrigating grassed areas of the city and for this a high quality was required.

DR I. VICKRIDGE, University of Manchester Institute of Science and Technology, UK
Does Dr Downing have any information on the use of reverse osmosis (or any other treatment process) for the treatment of radioactive wastes such as those currently being discharged from installations such as Sellafield nuclear reprocessing plant?

DR A. L. DOWNING
I have not been involved in the latest developments in the treatment of radioactive wastes, although I worked in the radiochemical field intermittently for several years. I do not know whether reverse osmosis is being much used in the UK for such duty but I doubt it. In principle the process would be expected to be as effective in removing radioisotopes as the corresponding non-radioactive species in the same chemical form and generally to effect removal of a wide range of cations and anions. If the process is not being much used then it is probably for good reasons such as that the radioisotopes of concern can be removed more economically in other ways from the particular types of waste water arising.

DR H. ABOUZAID, Office National de l'Eau Potable, Morocco
How is the rural water demand determined? Are steps taken to meet the final demand? If so, what are the stages?

MR M. WOOD, W. L. Wardrop and Associates, West Africa
What is Dr Padernal's experience of village level operation and maintenance of rural hand pumps as advocated by the World Bank? In the rural water supply programme that I am working on in Ghana, the hand pumps which have been selected are not ideally suited to this type of maintenance. What is the experience of this matter in the Philippines?

MR C. B. BUCKLEY, Welsh Water, UK
Concerning Dr Padernal's Paper, what analytical facilities are available in his country to determine the levels of pesticides and 'exotic' organic compounds? Is it the ability to analyse which is the driving force, or is it an overview of catchment control or management, e.g. the seasonal use of pesticides, which is directing the thrust of the analytical reserves?

Has Dr Downing also considered the general usefulness of the concept of catchment management as is now generally practised in the UK by regional water authorities in their role as integrated river basin managers? Sophisticated analytical facilities must be considerable to evaluate levels of complex organic compounds, but it is equally important to understand the risk assessment of storing and using such compounds and the likely effect of these compounds on the raw water resources of the district concerned.

MR J. MBUGUA, Church of the Province of Kenya, Kenya
Rivers may be 20 km or more away. Boreholes are poorly maintained and serve an 8-10 km radius, and if working fluoride and other undesirable species are high.

The development section of the Church is encouraging every home to dig a hemispherical pit, and to line it with black 1000 gauge ordinary paper. The paper is sandwiched between two mud layers. Locally available stones (saucer-shaped) are placed on the top layer and joints are finished with a weak mix lime cement sand (Fig. 1).

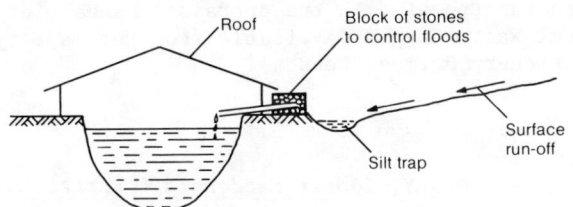

Fig. 1

MR B. LUNOE, Hifab International AS, Norway
Concerning the involvement of women in rural water supply and sanitation programmes

(a) village water committees comprise 50% women
(b) women are involved in the planning and implementation of the programmes through sociologists and anthropologists
(c) women are trained as hand pump caretakers
(d) women are trained as village health workers.

MR J. A. SAUDER, ACROSS, Kenya
Water committees made up of 50% women can still be dominated by men. In southern Sudan we have encouraged separate women's committees to obtain better feedback.

MR S. N. MAKHETHA, Urban Development Project, South Africa
Women are involved in water supply and all community development projects in Lesotho. The system works and is very successful. It can be used elsewhere and is very effective.

The Urban Sanitation Team also has women technologists, sociologists etc. and this also works well.

It is the women who are at home in the villages and they are the people who draw the water. They must be involved.

MS A. BINGHAM, Freelance Journalist, UK
Western specialists and advisors are often critical of the male-dominated attitudes of developing countries but from observation this attitude may often have been imported together with western advisors and consultants. Perhaps the consultants should set a better example by involving more women in their teams and by making more effort to make sure that their designs and solutions are truly helpful and appropriate to women's needs.

MR H. C. BALFOUR, Balfours, UK
Women do form a useful part of any study team but I would like to see an opportunity for those involved at the study stage to be able to assess the results and effects of the running of the realized project.

MR L. O. WILD, Laurie Wild Consultants, UK
There has been an interesting comparison in the technical notes brought about by Mr Balfour. Most speakers have concentrated on the minimum that can be provided whereas Mr Balfour has described a simple mechanical approach with piped water supplies to a number of villages - a system capable of expansion.

I have prepared a technical note which contends that providing local hand pumps can only marginally increase the lot of our less-advantaged fellow beings. What is needed is advanced technology designed to meet the needs which include piped supplies, sewage reservoirs, water for irrigation etc. If not it will need another influx of food to drought-affected countries soon. It is all very well to provide a few gallons of water but locally grown food is equally important as water as the key requirement.

The water decade is doing well as far as it goes, but the aid authorities must provide more money for modern schemes and improvement of training otherwise at the end of the decade most of the world will still be 'less advantaged'.

PROFESSOR M. J. HAMLIN, Birmingham University, UK
Dr Williams considers that an improvement of one or more of the following criteria represents an overall improvement in a rural water supply scheme: convenient access, reliable source, adequate quantity or safe quality.

About 12 years ago I examined a doctoral thesis at the University of Manchester Institute of Science and Technology where the candidate studied the needs of a village in Swaziland. He estimated that at the start of the new growing season the daily calorie intake of an adult in

PAPERS 5 AND 6: DISCUSSION

the population was 1400. Of this 555 calories were available for work and most of this was used by adult women carrying water for 2 miles. It was possible to pipe this water, untreated, into the village by gravity. Because the quality was poor the scheme was rejected by the donor agency although the possibility of an improved standard of living if the women weeded the crop instead of carrying water was acknowledged. This scheme made access more convenient and therefore met at least one of Dr Williams' criteria. Would he have recommended acceptance of the scheme?

DR P. G. WILLIAMS, National Institute for Water Research, South Africa
The implementation of rural water supply schemes is the main stumbling block, rather than the technology for rural successful projects.

Advantage should be taken of the successful experiments elsewhere, such as a drought relief scheme in part of South Africa, where the people are paid a low wage to provide labour for a rural water supply project, and then a charge is eventually made for the water. This is preferred by the people to providing full labour and receiving a free water supply.

MR C. RADHAKRISHNAN, Ministry of Water, Lands, Housing and Urban Development, Tanzania
With regard to community participation, the water supply has to be considered not only for social services but also for the indirect production sector for agriculture and also for industries. Regarding self-help the villagers take a more active part in grinding mills, tractors, lorries and buses because they buy and maintain them, whereas water is provided free by the government and donor agencies and the response to self-help from the villagers is therefore poor. It is advisable that villages should pay at least 1% or a small sum for water schemes for their effective operation.
Similarly villagers also can be paid as an incentive for self-help.

MR E. G. THOMAS, Sir William Halcrow & Partners, UK
Technical Note 15 covers just a few of the techniques which have been used by Oxfam and Register of Engineers for Disaster Relief engineers in the refugee camps of Somalia, Sudan and Ethiopia to provide basic reliable drinking water supplies for hundreds of thousands of refugees in communities ranging up to 50 000 people in size. The capital costs were £2 per refugee with an annual equipment replacement and maintenance cost of £0.5 per refugee.

Although these projects were low cost the technologies ranged from high technology solar pumping and saturated sand storage dams, built to improve the storage capacity and yield of the aquifer artificially, to lower technologies.

Low cost does not necessarily equate with low technology or no technology. Often low cost means a high standard of engineering design, in some instances of a higher standard than for conventional engineering.

These recent developments in refugee water supplies could be applied equally well in the whole rural sector if the responsible authorities including the engineering profession itself took them more seriously.

The aid agencies should provide more funds for this type of work.

The voluntary agencies, who fund much of the present work, should give more recognition to good engineering and specialist expertise when they receive it, and not accept lesser technical standards.

The profession should recognize that they have a major technical role to play in this low cost field which requires ingenuity and the sound application of both high as well as lower technologies as appropriate.

Finally I should like to endorse one point which Dr Wright of the World Bank has made. We should properly appraise the new technologies so that the better ones can be applied on a wider scale, to meet the water and sanitation needs of the rural areas, both in the second part of this water decade and beyond.

MR R. CADWALLADER, Overseas Development Administration, UK
The Overseas Development Administration has an evaluation department which commissions evaluations of past projects often from outsiders. The summaries of these evaluations (evsums) are freely available to outsiders and some of the detailed evaluations. Those on rural water supplies have been brought together in a grand summary. The lessons learned have been incorporated in the Appraisal Manual for Rural Water Supply (available from Her Majesty's Stationery Office, London).

MR L. M. SOLWAY, Cooper Macdonald & Partners, UK
Dr Padernal has clearly described the provision of rural water supplies in the Philippines and has highlighted the multitude of internal administrative agencies and international lending agencies. Could he say something about the benefits of providing the new facilities?

In paragraph 63 of the Paper he refers to the disparity in financing arrangements which leads to disparity in water rates charged but is it possible for the government to even out these disparities so that one village does not pay a different rate from another? Further on in paragraph 63 he writes that the effect of this disparity was that 'most if not all would like to get away from the responsibility of paying for water as a commodity'. This is a very depressing statement as it implies that the facility is looked on as an unnecessary burden and not as a benefit. I hope that this is not true and that there are social, medical, agricultural and industrial benefits to the provision of good water supplies in the Philippines and the villagers do appreciate the water but are like many other people in the world in being slow to pay their bills.

A. K. DJELI, Institut za naucna istrazivanja Inkos, Pristina, Yugoslavia, and A. H. IBAR and M. M. BERISHA, Prirodno-matematkicki fakultet, Pristina, Yugoslavia

TN11. The influence of surface exploitation of lignite on water pollution

The Kosovo coal basin is situated between the valleys of the rivers Sitnica and Drenica and contains an estimated 10 billion tons of lignite which, on the whole, can be surface exploited. The development programme foresees a large increase in production. From this, it can be assessed that, apart from other problems, an unavoidable problem is the purity and protection of water and river streams in the area of surface mining of lignite. One reason for this is that the area of exploitation is densely populated and water of unknown quality from the collectors of surface mining flows into the river streams, which is then used for various purposes (e.g. bathing, drinking water for livestock and irrigation) and also threatens to pollute underground waters.

AREA EXAMINED
2. Most results were obtained in intervals when there was high rainfall and drought in the region. During the winter, when there was little rainfall and frost, it was impossible to take representative samples.

SAMPLING
3. Sampling was carried out using 10 litre polyethylene bottles which were washed with HCl and rewashed with distilled and sampled water. The collected samples were treated at the same time in three ways: chemically, biologically and bacteriologically.

ANALYSIS
4. The metals were determined by AAS; other parameters were determined by the chemical method for water analysis which is comparable with the American standard methods for the analysis of water and sewage. The criterion for soluble material is the fraction which passes a 0.45 µm filter.

INTERPRETATION OF RESULTS
5. The simplest interpretation of the bulk of the results can be shown as an average value (this can be 50% of the arithmetic or the geometric average value). To know which average value is to be taken, the distribution type has to be found.

CONCLUSIONS
6. For suitable measurements of the control of water pollution to be carried out, it is necessary to consider and differentiate the main sources of pollution (anthropogenic, permanent and temporary). Sufficient data are needed for such an evaluation to give a better view using statistical methods.
7. The comparative analysis of two years showed that the quality of water from surface mines is not much more contaminated than water from conventional sources.

C. RADHAKRISHNAN, Ministry of Water, Lands, Housing and Urban Development, Dar es Salaam, Tanzania

TN12. Appropriate least cost reliable water supply technologies for developing countries

Due to prevailing economical conditions many countries cannot operate their existing diesel or electrically operated pumping water supply systems successfully. The people in rural areas are not getting potable drinking water; for survival, they fetch water from the old traditional sources which are highly polluted and unsuitable for drinking.

SIMPLE TECHNOLOGIES
2. The following simple methods can be easily constructed and maintained at village level and can also give a reliable water supply. They can also fulfil the target needs of the Drinking Water Supply and Sanitation Decade to a certain extent.
3. There are many renewable energy sources like biomass, biogas, charcoal and firewood, geothermal, desalination, ocean thermal energy conversion and tidal waves but only simple techniques appropriate for a village environment are described here.

Wooden lever system
4. Figure 1 shows an arrangement of a wooden lever with a rope and bucket lifting device supported on a wooden pillar with a pivot support and counterweight to lift water from a shallow well constructed with well rings. The well lining can be constructed with burnt mud bricks in mud mortar and the exposed surface plastered with cement. The lifting arrangement can be of hard local wood with a hardwood pivot support or GS pipes. Villagers can construct this system themselves and maintain it with the resources available. However, suitable training must be given to ensure the use of clean rope and buckets to avoid pollution.

Step wells with rope drawing method
5. Figure 2 shows a step well with brick or concrete walls with a rope drawing method using a bucket, rope and pulley. The details given for the wooden lever system are also applicable to step wells.

Tube wells with bailer
6. Figure 3 shows the borehole or hand augered well with a concrete platform and rope drawing method using a bailer, pulley rope and tripod. The details on pollution control given for the wooden lever system are also applicable to tube wells.

Bucket pump
7. Figure 4 shows a borehole or hand augered well with a concrete platform and a chain

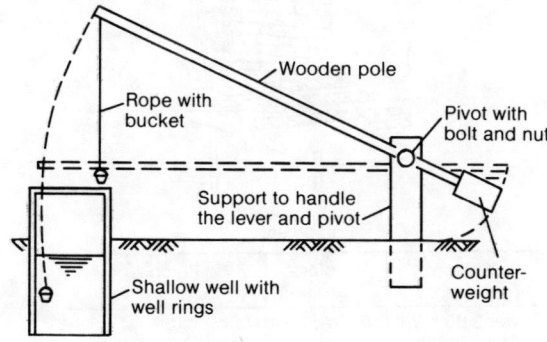

Fig 1. Wooden lever system

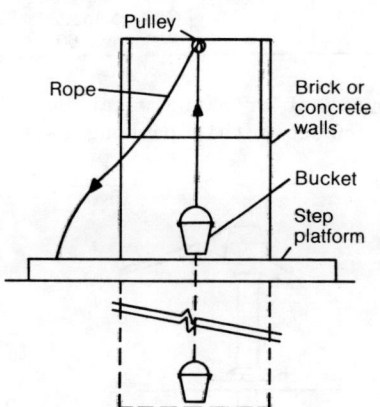

Fig. 2. Step well with rope drawing method

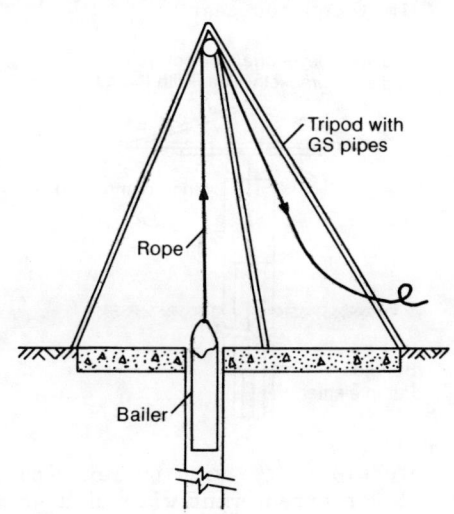

Fig. 3. Tube well with bailer

World Water '86. Thomas Telford Limited, London, 1987

drawing method using a bucket, disc pulley and hand crank. The pollution control of tube wells must be as described for the wooden lever system.

Animal-powered gear

8. Figure 5 shows an arrangement of a well with two buckets, rope and pulley drawing water with a pair of oxen. The animals move on a path between the well and the post supporting the reverse direction pulley. They are harnessed to the steel cable in such a way that as they move away from or towards the well, they raise the water.

Flywheel connected with mono pump

9. When a diesel or electrically driven engine is not operating due to unavoidable circumstances, the borehole mono pump can be operated manually if a mild steel bracket is welded over the pulley as shown in Fig. 6.

Windmills

10. Windmills can start to pump in winds of about 8 km/h and the governing or automatic furling speed is 32-40 km/h with a pumping capacity of 400-5500 l/h with a head of 3.6-258.0 m.

11. The renewable energy device of the windmill is directly by piston pump.

12. Windmills are also connected to generators and dynamos to produce electricity; the stored electricity is used for running pumps and for lighting.

13. The wind collected is stored directly in a compressor and operates a jet pump with pressure for pumping water.

Solar systems

14. Figure 7 shows a solar generator, solar venter and water pump for supplying water in a remote rural environment.

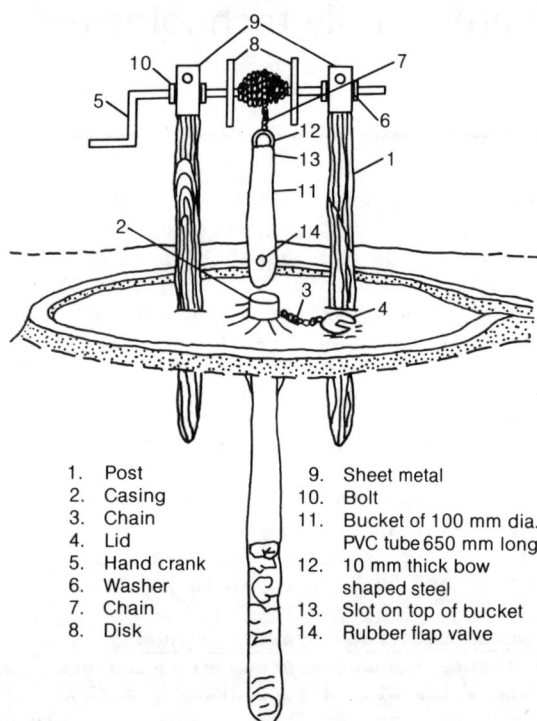

1. Post
2. Casing
3. Chain
4. Lid
5. Hand crank
6. Washer
7. Chain
8. Disk
9. Sheet metal
10. Bolt
11. Bucket of 100 mm dia. PVC tube 650 mm long
12. 10 mm thick bow shaped steel
13. Slot on top of bucket
14. Rubber flap valve

Fig. 4. Bucket pump

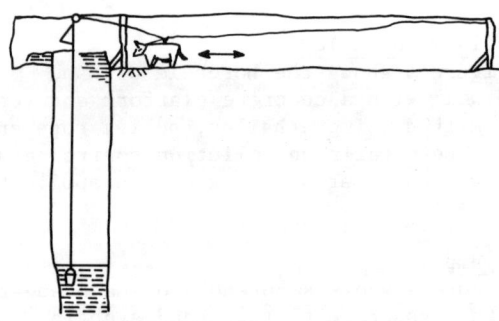

Fig. 5. Animal-powered gear

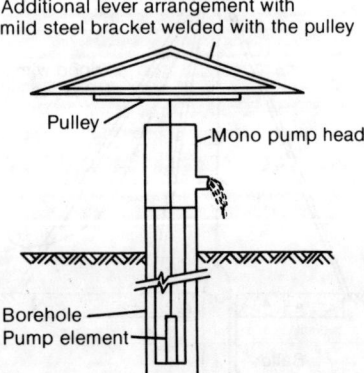

Fig. 6. Borehole fitted with pulley with additional lever arrangement with mild steel bracket for manual operation

Equipment		
Solar generator	Solar verter	Pump type
Power 3.65 kWp Voltage 300 V Modules 190	5 kVA DC/AC 3 phase 220/380 V	Subversible Power 2.2 kW Pumping height 76 m Possible average Water output 25 m³ Water tank 60 m³

Fig. 7. Pumping of drinking water by using solar energy (photovoltaic pumping system)

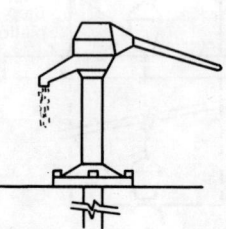

Fig. 8. Nira hand pump

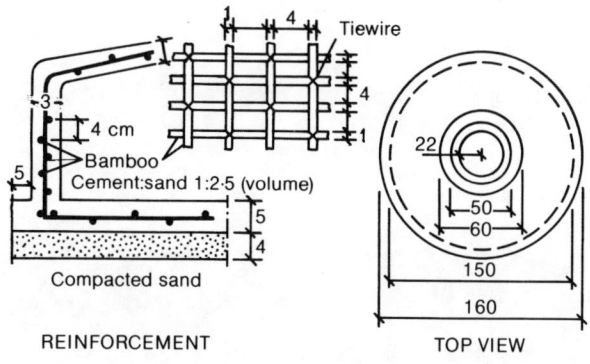

Fig. 9. Bamboo-cement storage tank 2500 litre

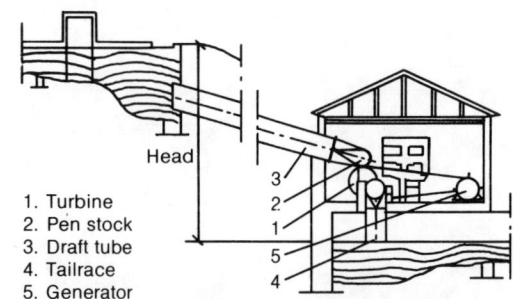

Fig. 10. Mini hydroelectric generator

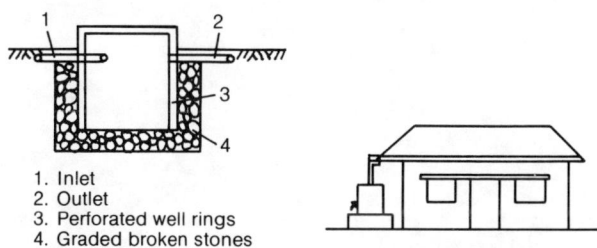

Fig. 11. Simple intake

Fig. 12. Rainwater collection

Hand pump system

15. Figure 8 shows a typical type of Nira hand pump used for lifting water from shallow wells.

Bamboo-reinforced concrete tanks

16. Some storage tanks have been built successfully with bamboo instead of mild steel bars (Fig. 9).

Mini hydroelectric generator

17. The principles of a mini hydroelectric generator are shown in Fig. 10. To reduce the cost of importation the penstock can be made with wooden stave pipes and the turbine with local hard wood.

Intakes

18. Figure 11 shows a simple intake with concrete perforated well rings sunk and filled with graded material.

Rainwater collection

19. Figure 12 shows the collection of pure rainwater from house roofs. Similarly ground catchments are used for collecting rainwater runoff.

WATER QUALITY

20. Normally water is mixed with various minerals, chemicals, organic matter, debris from vegetation and animals, soil particles and micro-organisms. It is always best to make physical, chemical and bacteriological analysis while investigating the source and also before supplying water to the public. All possible precautions have to be taken to supply clean and safe water.

TREATMENT OF WATER

21. As far as possible simple treatment methods have to be adopted which can be constructed and maintained at village level.

22. The water fetched from the source or from the water point has to be boiled to kill pathogenic organisms and stored in a clean vessel with a lid. The water can be drawn from a tap in the container.

23. Before boiled water is stored it can be filtered through double thickness cotton cloth to avoid the need to use expensive chemicals.

RECOMMENDATIONS

24. Simple technologies using local materials like burnt mud bricks, bamboo, wooden stave pipes and local wood with local manpower for construction and maintenance at village level have to be followed with minimum expense.

25. Local expertise has to be used to modify the standard of imported materials to local conditions to save foreign exchange.

BIBLIOGRAPHY

Gate health water and sanitation D-6236 Eschborn 1. Federal Republic of Germany magazine, 1985, June, 39, 41.

Radhakrishnan C. Rain water harvesting for drinking water supply. Regional Water Engineer's Conference, Arusha, Tanzania, 1985.

Van Markenlaan J.C. Rainwater harvesting for drinking water supply, p. 22. Training Modules Series, The Hague, Aug. 1981.

World Water 1981-1990 water decade first year review, p. 5. Thomas Telford, London, 1982.

R. D. H. TWIGG, MA, MSc(Econ), MICE, MIWES, Sir M. MacDonald & Partners, Cambridge, UK

TN13. The East Java water supply project

With a population of more than 160 million and a growth rate of about 2% per annum, provision of safe water supplies to the people of Indonesia represents a major challenge. The Government is meeting this through a substantial investment programme, and by the end of the third national plan period (1979-84) it was estimated that about 50% of the urban population and 20% of the rural population had access to clean water supplies.

2. Continuing progress towards higher coverage levels is being achieved through standardized planning approaches; during the third national plan two major programmes were adopted: a basic needs approach (BNA) for towns with populations of over 20 000 and the IKK programme (i.e. Ibu Kota Kecamatan - district towns) for communities with populations of between 3000 and 20 000. Further standardization has been established during the current national plan (1984-89), which stipulates standard design criteria for five categories of community size. These programmes are implemented by the Directorate of Water Supplies, which forms part of the Directorate General of Cipta Karya, within the Ministry of Public Works.

3. The East Java water supply project covers 25 towns in the BNA programme and 100 towns in the IKK programme, and is supported by two loans from the World Bank. The total population to be served by the project is 1.44 million, with individual community populations of 3000 and 160 000.

STANDARDS AND CRITERIA

4. The supply standards adopted for the project's BNA towns provide for piped water supplies to about 60% of the population. Of these, 25% receive a full house connection with a design supply of 120 litres/capita per day (lpcd); 25% receive a yard connection (i.e. a single tap outside the house) at 60 lpcd and 50% are provided with standpipe supplies at 30 lpcd. In determining total system capacity, allowances are made for non-domestic supplies, leakage losses and peaking factors. A conventional design philosophy is adopted for the BNA schemes with standard criteria laid down for items such as service reservoir capacity (about five hours' average demand), maximum and minimum working pressures in distribution systems and pipe materials.

5. For the IKK schemes, normally 50% of the served population is provided with a house connection (at 60 lpcd) and 50% with a standpipe supply (at 30 lpcd); these proportions can be adjusted in individual cases. The IKK programme design criteria have been developed nationally to provide limited supplies at low cost to as many people as possible. A modular design approach is used with standardized scheme capacities ranging from 2.5 l/s to 10 l/s. The systems are designed to run at constant output, using flow restrictors on consumer connections, so that no centralized storage is required. Householders are expected to provide their own small storage capacity on their premises, and standpipes are equipped with storage tanks to even out peak demands. Air/water pressure vessels are connected to pumping mains to maintain system pressures at steady levels when pumps switch on and off.

PROGRAMME AND PROGRESS

6. Initial studies and designs for the project began in late 1981, undertaken by teams of foreign and local consultants. The programme was drawn up with the aim of speeding implementation, by using a phased approach for studies and designs and by drawing up the procurement contracts for materials and equipment on the basis of preliminary design studies. It was thus possible to issue the first sets of tender documents in late 1982.

7. In order to phase the requirements for resources and funding, the 125 schemes were split into two groups: the first project for 50 towns, and the second project for 75 towns. Construction for the first project began in 1983. Some 14 schemes are now partially operational, with many others nearing completion. The second project was started in 1986.

WATER RESOURCES

8. The selection of suitable water sources was a challenging aspect of the studies. Surface waters are often heavily polluted and turbid with widely varying flows making intakes and treatment works expensive to build and operate. Springs exist in many locations but are often already committed to irrigation. Groundwater conditions in East Java are generally suitable for development but vary widely: a drilling programme was therefore undertaken at the study stage to confirm sources. Schemes were divided into three groups

(a) those where non-groundwater sources appeared to be the most attractive – no drilling was done
(b) those where good groundwater was known to exist, based on previous experience – no drilling was needed
(c) those where groundwater might be suitable but there was insufficient knowledge – test/production wells were completed as part of the studies.

9. Of the 25 BNA schemes, 12 were allocated to group (c); in eight cases groundwater was proved to be viable on the first drilling; in each of the remaining four cases, a second well (sometimes at a new location) established satisfactory quantity and quality. Of the 100 IKK schemes, 48 were considered to require drilling prior to source confirmation, and groundwater proved viable in 43 of these (for the 5 remaining schemes, a lower capacity was adopted or other arrangements were made). This drilling programme thus showed an excellent success rate; the ultimate cost to the project has been very low as most of the wells are being incorporated into the final schemes.

INSTITUTIONAL AND FINANCIAL ASPECTS

10. The 125 schemes will be administered by 20 semi-autonomous water enterprises; each of these enterprises covers a single kabupaten (an administrative district with an average population in East Java of about one million) or a municipality (for the two largest schemes). This consolidation into larger management units thus allows savings in staffing at the more senior levels and unified tariff levels which can be applied throughout the enterprise.

11. Tariffs for project schemes will be in accordance with the national water rate structure which sets out the relationship between tariffs for each consumer category and each of four consumption bands. Overall tariff levels will be set to cover operating and maintenance costs plus depreciation on net revalued assets, where possible; the project studies included evaluations of consumers' ability and willingness to pay for the supplies.

P. G. WILLIAMS and L. R. J. VAN VUUREN,
National Institute for Water Research,
Pretoria, South Africa

TN14. Appropriate technology for rural water supplies

Water supplies for rural communities in developing countries, including South Africa, are often unsatisfactory because of quantity, quality and accessibility considerations, thus posing serious health risks to the people who use them. The water is usually drawn from individual sources (e.g. streams, springs and boreholes) which are often polluted. Water supply improvements generally rely on self-help by the rural people themselves, possibly with a contribution of materials and equipment (e.g. hand pumps) by the regional Department of Works or other authority.

2. The uplifting of a rural community involves the staged upgrading of all of the community's basic needs, which includes water supply as well as housing, medical services, schools and so on. The amount of funds and other resources which are made available for improving a rural water supply depends largely on the priority which the rural people themselves give to this particular need. This, in turn, depends on the benefits which they perceive will come from water supply improvements. These benefits may relate to convenience of access and adequate quantities of water rather than improved bacteriological quality, which is likely to be more of a concern to the local health authorities.

3. In rural societies the men are usually the spokesmen for the communities, but the women are responsible for collection of the water. Therefore special effort must be made to ensure that the views of the women are considered in planning water supply projects.

4. The challenge of appropriate technology, as far as a rural water supply is concerned, is to make the best use of the available resources to improve all aspects of the supply. However, it must be realized that, with limited resources, it will probably not be possible to achieve the ultimate standard, i.e. individual household reticulation of water to WHO standards. Nevertheless, if a significant improvement can be made to the original water supply situation, resulting in better health and hygiene conditions for rural people, then it should be seen as a success for the appropriate technology.

RESEARCH, DEVELOPMENT AND TESTING

5. Appropriate technology in rural situations must be affordable, effective, reliable and acceptable to the users. It is likely to be something which the rural people use themselves (e.g. hand pumps) or else is operated by someone from the community (e.g. slow sand filters).

6. There is a need for trained people who have a sympathetic understanding of the people's needs in rural areas, and can translate these needs into research and development projects which scientists and engineers, in both developing and developed countries, can tackle. Appropriate technology research and development can be a challenging career requiring fundamental scientific knowledge and the skills of competent engineers.

7. Appropriate technology should make use of new technologies, as well as adapt existing technologies. For instance, new plastics can be used in hand pumps to eliminate wear of bearings. In addition, research should be carried out into unconventional technologies which could be appropriate for rural areas. The worldwide designs of VLOM pumps is an example. The National Institute for Water Research (NIWR) of the Council for Scientific and Industrial Research (CSIR) in South Africa is currently investigating alternatives to chlorine for disinfecting small water supplies.

8. There are many items of equipment already on the market which are claimed to be suitable for small water supplies. The NIWR has surveyed this equipment and produced a catalogue with information on almost 200 different items.

9. This type of equipment needs to be tested objectively and evaluated for suitability in rural conditions of developing countries (similar to the hand pump testing by the Consumer Council in the UK), and the results need to be made available to the general public. Appropriate equipment should be manufactured locally in developing countries, if possible, and marketed at a realistic price in order to encourage a large volume of sales.

10. There is also a need to develop technologies which the local people can make and operate themselves.

PLANNING, CONSTRUCTION, OPERATION AND MAINTENANCE

11. Community involvement in the planning and construction of water supply systems is important if projects are to be successful. This can be done through local water committees which are set up in a way which does not usurp the authority of the traditional chiefs.

World Water '86. Thomas Telford Limited, London, 1987

However, committees must be organized competently and run on businesslike lines.

12. The operation and maintenance of water supply systems is of vital importance and must be organized properly by the water committees. Equipment operators must be well trained, and given reasonable remuneration. They must also be able to get necessary chemicals and items required to maintain the equipment.

TECHNOLOGY TRANSFER

13. Technology is successfully transferred only once it is in widespread use. Transferring new technologies to rural communities must be done carefully and with sensitivity to the needs and cultures of the users.

14. There is already a wide variety of appropriate water supply technologies in use in developing countries around the world. A number of international organizations are making effective contributions to transferring these technologies between countries through their publications (e.g. WHO, IRC, Waterlines and World Water). However, these publications generally do not reach the rural people directly, and therefore each country needs to make relevant information widely available locally by means such as information sheets in local languages, demonstrations and video films.

15. There is also a need to make engineers and government authorities aware of appropriate technology and its benefits. The NIWR and National Building Research Institute of CSIR have helped to organize several seminars in southern Africa for this purpose over the past few years.

E. G. THOMAS, BSc(Eng), MICE, MIWES, Halcrow Water and Management Committee of Register of Engineers for Disaster Relief, UK

TN15. Recent developments in refugee water supplies

The water supply and sanitation programmes undertaken over the past five years by international relief agencies such as Oxfam, the Save the Children Fund, War on Want and UNHCR have contributed greatly to the survival and health of many hundreds of thousands of people. For the most part these agencies have been assisted in this task by over 70 professional engineers on the Register of Engineers for Disaster Relief (REDR) and specialists who have undertaken three-month voluntary assignments in ten countries.

2. The refugee water supply programmes in the semi-arid zones of north-western Somalia and eastern Sudan were just two of these schemes conceived and implemented with REDR help (ref. 1). These were large-scale programmes serving 400 000 and 200 000 refugees respectively. They had to be implemented immediately and developed progressively, requiring a high degree of improvization and the sound application of both high and low technology.

3. The appropriate technology ranged from hand-dug wells sited from information obtained using a modern lightweight water proving rig, to boreholes drilled using large commercial rigs. Wells were variously equipped with diesel, solar and hand pumps and the water was filtered and chlorinated by simple methods as required. Aquifers were tapped using infiltration galleries, and were enlarged artificially through the construction of saturated sand storage dams.

SAND RIVERS AS A SOURCE

4. The most obvious water source in much of semi-arid Africa, where the basement rocks are exposed and eroded into coarse sands and gravel, is 'sand rivers'. In north-west Somalia and eastern Sudan bordering the Red Sea and Ethiopia these are the traditional sources for water-holes.

5. Although the saturated sand in the ribbon aquifers of north-west Somalia was shallow (typically 1-3 m in depth), with consequent low unit water storage capacity and subject to high evaporation loss, this disadvantage was compensated by their relatively frequent recharge. In the region of the refugee camps, where catchment areas were several hundred square kilometres in extent, studies revealed that major floods were generated by rainfall thresholds of intensity equal to or exceeding 24 mm/day. On 20 years of rainfall records this indicated that there were likely to be at least four major floods in a year, and only once in 40 years would the aquifer not receive any usable replenishment. Furthermore these aquifers were recharged over the greater part of the dry season from subsurface outflows from extensive ribbon aquifers upstream.

6. It was concluded that a system of development based on the construction of shallow wells and infiltration galleries sited along the sand rivers could supply drinking water to most camps at a rate of 5 litres per refugee per day during the most critical drought in years of normal rainfall.

SHALLOW WELL DEVELOPMENT

7. The shallow well development in north-west Somalia was essentially the development of the traditional sources using a modern approach (ref. 2). Water-holes directly sited in the river sands were easily polluted, and were replaced by fully lined open wells sited in the adjacent river terraces. Refugee housing was moved at least 0.5 km from the sand river source to avoid its pollution, and refugees were directed to use designated defaecation fields or to build pit latrines on a family basis. Water was supplied to camps with populations of around 30 000 from 15 wells of 2-3.5 m diameter and 3.5-6 m depth, spaced along a 3 km river frontage. Because of the basic shortage of water during the dry season it was initially necessary to limit the consumption to the maximum sustainable yield of around 5 litres per refugee per day by supplying water at stand-pipes near each well with a policy of minimum distribution within each camp. Infiltration galleries were built to improve the flow of water to the wells.

SATURATED SAND STORAGE DAMS

8. The idea of using sunk structures to enlarge artificially the natural ribbon aquifers is not new and had for years been used in southern Africa (refs 3 and 4). The 50 m wide gabion structures used in Somalia were impermeable as they were faced with cement and were keyed into the natural rock barriers; their immediate effect was to divert the whole subsurface flow to the open well system in the river terraces upstream. Later on during the successive flood seasons the coarse sand and gravel river bed loads built up to a sill level progressively raised in steps of 1 m, thus

World Water '86. Thomas Telford Limited, London, 1987

improving the reliability of the aquifer dramatically (ref. 2).

SOLAR POWERED PUMPING

9. Although much of the pumping capacity at the wells was diesel-powered, the system in Somalia was supplemented by 25 solar pumps each able to deliver water at 2 l/s through a lift of 5 m. Full technical evaluations of solar pumps have been undertaken for irrigation and water supply (ref. 5) but not in refugee camp situations where they have obvious potential.

CONCLUSIONS

10. The north-west Somalia refugee water supply programme confirmed that sand rivers can provide a long-term water supply for large communities, with good quality and moderately reliable quantity, at a capital cost of £2 per refugee. Over the five-year period for which the scheme was originally used, there has additionally been an annual equipment replacement and maintenance cost of £0.5 per refugee. As an indirect result of the success of this project, Oxfam has developed water kits (ref. 6) for use in relief situations where the sources are either open wells, springs, simple boreholes or rivers.

11. Saturated sand storage dams have exceeded expectations. Solar pumping despite high capital costs and earlier technical faults with the pump motors has provided a valuable contribution to the Somalia programme and has future potential.

12. If the developments described are to be accepted more widely in the future, it will be necessary for the existing refugee water schemes to be monitored continuously and evaluated professionally in all respects. If this is not done the impact of appropriate technologies on water supplies for refugees and rural communities will not be fully appreciated by development planners and funding agencies.

REFERENCES

1. THOMAS E.G. et al. Refugee water supplies in Somalia and Sudan. WEDC 12th conference, Calcutta, 1986.
2. SIR WILLIAM HALCROW & PARTNERS and OXFAM. NW Somalia refugee water supplies interim report, 1982.
3. WIPPLINGER O. Storage of water in sand. Namibian Department of Water Affairs, 1958.
4. BURGER S.W. and BEAUMONT R.D. Sand storage dams for water conservation. Southern African Water Year Conference, 1970.
5. KENNA J. and GILLETT B. Solar water pumping handbook. IT Publications, 1985.
6. GRAHAM N.J.D. and TOWNSEND G.H. Appropriate water supply systems for disaster relief. Public Health Engineer, 1983, Oct.

H. C. BALFOUR, FICE, and J. M. WENN,
BSc, PhD, Balfours, UK

TN16. Trends and issues in rural water supply in the Sahel area

In the past the development of the Sahel area was assisted by the construction of power supplies, transport and communication systems. Such development was of interest to European countries because it involved them in the design and construction of these substantial schemes. Such projects did not provide training for the local people, and did not prevent movement of population away from the rural areas. As a result, agricultural and industrial output declined in these areas.

2. In the Sahel desert area the value and importance of water is fundamental. The problem, therefore, is to indentify areas which, if supplied with a basic infrastructure, can develop agriculture or industry based on natural resources in order to provide a marketable surplus as well as to sustain the community.

3. Two projects carried out in Mali and Senegal are of interest. Both these contracts were designed with the assistance of local counterpart staff from Water Departments. The construction was carried out by using local labour.

4. We provided the construction equipment, materials and engineering expertise. The implementation of projects in this way provides basic training of local administrative, technical and manual personnel and provides experience in the running and maintenance of the equipment. Thus the developing countries benefit by their own efforts, obtain experience in both design and construction which should be useful in the future, and at the same time the manufacturing interests of the West are supported.

SUPPLEMENTARY SOURCE FOR SIKASSO, MALI (75 000 POPULATION)

5. In 1974 our firm completed reports for the WHO for a 'sector study' of water supply for the whole of the Republic of Mali. The report included a study for the town of Sikasso, situated near the border with Ivory Coast.

6. Sikasso had a water supply which comprised a small concrete dam, a treatment plant and a distribution network over part of the town. Unfortunately, the water catchment area for the dam was too small in relation to the town. Thus the necessary supply to the town during the dry season was by tankers provided by private enterprise at high cost.

7. The solution was to provide a supplementary supply from an adjacent larger catchment area, upstream of the dam. This involved the construction of a new river intake, a pumping station nearly 8 m deep and 16 km of concrete-lined iron pipe pumping main. Water was also supplied to three small villages along the route of the pumping main by the provision of small water towers, package treatment plants or chlorinators. The estimated cost of the work was £500 000.

8. Due to the remoteness of the area, it was not possible to find local contractors able to undertake any major part of the work, and no European contractors were interested in such a small scheme. Thus in 1976 we provided two of our engineers to the local Water Department who employed up to 300 locals to construct and subsequently to operate the works. Each engineer had a local counterpart, who received training in all aspects of the construction and commissioning of the works.

9. During a visit to Mali in December 1985, a local counterpart engineer, who was then well placed in the Water Department, was proud to say that the only part of the equipment which was not functioning properly was the small chlorinator. Thus the plant is still in operation ten years after construction - a considerable achievement in such a remote area.

10. We believe that the project is successful because of the simplicity of construction and because of the expertise of the local engineers, resulting from the extensive practical training throughout the construction and commissioning periods.

WATER SUPPLY FOR 18 VILLAGES IN SENEGAL (85 000 POPULATION)

11. In 1983 we studied the water supply possibilities for 18 rural villages in Senegal. It is to be hoped that such a project will encourage a more settled lifestyle with more time available for agriculture, and will reduce the need for large numbers of people (and cattle) to be nomadic. The construction is now being carried out with local labour, as neither local nor international contractors were interested.

12. Nearly all water is from deep wells, most of which are under artesian pressure. A village area includes everybody housed within a 5 km radius of the well. Some of the 18 villages already have wells which have never been used. The remainder are being provided with new wells, eight of which have already

been satisfactorily completed.

13. Each well is to be provided with a pump, driven by a diesel engine which is housed in a prefabricated 'motor house'. The water is then pumped into a prefabricated steel water tower, and into a distribution system of about 8 km average length per village. The system is constructed complete with cattle troughs, water stand-pipes and small tanker filling points. All this work is undertaken by our engineers, assisted by local counterpart staff, craftsmen and technicians.

14. This project is quite sophisticated in comparison with most rural projects, and can form a useful basis for larger projects in the future. Its running and maintenance cost will be paid for largely by local subscription.

15. This is a large self-help project, with the villages widely spread out, both in the northern and southern parts of the country. The construction will cost in excess of £5 million and will comprise 18 water towers, 18 pump sets and over 250 km of distribution pipelines.

16. The original study involved civil and mechanical engineers, a geologist, a geophysicist, an economist and a social anthropologist. The construction phase is being supervised by the same engineering team and the geologist, supplemented by two experienced mechanics and a specialist in health education.

17. The work is part of a trend for rural development in the Sahel area. Together with the smaller works being carried out by the funding agencies, WaterAid, and many voluntary and charitable organizations working in the area, it is providing an improved water supply in this arid area, aimed at making life there more attractive.

ACKNOWLEDGEMENTS

18. We would like to express our thanks to the WHO and the Ministry of Overseas Development who funded the work described, and to their representatives and engineers who have on many occasions assisted the projects with their knowledge and experience.

P. J. HUISWAARD, MSc, CE, PhD, and
J. G. BRUINS, MSc, AE, BKH Consulting Engineers, Bongaerts, Kuyper and Huiswaard, The Hague, The Netherlands

7 Recent developments in wastewater treatment

The most generally used wastewater treatment methods are discussed, whereby the state of the art of each method is updated. Attention has been given to pond systems, anaerobic treatment, aerated systems and ponds with aquatic plants. Several methods for disposal of treated waste water have been described, such as irrigation, land disposal and fish ponds.
Then some tertiary treatment methods for removal of nitrogen, phosphorus, pathogen micro-organisms and toxic substances out of wastewater treatment plant effluents have been described.

INTRODUCTION
1. Many developments have taken place in wastewater treatment technology during recent years. In conventional, aerobic treatment special attention has been given to reducing power consumption by aeration systems. Some new aeration methods have been introduced in which much attention is paid to the removal of nitrogen and phosphorus. The loading factors of existing pond systems have been investigated in order to arrive at more universal design criteria.
Algae ponds and aquatic plant systems are examples of recently developed wastewater treatment systems which have more and more application. In the last 10 years anaerobic digestion has become a widely accepted method, in particular, for industrial waste water. Re-use of waste water, for example, for irrigation or groundwater recharge has been found in many cases to be a feasible method of wastewater disposal.

Wastewater treatment methods
2. The ideal wastewater treatment system is characterized by low investment costs, low energy consumption, low operational costs, and limited land requirements, with simple operation and maintenance, and a high treatment efficiency. Some systems, which have been the subject of recent developments and studies, are described briefly below.

Pond systems
3. A pond system, especially in hot climates, is an efficient method for waste water treatment. Investment costs depend largely on the price of land. Operational, maintenance and energy costs are low. A large land area is, however, required. Treatment efficiency with respect to the removal of organics, nutrients and pathogens is high. A pond system usually consists of a chain of anaerobic, facultative and maturation ponds.

New developments include high rate algae ponds and aquatic plant ponds. The various types of ponds are discussed below.

Anaerobic ponds
4. In this system waste water is stored in a pond with a depth of 2.5 - 4.0 m. In the pond anaerobic micro-organisms digest organic matter causing the formation of methane, hydrogensulphide, ammonia and carbondioxide gases.

Design criteria for anaerobic ponds
5. In recent years the following design criteria have been developed for anaerobic ponds:
Loading rate   : 0.1-0.4 kg BOD_5/m^3/day
Residence time   : 1 - 5 days
Minimum temperature   : 15°C
Depth   : 2.5 - 4.0 m
pH   : about 7
Usually a parallel pond system is operated. The flow pattern in the ponds is completely mixed. Anaerobic ponds are used for both domestic and industrial waste water. Some data on BOD_5 reduction by biodegradation and by sedimentation in anaerobic ponds are given below:

temperature °C	residence time days	BOD_5 reduction %
< 15	4 - 5	30 - 40
15 - 20	2 - 3	40 - 50
20 - 25	1 - 2	40 - 60
25 - 30	1 - 2	60 - 80

Start-up of anaerobic ponds
6. When starting up an anaerobic pond system, the ponds have to be inoculated with well fermenting sludge.

Anaerobic pond odour problems
7. Anaerobic ponds may develop unpleasant odours and should, therefore, be built 0.5 - 1.0 km from settlements.

World Water '86. Thomas Telford Limited, London, 1986

Facultative ponds

8. After passing through the anaerobic pond the waste water undergoes further treatment in a facultative pond.
In facultative ponds anaerobic degradation takes place at the bottom of the pond and aerobic degradation in the top layers. Growth of algae is induced in the pond by the penetration of sunlight. The algae are responsible for most of the oxygen input.

Facultative pond design

9. Many different formulae exist for design facultative ponds. One formula is based on a completely mixed flow pattern:

$$L_e = \frac{L_i}{1 + K_T \cdot t}$$

L_i/L_e = BOD_5 of influent/effluent (mg/l)
K_T = reaction constant (day^{-1}), depending on the temperature
t = retention time (days)

For ideal plug flow the formula is:

$$L_e = L_i \times e^{-K_T \cdot t}$$

In practice there is a dispersed flow pattern, a mixture of completely mixed and ideal plug flow.
Empirical formulas have been developed in many places in the world for the design of facultative ponds. The depth is usually about 1.2 m for (sub-) tropical climates, but in colder climates, deeper ponds are more suitable. Design formulas are mostly based on the temperature and organic loading rate. The formulas mostly apply to local circumstances. Depending on temperature the residence time in the facultative pond can vary between 7 and 30 days, while the organic loading rate can vary between 100 and 400 kg BOD_5/ha/day. In facultative pond design the following aspects have to be considered:
- length/width ratio = 2/1 - 3/1
- length of pond in the prevailing wind direction
- installation of freeboard because of wind -induced wave action

It is advantageous to design a number of ponds in series instead of one pond with the same surface area, since this promotes plugflow, and, therefore, a higher BOD_5 removal efficiency.

Application and costs of pond systems

10. For strong waste water the best system is an anaerobic pond in series with facultative and maturation ponds. For weaker waste water (BOD_5 < 500 mg/l) the anaerobic pond may be omitted. In this case a pit can be constructed in the first pond for sedimentation of the solids. Pond systems are a good method of sewage treatment especially in warm climates where plenty of land is available.
For Quetta (Pakistan) a wastewater treatment system, consisting of anaerobic, facultative and maturation ponds was designed.

Design flow is 25,000 m^3/day. The waste water has a BOD_5 = about 500 mg/l.
The total area of the ponds is 20 ha. Total investment costs of the system are estimated at about Dfl. 3,200,000.-, at a local price of about Dfl. 31,000.- per ha.

Performance of facultative ponds

11. A combination of anaerobic and facultative ponds can give a total BOD_5 reduction of more than 90%. Residence time and loading rate depend on the local circumstances and on the type of waste water. Effluent from facultative ponds generally contains a considerable quantity of algae.

Maturation ponds

12. The effluent of facultative ponds should flow into a maturation pond. The purpose of maturation ponds is to achieve further die-off of fecal bacteria. A further reduction of BOD_5 and algae removal also takes place. Maturation ponds are completely aerobic, with depths in the range 1.2 - 1.5 m.

Design of maturation ponds

13. The following formula is used for the design of maturation ponds:

$$B_e = \frac{B_i}{1 + K_{B(T)} \times t^*}$$

in which:
B_e = number of fecal bacteria/100 ml in effluent
B_i = number of fecal bacteria/100 ml in influent
t^* = residence time in days
$K_{B(T)}$ = first order fecal bacteria removal

Maturation ponds deeper than 1.0 - 1.5 m may be designed but the BOD_5-loading should remain less than the loading rates for facultative ponds, in order to maintain aerobic conditions in the pond.

Artificially aerated ponds

14. In artificially aerated ponds the oxygen is supplied by mechanical aerators instead of algae.
Retention time in the ponds is usually about 4 days. Depth of the ponds may vary between 2.5 and 5 m.
Basically two different systems may be marked out:
- facultative aerated ponds
  In these ponds oxygen is supplied to the top layers of the pond, while the deeper layers are anaerobic.
- completely mixed ponds
  The whole volume of the pond is kept aerobic. There are systems with and without recirculation of sludge.

The design aeration capacity depends on which of above mentioned system will be applied, and by the quantity of oxygen that has to be supplied for oxidation of organic material so that the desired BOD_5 of the effluent will be reached.
According to Mara a power of 5 W/m^3 is needed for creation of a completely mixed pond. Extra power is needed for the supply of oxygen. Other researchers give different numbers for the power that is needed for complete mixing, lower as well as higher than 5 W/m^3.

High rate algae ponds

15. A recent development in sewage treatment is the high rate algae pond system in which the waste water, mostly presettled municipal sewage, is circulated in a shallow basin. Algae growth is enhanced by light radiation into the water surface. The algae produce oxygen. Because of the relatively high liquid velocity algae and oxygen are well mixed with the water in the basin and as a result aerobic biodegradation is promoted. The effluent of the circulation basin, which is rich of algae, is conveyed into a separation unit, where the algae are removed from the mixed liquor. This separation unit may be a filter system or a coagulation - flocculation system. The algae can be used as a protein source, for example, for animal feed.

Design criteria for high rate algae ponds

16. The following design criteria apply to high rate algae ponds:

temperature	> about 15°C
irradiance	> 160 cal/cm^2.day
residence time	= 2 - 7 days
depth	= about 0.5 m
surface loading rate	= 220-800 kg BOD_5/ha/day

Performance of high rate algae ponds

17. High rate algae ponds have a BOD_5 removal efficiency of more than 90%. Mean productivity of biomass is about 0.034 kg dry solids/m^2/day. About 30 - 90% of the dry biomass is algae. Separation of algae out of the effluent is generally a difficult process.

Wastewater treatment by means of aquatic plants

18. Recently more and more attention has been paid to the ability of aquatic plants to purify water. The following mechanisms play a role in the process:
- adsorption of volatile matter onto the plants
- biodegradation of organic matter by bacteria adsorbed onto the aquatic plants
- nutrient uptake by the aquatic plants
- algae growth due to irradiation
- oxygen input by algae

Two different systems can be applied:
- ponds with floating aquatic plants
- basins with emergent aquatic plants

The aquatic plants have to be harvested periodically.

Ponds with floating aquatic plants

19. These ponds are comparable to facultative ponds. The presence of the aquatic plants, mostly water hyacinths, makes higher loading rates possible. Efficiency of BOD-removal and removal of nutrients is high. The very high evaporation may be a disadvantage. This system is only possible in warm climates because of the type of plants required.

Basins with emergent aquatic plants

20. The waste water flows through shallow basins in which, for example, reeds or bullrushes are growing. Only the roots and the lowest part of the stalks are submerged. The purifying mechanism consists of biodegradation, adsorption, nutrient uptake and sedimentation. In temperate climates the systems loses its efficiency in winter time. A new development in this treatment system is a system in which the waste water flows through the root zone, so that the roots also participate in the purifying mechanism.

Application of aquatic plants in wastewater treatment

21. Aquatic plants can be applied for treating waste water which is not too strong, for example, pre-settled domestic waste water. Aquatic plants are also applied for the further treatment of secondary effluent for the removal of nutrients and fecal bacteria. Investment costs for aquatic plants systems are comparable to those of ponds systems, but operational costs are much higer since the aquatic plants have to be harvested. After harvesting, however, the aquatic plants may be used, for example, for soil conditioning, pulp making for the paper industry, for cattle feed or for production of biogas by anaerobic digestion.

Anaerobic wastewater treatment

22. Anaerobic digestion is a well known technique for the treatment of sludge from primary and biological wastewater treatment plants. During the last 10 years the technique has also found application in the direct treatment of waste water, mostly from agricultural and food industries. Different processes have been developed, such as the anaerobic filter, the anaerobic contact process and upflow anaerobic sludge blanket process (UASB). Of these processes the UASB process is considered to be one of the most promising and has already been applied in full scale installations in the following industries: starch, sugar, candy, paper, potato processing, pharmaceutics, alcohol, brewing and dairy. Treatment of waste water in UASB reactors is based on the two following principles:
- the anaerobic sludge has good settling properties making it possible to operate a reactor in which the waste water flows upwards through a sludge blanket
- loss of sludge with the effluent is prevented by an early separation of the biogas out of the sludge/water mixture.

In properly controlled UASB reactors formation of granular sludge takes place. This type of sludge has superior settling properties.

Design of UASB reactors

23. The following process parameters, amongst others, have to be considered in the design of a UASB reactor:
- organic load
  Maximum organic load is about 15 kg/COD/m^3.day, the design load depending on the strength of the waste water and the retention time at this load, and on the quantity of sludge in the reactor.
- retention time:
  Minimum retention time is about 8 hours
- process temperature:
  Optimal temperature in the reactor is about 35°C.
- composition of waste water:
  Important factors here are the absence of suspended solids, the presence of nutrients, the absence of toxics and a neutral pH.

Application and costs UASB systems

24. A UASB reactor which is functioning well will remove usually about 80% of the influent COD, at influent COD's between 4,000 and 5,000 mg/l. Biogas production is about 0.4 Nm^3 per kg COD removed. The biogas contains about 80% methane. Excess sludge production is small.
Investment cost of UASB installation amount to about Dfl. 1,000.- per m^3 reactor volume for medium sized reactors of about 2,000 m^3.

Recent developments in anaerobic waste water treatment

25. The fluidized bed reactor was recently introduced into anaerobic wastewater treatment. Hereby the anaerobic sludge in the reactor is adsorbed onto a bearer material, for example, sand granules. This improves the settling properties of the sludge considerably and the dry solids concentration of the sludge in the reactor can be much higher. As a result much higher loading rates and shorter retention times become possible.

Aerobic wastewater treatment

26. With artificial aeration wastewater treatment systems special attention is now being given to the development of systems that consume less energy for aeration. In small extended aeration plants the energy consumption can be reduced by installation of a time mechanism in the aerator, which adjusts the oxygen supply to the daily variation of the wastewater flow to the treatment plant. In large extended aeration plants the oxygen supply can be controlled by means of oxygen electrodes, to ensure that the dissolved oxygen concentration in the aeration basin does not exceed its design concentration. A denitrification compartment in the treatment plant is also used to lower the oxygen and energy input, since in this case nitrate is used for oxidation of organic matter.
Another development in this field is the Schreiber system.

With this system the sludge/water mixture is aerated in a circular tank by means of aerators at the bottom, connected to a rotating bridge. The oxygenation efficiency is higher than, for example, in an oxidation ditch with surface aeration.

New developments in aerobic waste water treatment systems

27. With the growing need to reduce the power consumption the "bio-disc" has again become more interesting. The biodisc consists of a number of vertical circular plates attached to a horizontal axis, the whole rotating vertically. The plates rotate in a basin containing waste water. The plates are covered with biologically active material. For high BOD-reduction a number of discs should be operated in series. Another recent development is the "tower reactor", developed for industrial waste water treatment. Advantages of this system are:
- installation occupies little space
- very efficient use of oxygen

Wastewater disposal

28. Often the effluent of a wastewater treatment plant is discharged into surface water. In many cases however, it may be useful to consider whether the effluent can be used directly for some purpose or other. Examples of this are:
- irrigation
- land disposal
- fish ponds
- recirculation in factories

Irrigation

29. It may be possible to use wastewater treatment plant effluent for irrigation. There are many restrictions however, connected with the use of effluent for irrigation. Important factors are:
- hygienic quality of the effluent. Different crops put different demands in the hygienic quality of the effluent. For instance, some fodder crops may be irrigated with primary effluent, while vegetables which are eaten fresh, like lettuce, need completely treated and disinfected effluent.
- chemical quality of the effluent
  Most important restricting factors are the salinity of the water, the ratio between sodium and calcium + magnesium, and the boron concentration.
- type of irrigation. Sprinkler irrigation demands a very high hygienic quality because of the danger of spreading pathogenic micro- organisms by aerosols.
With ridge and furrow irrigation the demands on the hygienic water quality are less strict.

Hereunder some criteria for use of treated waste water for irrigation are summarized:

	Crops not for direct human consumption	Crops eaten cooked; fish culture	Crops eaten raw
Health criteria (see below for explanation of symbols)	A + D	B + D or C + D	C + D
Primary treatment	**	**	**
Secondary treatment		**	**
Sand filtration or equivalent polishing methods		*	*
Disinfection		*	**

Health criteria:

A Freedom from gross solids; significant removal of parasite eggs.
B As A, plus significant removal of bacteria
C Not more than 100 coliform organisms per 100 ml in 80% samples.
D No chemicals that lead to undesirable residues in crops or fish.

In order to meet the given health criteria, processes marked ** will be essential. In addition, one or more processes marked * may also be required.

Land disposal
30. Disposal of treated waste water on land, by means of infiltration basins or by pits, can serve two possible purposes.
The first possibility is that the water, by passing through the soil layers, undergoes further purification. The water can be collected in an underground drainage or wells system, and then re-used, for example, for irrigation. Usually in such a system there is an effective removal of organics and phosphate. Often the infiltration schedule has to be adapted to the nitrification and denitrification mechanisms in order to obtain an optimal nitrogen removal.
The second possibility is to use the treated waste water for replenishing a groundwater aquifer. Basically the criteria used for the first possibility are applicable here. Replenishing of aquifers is necessary when the aquifer has been overdrawn and, for example, salt water intrusion, takes place.

Fish ponds
31. Waste water treatment plant effluent can be used very effectively for fish ponds. In the fish ponds the effluent will undergo further purification, removal of BOD_5 and removal of algae. Dissolved oxygen concentration of the water should be at least about 6 mg O_2/l. No toxic substances should be present in the water. For instance the ammonium concentration should be under 5 mg N/l. If the effluent does not meet the fish pond standards, it might be possible to reach the standards by mixing it with surface water.

Tertiary treatment
32. Tertiary treatment is the polishing of the effluent of a secondary or biological waste water treatment plant. Thus, maturation ponds, water plants, irrigation, land disposal and fish ponds may be considered as tertiary treatment, but in some cases aquatic plants, irrigation and land disposal may replace secondary treatment. The purpose of tertiary treatment is further removal of BOD, suspended solids, nitrogen, phosphorus, heavy metals, toxic organics and pathogenic micro-organisms. In addition to the methods already mentioned there are several other physical and/or chemical methods for tertiary treatment, such as:
- coagulation/flocculation
- sand filtration
- magnetic separation
- granular activated carbon adsorption
- membrane processes
- disinfection.

Coagulation/flocculation
33. By addition to waste water of certain chemicals, for example, aluminium salts, flocs are formed by colloidally dissolved material. Volatile particles are adsorbed on to the flocs. In a next step the flocs are separated out of the water by filtration or sedimentation.
Phosphates, pathogen bacteria and viruses are well removed by this method. Total costs of this method are about Dfl. 0.40 per m^3.

Sand filtration
34. The water is filtered through a sand bed. Hereby volatile matter is well removed out of the water, but dissolved materials are very little removed. Total costs of this method are about Dfl. 0.10 per m^3.

Magnetic separation
35. Particles in the water are magnetized by addition of, for example, ferrite. Then the water is sent through a magnetic field, whereby the magnetic material is separated out of the water.
This is especially a good method for regaining heavy metals out of waste water.

Granular activated carbon adsorption
36. The water is led through columns that have been filled with granular activated carbon.

Particles out of the water are then adsorbed onto the activated carbon granules. This method is especially effective for removal of complex organic micro pollutants. Total costs of this method amount to about Dfl. 0.90 per m^3.

Membrane processes

37. Membrane processes, such as reverse osmosis and microfiltration, are quite efficient for removal of phosphates, micro-organisms and toxic organics.
Average costs of these processes are about Dfl. 0.50 - 1.50 per m^3.

Disinfection

38. Disinfection is applied for killing pathogen bacteria and viruses in the water. Most applied chemical for disinfection is sodium hypochlorite (NaOCl), which is very effective for killing bacteria, but less effective for killing viruses. For killing viruses ozone (O_3) and chlorodioxide (ClO_2) are more effective.
Costs of disinfection with Na OCl are about Dfl. 0.023 per m^3, with O_3 about Dfl. 0.08 per m^3 and with ClO_2 about Dfl. 0.036 per m^3.

REFERENCES

1. MARA D.D. Sewage treatment in hot climates, Wiley, London, 1976.
2. ARTHUR J.P. Notes on the design and operation of waste stabilization ponds in warm climates of developing countries, World Bank
Technical Paper No. 7.
3. GLOYNA E.F. Waste stabilization ponds, WHO, Geneva, 1971.
4. CORNWELL D.A. et al. Nutrient removal by water hyacinths, JWPCF, 49 (1977), 1, 57-65.
5. DINGES R. Upgrading stabilization pond effluent by water hyacinth culture, JWPCF, 50 (1978), 5, 833-839.
6. WOELDERS J.A. et al. Anaerobic reactors and biodiscs: a promising combination for treatment of industrial waste water (Dutch) H_2O, 15 (1982), 22, 608-612.
7. HABETS L.H.A. et al. Anaerobic wastewater treatment at paper factory Roermond successful and economic (Dutch) H_2O, 17 (1984), 26, 626-629.
8. PETTE K.C. Anaerobic wastewater treatment at CSM sugar factories. La Sucrerie Belge, 99, December 1980, 473-479.
9. SAYED S. et al. Anaerobic treatment of slaughterhouse waste using a flocculant sludge UASB reactor. Agricultural wastes, 11 (1984), 197-226.
10. WOUDA T.W.M. Nitrogen removal with biodiscs (Dutch) H_2O, 15 (1982) 22, 595.
11. Energy conservation at sewage treatment plants. BKH Consulting Engineers Bongaerts, Kuyper and Huiswaard.
12. ENGWIRDA S. Underground wastewater treatment according to multireactor systems. (Dutch) Land en Water Nu, 24 (1984), 11.
13. Reuse of effluents: methods of wastewater treatment and health safeguards, WHO, Geneva, 1971.
14. DAYDEN F.D. et al. Virus removal in advanced wastewater treatment systems. JWPCF, 51 (1979), 8, 2098-2109.
15. VAN DALEN R. et al. Removal of phosphates from dephosphatized waste water by means of a cober filter. (Dutch) H_2O, 10 (1977), 21, 480-485.
16. GROS H. Reducing the phosphate concentration by wastewater filtration (German) Wasserwirtschaft, 73 (1983), 12, 528-529.
17. VAN HAUTE A. Tertiary treatment, lagoons and reuse of waste water (Dutch). Leefmilieu Dossier No. 8. Antwerp, 1983.
18. MITCHELL R. et al. Magnetic separation: a new approach to water and waste treatment. Progr. in Water Technology, 7, 3/4, 349-355.
19. TOFFLEMIRE T.J. et al. Activated carbon adsorption and polishing of strong waste water. JWPCF, 45 (1973), 10, 2166-2179.
20. APPLEGATE L.E. Membrane separation processes Chem. Engineering, (1984), 6, 64-89.

G. A. THOMAS, FIEE, FIGasE, MIEEIE, MIWES, and A. J. RACHWAL, BSc, Thames Water, UK

8 The activated sludge process and energy efficiency for the 80s

The use of modern design and process control technology for optimising activated sludge treatment of domestic and industrial wastewater in Thames Water is reviewed. Transfer of appropriate technology for Third World use is discussed. Experience in reducing energy consumption via integration of aerator efficiency trials, dissolved oxygen control and use of anoxic zones for nitrate removal is presented for full scale plants. Cases for employing conventional activated sludge plant, with either diffused air or surface aerators are compared with oxidation ditches and package plants.

INTRODUCTION
1. In 1986 the activated sludge process can celebrate seventy years of continuous employment for treating wastewater in operational scale plants. Ardern and Lockett's original pilot plant work at Manchester in 1914 was followed two years later by the first purpose designed plant at Worcester (1). Today the activated sludge process and its many developments and variants provides a means for biological treatment of sewage worldwide. Hundreds of thousands of plants are operating in widely varying climates and serving populations from a few hundred people in a village to many millions in the large cities. The original purpose of preventing the pollution of rivers has extended equally widely to cover indirect reuse for potable supply and direct reclamation of valuable water resources for irrigation purposes in arid areas of the world.
2. For many years development work was mainly carried out by manufacturers and contractors and tended to concentrate on the improvement of mechanical devices and methods of construction rather than on the process aspects of the system which in the case of the biology have never been fully explained or expoited. The organisations requiring activated sludge plants attached primary importance to removing polluting load, achieving good quality effluents and providing adequate capacity for future growth and stricter effluent standards. To remain successful the plants had to be reliable and simple to operate.
3. Over the past ten to fifteen years these aspects of the activated sludge process have been taken for granted and there has been an increasing demand for better process optimisation. The same period has seen re-organisation of the water industries in some countries of the developed world, which has created much larger units, with both the resources to carry out in-house design and to support research work together with the economic need to improve efficiency. In the UK these Water Authorities have been required to minimise energy and operating costs, whilst often at the same time to improve effluent quality with the minimum of capital investment. This has required application of new technology in the fields of biology, process design, sensors and computer control systems as well as developments in traditional engineering and operational methods.
4. This paper presents Thames Water's experience in applying modern design and process control technology for optimising activated sludge treatment of domestic and industrial wastewater. The work includes both in-house developments and the application of research carried out on behalf of the UK water industry by the Water Research Centre.

ACTIVATED SLUDGE PLANTS IN THAMES WATER
Domestic and Industrial Wastewater Treatment
5. In Thames Water more than 90% of the 4 million cubic metres of sewage treated each day involves some form of the activated sludge process. The fifty plants involved range from coarse bubble aeration package plants for 100 population, through oxidation ditches for 1000-100,000 population to conventional surface and diffused air plants serving populations up to 2.5 million (Table 1). Industrial wastewater is normally treated at these plants in admixture with domestic sewage, typically representing 25% of the total load. In most cases this is preferable to separate treatment plants for industrial waste as these can experience significant problems with nutrient inbalances excessive flow variation and toxicity.

Industrial wastewater treatment
6. In a few cases, where industrial wastewater represents more than 50% of the applied load or it is known to cause treatment problems, Thames Water has constructed activated sludge plants specifically for treating trade wastes. At Reading a pure oxygen activated sludge plant of 200,000 population equivalent was constructed in parallel with an existing biological filter plant to treat waste from a new brewery. Pure oxygen was selected after pilot plant trials indicated improved sludge settlement characteristics and a rapid response to large variations in load. The pilot trials also showed that the brewery waste was deficient in nitrogen and

phosphorus. This was solved by including some domestic wastewater in the pure oxygen plant feed. The full scale plant has easily achieved its design performance of 95 percent compliance with a 50 SS, 50 BOD standard since commissioning in 1980 even when treating a peak BOD concentration of 2500 mg/l. The non-nitrifying pure oxygen plant is followed by a nitrifying activated sludge system which enables the domestic and industrial wastewater treatment plants at Reading to meet their combined effluent to river standard of 95 percent compliance with a 20 BOD 4 NH_3-N standard.

7. At Riverside sewage treatment works, a different industrial wastewater treatment problem was also solved by the construction of a trade waste activated sludge plant designed to operate in parallel with a domestic sewage treatment plant. The industrial wastewater came from two major industries, motor vehicle manufacture and chemical manufacture. This mixture of industrial wastewaters presented inhibition sludge treatment and disposal problems when treated with domestic wastewater. The least cost solution was to design and construct a separate surface aeration activated sludge plant operating at reduced loadings and with an activated sludge that could become conditioned by frequent contact with inhibiting chemicals. A separate sludge treatment and disposal arrangement was also necessary for the industrial wastewater sludge produced.

8. Apart from these two major industrial wastewater treatment plants, Thames Water's proven treatment practice is to treat domestic and industrial wastewater in the same plant. To reliably achieve this it is vital to have a trade effluent monitoring and control policy together with industrial waste discharge consents to ensure that activated sludge plants treating domestic and industrial wastewater are not subjected to toxic inhibition or shock overloads. This also provides a means of charging industry a fair price for treating their polluting load.

ACTIVATED SLUDGE OPTIMISATION

9. Financing and depreciation charges represent between one third and one half of the annual cost of a typical activated sludge plant. This illustrated the importance that must be attached to minimising initial capital cost. Direct operating costs can generally be allocated to three headings:-
   i) energy
   ii) employee and related costs
   iii) sludge treatment and disposal

Each heading represents between 20 and 40% of direct operating costs depending on size, location, treatment, process options and sludge disposal routes.

Areas in which Thames Water has applied new technology to minimise activated sludge treatment costs are numerous.

Energy consumption

10. The UK Water Industry is not a major user of energy in National terms - its overall consumption is less than 0.5% of total industrial and domestic consumption. Nevertheless the water industry spends more than £150 million on energy each year. Over 80% of the energy is supplied as electricity with the majority used for two purposes:-

   i) pumping water and sewage
   ii) aeration in the activated sludge process

An attempt has been made by the authors to calculate annual energy consumption per head of population for water supply, sewerage and sewage treatment (Table 2)(2).

11. Caution should be used when applying these derived data, but for example an average water supply uses 30 kWh/head/yr, sewage pumping 9kWh and activated sludge treatment 17kWh. On site combined heat and power system (CHP), where the methane gas developed by anaerobic digestion of sewage sludges is used to generate electrical power with principally gas fired dual fuel engines, can provide 80% of treatment plant energy needs (13kWh/head/year). This can reduce the amount of power required to be bought from Electrical Supply Undertakings to 15 kWh/head/year.

Combined Head and Power Generation (CHP

12. The majority of activated sludge plant energy demand is for aeration, 60-80%, with return sludge pumping consuming up to 10% and tertiary treatment 10-15%. It is estimated

Table 1. The 4 largest activated sludge plants in Thames Water

Plant	Population (millions)	Flow $10^3 m^3$/d	Type[1]	IDT^2 (hrs)	Effluent Quality BOD	NH_3-N
Beckton	2.4	1030	DA	7	9	2
Crossness	1.7	620	SA	6	7	5
Mogden	1.4	480	DA	8	13	2
Deephams	0.7	230	DA	12	3	2

1. DA - Diffused Air, SA - Surface Aeration
2. IDT - Influent Detention Time

that aeration would cost Thames Water in excess of £9 million per annum if all the electricity required was purchased from the Electricity Supply Undertaking.

In practice Thames Water generates the equivalent of 180 million kWh of electricity in CHP units each year burning 100 million cubic metres of sludge gas and 15,000m of diesel (for engine starting and gas make-up) in the process. CHP using dual fuel engines is only economic at Thames Water's largest works serving populations of 0.25 million and upwards. This is due to high capital and maintenance costs together with diesel requirements and a significant labour cost for shift manning. Automatic and computer control of generating plant as presently under design for Mogden (1.4 million population) together with the gradual introduction of package spark-ignition engines may reduce the size of plant where CHP is economic to the 10,000 to 100,000 population range but the case has yet to be proven.

13. In countries where power supplies are limited and unreliable there may be a stronger case for CHP at small plants. However the problems of maintenance and skilled operation must be taken into account. For small and medium sized plants it may be wiser to design plants that can accept interruption of power supplies for short periods or to provide simple diesel powered standby generators. Coarse bubble aeration package plants and oxidation ditches are the most suitable for this application.

Automation to Improve Efficiency

14. Next to financing and depreciation charges manpower is the highest revenue cost in most activated sludge treatment plants. Thames Water in common with WRC and the rest of the UK water industry has directed a major effort in the 1980's towards developing and applying instrumentation, control and automation (ICA) systems for the management and control of water and sewage treatment processes in order to:

i) reduce operating costs (energy, chemicals manpower, maintenance)
ii) defer capital expenditure by increasing throughput of existing plants
iii) enable plant control to be more adaptable to changing operational needs.

15. In the field of activated sludge treatment Thames Water and WRC have established a computer monitored and controlled Evaluation and Demonstration Facility (EDF) at the Witney activated sludge plant. This facility is utilised for the evaluation and demonstration of sensors and control equipment. Strategies and reliable methods for automatic process control and optimisation are being evaluated with WRC as part of a long term research programme for all of the UK water industry.

16. Thames Water has also installed fully operational minicomputer monitored and controlled ICA systems at High Wycombe and Mogden serving a total population in excess of 1.5 million. Application of low cost microcomputer based ICA, telemetry and alarm call out systems is also becoming well established. In addition to optimising process performance these systems linked via telemetry to remote control centres enable significant reductions to be made in the need for high cost shift manning of these plants.

Table 2. Typical energy usage per head for population served for water supply, sewerage and sewage treatment

Process stage	Energy Consumption kWh/m^3	Energy Consumption kWh/head/year	Energy Cost £/head/yr*
Water abstraction, treatment and supply (Pumping)	0.3	30	1.20
Sewerage and sewage pumping	0.1	9	0.36
Sewage preliminary and primary treatment	"	Negligible	"
Secondary treatment			
Activated sludge aeration	0.2	16	0.64
Return sludge pumping	0.01	1.1	0.04
Tertiary treatment			
Microstrainers or filters	0.02	2.2	0.09
Sludge treatment and disposal			
Digestion + power generation	-0.16	-13	-0.52
Tankering to land	1 litre fuel/m^3	0.5l/hd	0.15

* 1 kWh costed at 4p

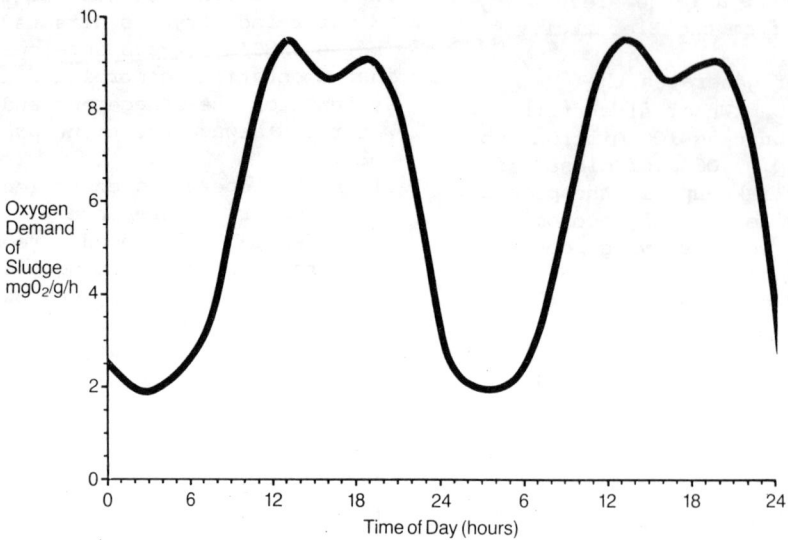

Fig. 1. DO control: typical diurnal variation in oxygen demand of a ditch activated sludge

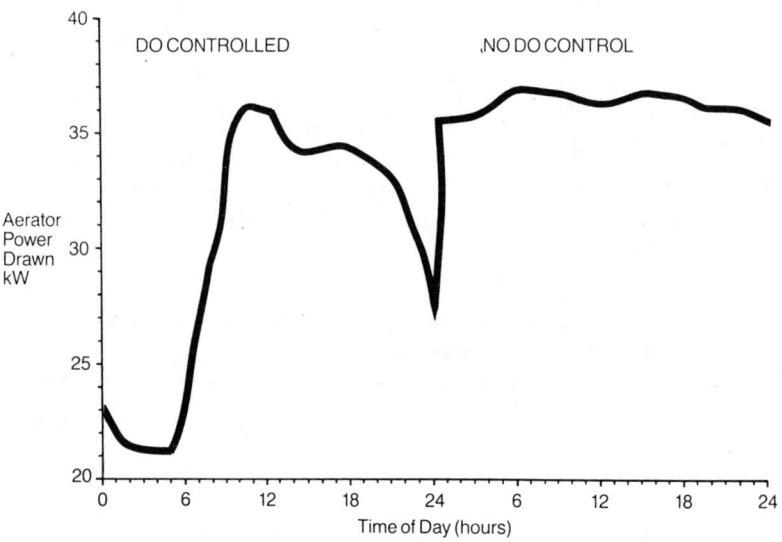

Fig. 2. DO control: surface aerator power consumption with and without DO control

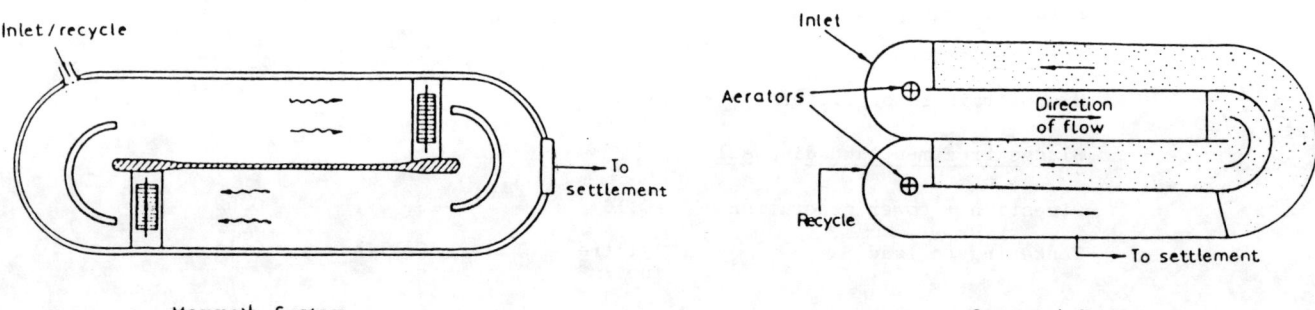

Fig. 3

MINIMUM OPERATING COST ACTIVATED SLUDGE PLANTS - OXIDATION DITCHES

17. Retrofitting ICA systems to conventional primary sedimentation, activated sludge and anaerobic digestion plants permits reduced labour costs at existing works. Ideally new plants should be designed which are relatively simple to automate and require the minimum of maintenance and manning. Traditionally whole sewage package plants and Pasveer type oxidation ditches have been used for activated sludge treatment at small and infrequently visited sites.

18. Oxidation ditches can be considered as low-intensity activated sludge systems operating in the extended aeration range of sludge loading. Usually they treat raw sewage, without the use of primary sedimentation, using an "endless loop" or closed circuit aeration channel with an influent detention time of 1 to 3 days. Typically a surface aeration system is used to provide oxygenation, mixing, and horizontal impetus to oxidation ditch aeration tank contents. Final settlement may be achieved within the aeration tank or, more commonly, by utilising conventional final settlement tanks.

19. This basic or traditional oxidation ditch is often known as the Pasveer Ditch owing its name to Dr. A. Pasveer, of the TNO Research Institute in Holland, who developed the first full scale ditch in 1953(3). Many hundreds of small Pasveer oxidation ditches (<10,000 p.e.) have been installed in the UK during the last 20 years. Most have been cheap to build and have reliably produced high quality effluents, at low manning levels, with claims of low "stable" sludge production and minimal sludge disposal problems.

20. That this is so gives credit to Pasveer's original philosophy which was to provide simple systems that would treat sewage and aerobically digest sludge in one tank.

21. However, although large in number such ditches do not represent a very significant proportion of the total treatment capacity in the UK mainly because the traditional oxidation ditch extended aeration process has been considered too extravagant in aeration tank size and aerator energy consumption to use for larger treatment plants.

22. The development of more efficient aeration and mixing systems in the decade 1960-70 permitted the construction of deeper ditches operating at higher volumetric BOD loadings, thereby reducing aeration tank sizes and capital costs. The more efficient aerators, operating with close limit DO control, also permitted a considerable reduction in aeration energy consumption with only a small penalty in increased sludge production.

23. Where the option of direct land disposal of gravity thickened oxidation ditch sludge was available the overall cost of sewage treatment by second generation ditch systems was often reported to be significantly lower than that for conventional treatment options. This cost advantage was particularly enhanced wherever high quality effluents (better than 15 SS: 10 BOD: 5 NH3 at 95 percentile compliance) were required, as conventional plants needed a tertiary treatment stage, whereas efficiently designed ditches did not. These developments led to the increasing adoption of second generation ditch systems for sewage treatment works, serving populations of 10,000 to 100,000 in the UK, and sometimes more than 500,000 elsewhere in the world.

24. More than 30 second generation ditches have now been installed in the UK, serving a total design population of 1 million. Together with 240 Pasveer ditches, the oxidation ditch process now serves some 2 million people representing nearly 20% of all UK activated sludge plant treatment capacity. Oxidation ditches currently appear to be taking 30 to 50% of the UK market for new biological treatment plants in the population range 1000 - 100,000 (4).

25. Modern oxidation ditches provide appropriate technology for activated sludge treatment in much of the third world. They employ advanced process concepts but require simple low cost construction techniques and can be operated efficiently with simple and robust process control systems.

DISSOLVED OXYGEN CONTROL TO SAVE ENERGY

26. Bacteria in the activated sludge treatment process require no more than about 2 mg/l of DO in their surrounding water, to treat sewage and oxidise ammonia at an acceptable rate. Supply too little oxygen to the system and the bacteria will reduce the dissolved oxygen (DO) levels compromising treatment performance. Conversely, supplying more oxygen than the bacteria need, results in a rise in DO levels to unnecessarily high and inefficient concentrations.

27. The oxygen demand in an activated sludge plant varies with diurnal changes in load. A typical Thames Water oxidation ditch plant has been found to require four times as much oxygen at mid-day as in the early hours of the morning Figure 1. This oxygen demand variation can be met by continuously measuring DO levels, establishing the optimum DO concentration and then maintaining the plant between two closely set DO limits by varying aerator oxygen input. Conventional treatment plants have generally employed fixed input aeration systems. These were designed to satisfy peak load demands, usually 1.3 to 1.5 times average load and hence wasted energy at times of low load, usually the early hours of the morning.

28. Some managers in the authority have modified their installed plant, fitting time switches or dissolved oxygen meters to switch off one or more aeration devices during low load. This has resulted in up to 10% savings in energy costs. The greatest savings were obtained when dissolved oxygen control was included in the plant design at the outset. A good DO control system can save 25% of the energy consumption of a conventional plant. This saving was shown as Ash Vale (25,000 population) in 1976 when the plant was operated first under DO control and then on full aerator immersion, Figure 2.

29. Most of the new works constructed by Thames Water since its formation in 1974 have been in the size range 10,000 to 100,000 population equivalent. Many of these plants have employed the oxidation ditch principle

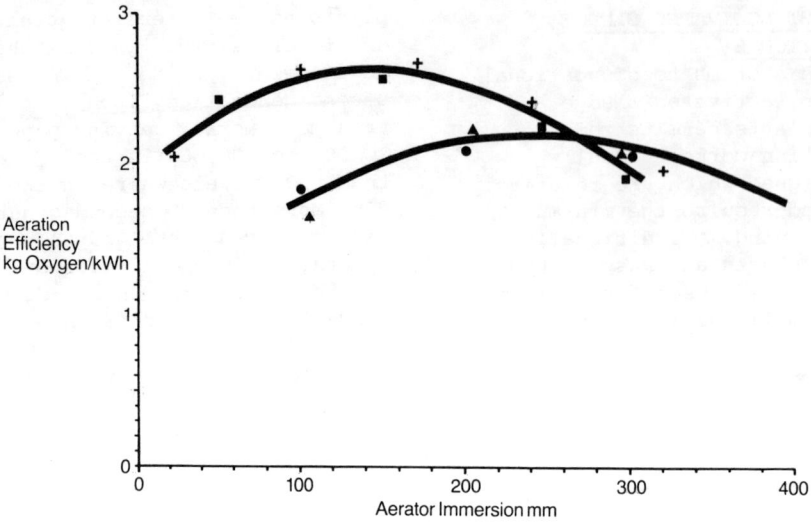

Fig. 4. Aerator optimising: aerator efficiency curves of some recently completed ditches

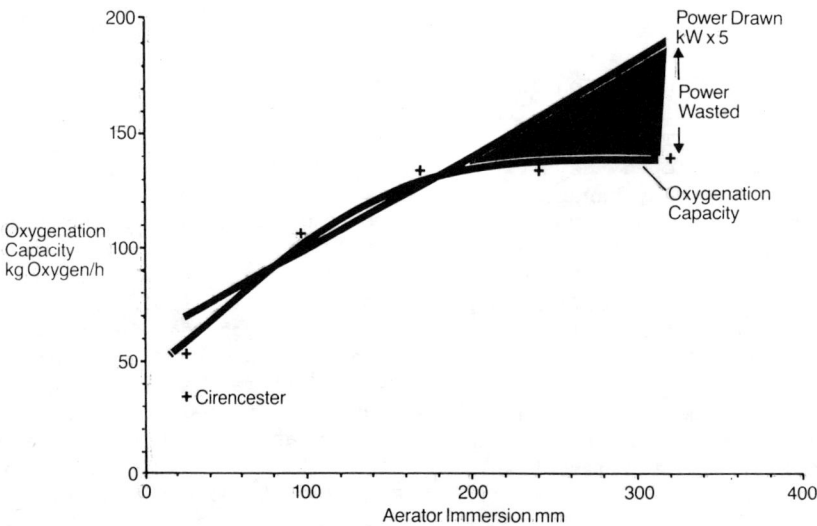

Fig. 5. Aerator optimising: oxygenation capacity and power drawn by Cirencester cone aerators

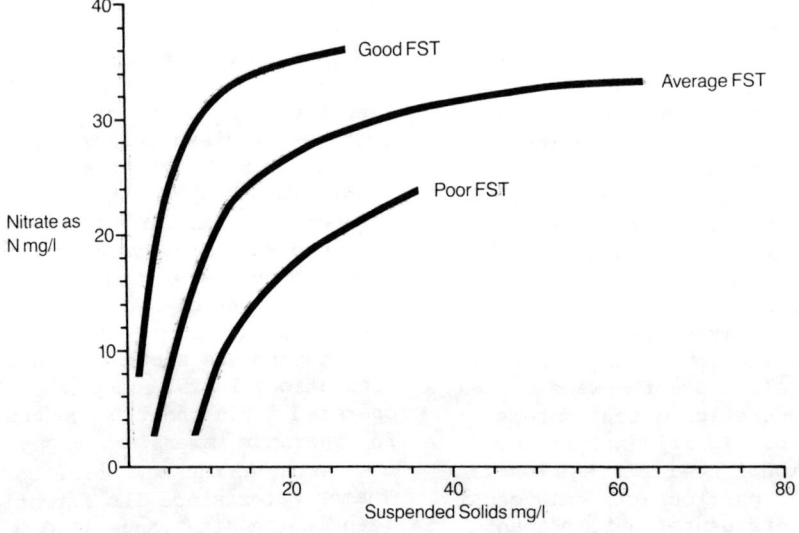

Fig. 6. Effect of nitrate on final settlement tank (FST) performance: effluent nitrate - effluent suspended solids

typified by the Carrousel at Ash Vale and the Mammoth Rotor type of plant recently completed at Esher (Figure 3). These plants were simple to construct and operate but employed a relatively advanced concept, for the sewage treatment field, in that they utilised variable output aerators operating under "close limit" dissolved oxygen (DO) control. The output of such aerators is usually varied by their depth of immersion in the aeration tank contents.

30. The DO control systems used with the variable immersion aerators have been relatively low in cost, < £1,000. They consisted of Makereth type electrodes, angled at 45 degrees with direction of flow, immersed by 0.5m and with all protective shielding removed The electrode monitors had upper and lower limit switches which sent a signal to the aerator process controller. The latter gave the operator a degree of flexibility by providing dials for frequency of DO interogation and % aerator immersion change whenever outside operation selectable limits. Typically a 0.5 mg/l DO spread was set, with interogation every 10 minutes and 5% aerator immersion increments. Such systems have provided reliable service over 10 years of operation. Maintenance has consisted of weekly checking DO sensor readings in the tank by means of a portable DO meter and replacement of the electrode assembly (£20) every 18 months to 2 years.

31. Similar DO control systems have also been installed at a number of diffused air plants, for example at Basingstoke (100,000 population) where the variable input aeration device is a thyristor controlled DC motor blower system. Recently a DO control system was retrofitted, for a total cost of £1,200, at Riverside trade wastes surface aeration plant where a 38% reduction in aerator energy consumption has been reported, representing an annual saving of £10,000 on the electricity bill (at 4p/kWh).

32. More sophisticated computer controlled air flow systems have been under investigation at Rye Meads as part of the WRC managed project 1815 "Energy Saving: Optimisation of Fine-Bubble Aeration"(5). This study indicates greater predicted energy savings using computerised DO control than simple single point monitors. Thames Water is now adopting this technology at some of its larger plants.

AERATOR EFFICIENCY CURVES

33. The ideal variable output aerator would have a linear relationship between oxygenation capacity (OC) and power consumption i.e. be equally efficient over its operational range. In practice there is usually a curved relationship with an optimum aerator efficiency which can only be found by full-scale aerator efficiency tests. Figs. 4 & 5. The shape of these curves, for a given aerator design can be influenced by the size and shape of the aeration tank in which it is installed. It is therefore desirable to carry out aerator trials on every new plant completed until a sufficiently large database is established to identify fully the effect of tank dimensions, aerator position and aerator design on performance characteristics.

Aerator efficiency trials: theory

34. A standardised test procedure is used to measure aerator efficiency employing the so called "unsteady state" method. This involves de-oxygenating clean water in the test tank by addition of the chemical sodium sulphite. The aerator is adjusted to its desired setting and switched on. Dissolved oxygen levels rise and are monitored chemically or by instrumentation until the tank contents are saturated. Calculations based on the rate of rise or oxygen levels enable the aerator oxygenation capacity, OC, to be determined. This is usually expressed as Kg oxygen/hour. Dividing this figure by the power consumption of the aerator gives its efficiency, quoted as Kg oxygen/kWh. A typical surface aerator will provide $2KgO_2$/kWh into clean water. This may sound a simple procedure, but in full-scale practice the problems involved in obtaining meaningful results can be considerable.

Aerator efficiency trials: practice

35. The major problem is one of scale. Aerator trials are amongst the largest form of "bucket chemistry" in the chemical world. A typical oxidation ditch aeration tank may have a volume of 7000m^3 (1½ million gallons) and require 10 tonnes of concentrated sulphite solution to deoxygenate it. Preparing such a solution and then evenly distributing it throughout the aeration tank is a considerable logistical problem, particularly when it has to be repeated daily for six separate tests.

36. Monitoring of dissolved oxygen levels can also place a large demand on staff time. It is usual to employ 6 recording dissolved oxygen meters each of which must be continuously manned, to provide accurate timing of readings. The large number of data generated require considerable processing and computing time before the final aerator efficiency curve can be produced.

Aeration efficiency specification

37. Thames Water has designed and constructed a computer controlled mobile aerator testing unit for this purpose. This system is used for performance assessment of aerators against energy efficiency clauses included in the tender specification for all new plants installed in Thames Water.

The clauses require manufacturers to state the performance of their aerators at several points on the efficiency curve. The tender documents state that aerator trials will be carried out on completion and that aerators that fail to meet claimed specifications will be rejected or the difference in operating cost over 15 years be made up by the manufacturer. This can only benefit the industry, weeding out unsuccessful designs, encouraging manufacturers to develop more efficient aerators perhaps with more linear oxygen input/power relationship that would be ideal for DO control systems. It should be noted that an aerator consumes electricity to the value of the original equipment every one to three years emphasising that energy is the major cost in owning an aeration system!

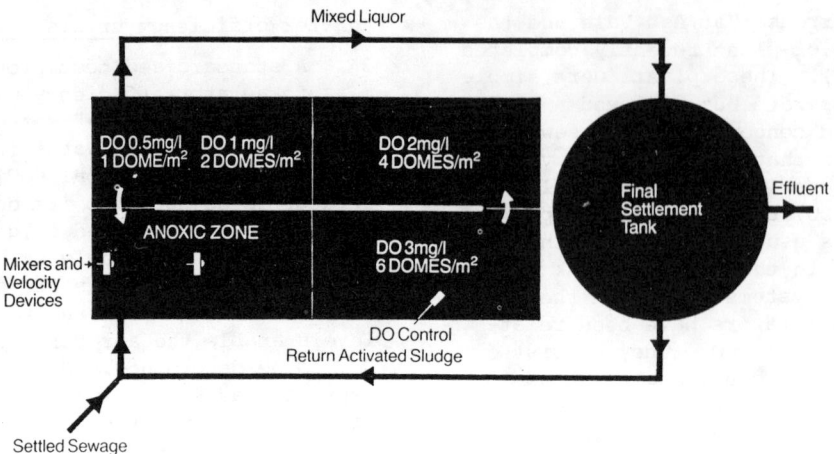

Fig. 7. Schematic of Rotanox experimental activated sludge plant at Basingstoke STW

Modification of dome layouts in diffused air plants and use of anoxic zones

38. Recent studies by Boone[5] and Houck [6] have shown that there is a considerable variation in oxygen transfer efficiencies (1.0-2.2 KgO_2/kWh) between the many fine-bubble diffused air activated sludge plants in the UK. They attributed this variation to a number of reasons, the major ones being lack of DO control and poor dome layout. Completely mixed aeration tanks are ideal for DO control and design of dome layout but present other process operation problems including a tendency to produce poor settling sludges (bulking). A large number of diffused air plants in the UK have length to width ratios of 10:1 or greater and can be considered to have considerable plug flow characteristics.

39. In an activated sludge aeration tank with high length to width ratio the uptake of dissolved oxygen by the biomass is much greater at the beginning of such tanks and gradually decreases as the contents flow along the length of the tank. Consequently in tanks with even dome distribution the concentrations of DO tends to rise to levels approaching saturation (approximately 10 mg/l) at the end of the aeration tank. This situation is wasteful of energy as the rate of oxygen transfer from gas to liquid is proportioned to the oxygen deficit from saturation. DO control alone will not solve this problem as the oxygen demand at the end of an aeration lane can be much lower than the minimum recommended air flow per dome of 0.85 m^3/[7] with conventional dome densities.

40. Tapering of dome densities and air distribution has become more common practice at many works but rarely has a taper of more than 2:1 been employed. Recent development by Thames Water at Basingstoke[8] and WRC at Rye Meads[5] have suggested that tapers of between 3 and 6:1, linked with automatic DO control, can be employed to provide better overall control of DO levels.

Anoxic Zones and Denitrification

41. In conjunction with DO control and tapered aeration, there have been moves to install stirred anoxic zones at the inlet end of many activated sludge aeration tanks. These zones allow the bacteria responsible for BOD removal to utilise oxygen from nitrate, present in return sludge, instead of requiring energy intensive aeration. In theory 67% of the oxygen used for nitrification can be reutilised for BOD removal if all nitrate formed is removed by denitrification in an anoxic zone. In practice, single pass anoxic zones achieve on average a 40% reduction in final effluent nitrate levels. Such a denitrifying system was installed during the mid '70s in the 95,000 m^3/d (DWF) plant at Rye Meads[9]. The main reason for this was to reduce effluent nitrate levels going into the River Lea. Subsequent studies showed that aeration energy consumption had been reduced by up to 15% after the installation of the anoxic zones.

42. Anoxic zones have now been retrofitted to 5 more works in Thames Water and retrofits are under consideration at a number of other works. This entails baffling off 10 to 12% of existing aeration tanks, removing all aerators and installing low power mixers. All the 15 new oxidation ditches installed by Thames Water over the past 10 years have also included anoxic zones.

The justification for installing anoxic zones does not solely depend on energy saving but rather on a number of aims:
1. Energy saving by utilisation of nitrate oxygen
2. Improvement of effluent quality with respect to suspended solids and associated BOD by prevention of denitrification problems in final settlement tanks (Figure 6)
3. Reduction in effluent nitrate levels
4. Reduction in the need to construct and operate tertiary treatment plants because of improved effluent quality (Figure 6)

ROTANOX

42. The denitrifying system installed on an experimental plant at Basingstoke was novel in that it employs rotary flow through an anoxic zone, hence the trademarked name of ROTANOX. By inter-connecting two parallel plug flow aeration lanes and installing horizontal axis impellers in an anoxic portion of the tank it was possible to achieve a flow pattern and oxygen profile similar to that found in the latest generation of oxidation ditches (Figure 7)

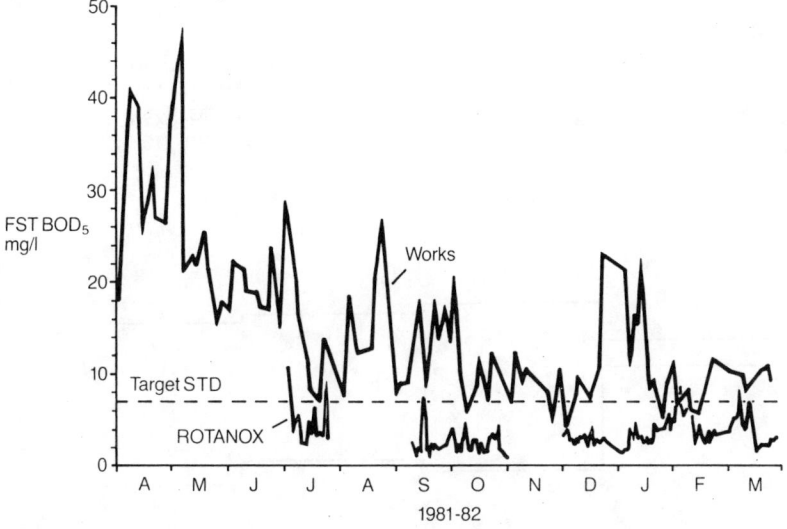

Fig. 8. Rotanox effluent BOD: Rotanox final settlement tank BOD_5 atu compared to main works FST's

The initial 25% of the aeration tank had all domes removed and low energy mixers installed to create the anoxic zone. The following aerobic zones were arranged with the domes tapered from 6 domes/m² to 1 dome/m², and was operated under close limit DO control.

43. The experimental ROTANOX plant treated 4000 m³/d of settled sewage for a three year period between 1981 and 1984. Operating at a loading (F/M) of 0.11 gave excellent quality effluent with a mean of 11 SS, 4 BOD and 2 NH_3-N (mg/l) (Figure 8). A high degree of denitrification was achieved, 76% total N removal, maintaining nitrate levels at between 7 and 10 mg/l NO_3-N (Figure 9). Energy consumption for aeration averaged 0.9 kWh/Kg BOD removed at a calculated oxygen transfer efficiency of up to 2.4 KgO_2/kWh. These figures are between 15% and 50% better than the majority of conventional nitrifying activated plants which have not employed energy saving techniques (Figure 10) (8)

ROTANOX Package Plant

44. The ROTANOX concept has now been developed and installed in four whole sewage package plants which initially utilised coarse bubble aeration. These plants can be used for populations ranging from 100's up to more than 10,000 in one prefabricated unit. Package plants have the reputation of low aeration energy efficiency, mainly due to the use of coarse bubble aerators. In many cases such aerators were justifiable in that they were simple to maintain and were not seriously effected by power failures causing blockage of aeration pores. The advanced ROTANOX package plants that Thames Water has now developed include DO control, anoxic zones, tapered aeration and the option of coarse bubble or fine bubble aeration. The latter system uses plastic disc aerators which can periodically be treated with formic acid to remove dome sliming and sludge blockages whilst still in operation. Energy consumption is reduced to 40-60% of that for unmodified package plants.

45. Thames Water has assisted in the design of ROTANOX plants for overseas installation in

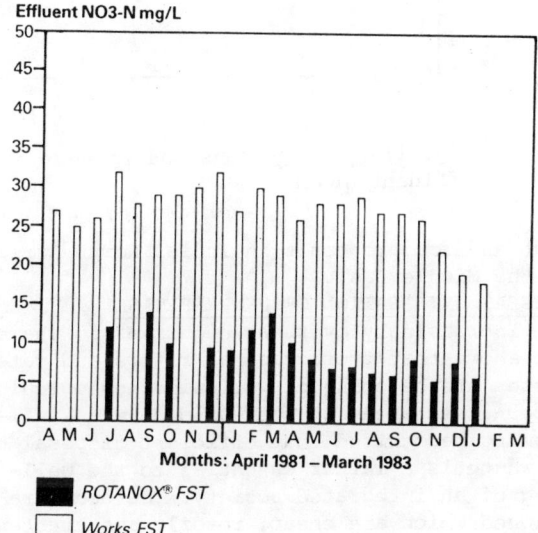

Fig. 9

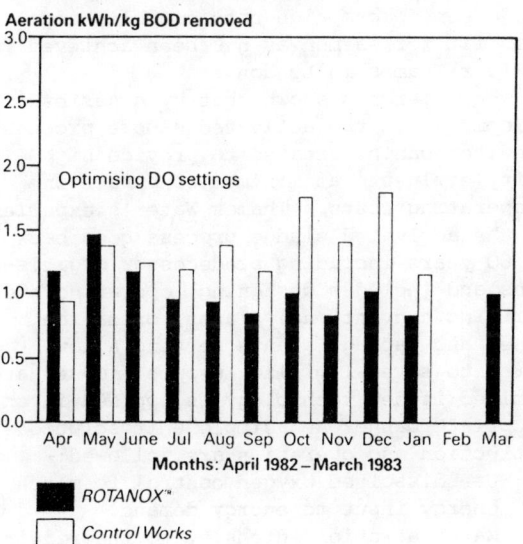

Fig. 10

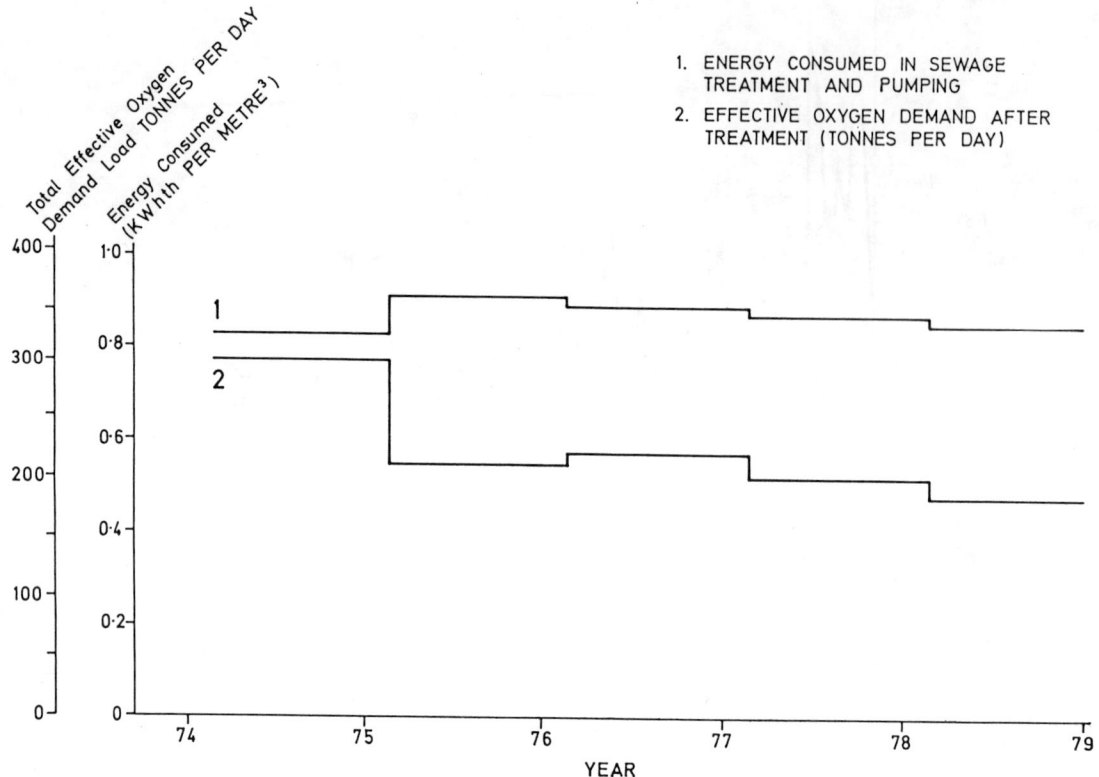

Fig. 11. Energy consumed in sewage treatment and pumping compared with final effluent quality

Egypt and for proposals in India, North Africa and the Middle East.

OVERSEAS UTILISATION OF UK EXPERTISE

46. Increasingly overseas enquiries to Thames have shown that protection of rivers and watercourses are becoming of worldwide ecological concern. As areas of the world develop it is becoming recognised that efficient disposal of all waste is a matter essential to the well-being of an integrated society. Methods are designed which are cheap, readily constructed and offer unit solutions to disposal problems. In sewage treatment this means the provision of plant which will provide the required effluent standard to suit the planned usage of the receiving river or watercourse into which it is designed to discharge. This will vary from a basic clean up campaign upto establishing rivers fit for salmon as has been achieved for the River Thames in London.

47. The paper has shown that by a series of developments in the activated sludge process wastewater can be treated to provide high quality effluents at minimum construction and operating costs. Thames Water's experience with the activated sludge process goes back over 60 years (including predecessor organisations) and includes designing, operating and optimising conventional plants, oxidation ditches and package plants serving a few hundred to several million people. Costs are minimised if new technology is applied in an appropriate manner and simple unit solutions of construction and operation are followed:-

(i) Use dissolved oxygen control to match energy input to energy demand.
(ii) Match aeration intensity to the requirements of the plant layout using tapered aeration where appropriate.
(iii) Use anoxic zones wherever nitrification is required or may occur. Improvements in both effluent quality and energy consumption should result.
(iv) Adopt new technology but use simple treatment processes such as modern oxidation ditches for cities and towns and advanced package plants for villages. Benefits result in both construction and operation.

48. British Manufacturers have risen to this need and are able to quote plant which can follow and utilise these basic principals. Thames have via their Consultancy Service the means to advise on plant suitability to meet many influent and effluent treatment demands which is of course a vital first step. The final figure (Figure 11) illustrates that efforts to minimise energy use whilst reducing pollution load can be successful even at the largest of plants. Over a seven year period polluting load was halved and energy efficiency doubled. Improved efficiency is an ongoing process but it is hoped that the paper indicates areas of potential and examples for those about to become involved in the design and operation of activated sludge treatment plants.

REFERENCES

1. Ardern, E., and Lockett, W.T., "Experiments of the Oxidation of Sewage Without the Aid of Filters, Parts I, II and III" J.Soc.Chem.Ind.1914
2. Rachwal, A.J., and Hurley, B.J.E., "Energy Savings in Wastewater Treatment Processes" J.Inst. Public Health Engineers, 11, 3, 1983.
3. Pasveer, A., "A Contribution to the Development in Activated Sludge Treatment"

J. Proc. Inst. Sew. Purif., 4, 436 (1959).

4. Rachwal, A.J., Oxidation ditches the up and coming solution for the UK?: Presented IPHE/BWETPA Seminar on Oxidation Ditch Technology London March 1984. Awaiting publication J.Inst. Public Health Engineers.

5. Boon, A.G. (1981). Project 1815 - Energy Saving: Optimisation of fine-bubble aeration. Report submitted by WRC to government under the Energy Conservation Demonstration Projects Scheme.

6. Houck, D.H. (1980). Survey and evaluation of dome/disc fine bubble aeration equipment for activated sludge treatment. U.S.E.P.A. Report, Contract No. R806-990-010.

7. Templewood Hawksley Activated Sludge Ltd., Fine bubble dome diffuser activated sludge system. Publication No. FBDD 675.

8. Best, G.A., Hatton, C.J., Rachwal, A.J., and Hurley, B.J.E., 1984. "Biological phosphorus and nitrogen removal at an experimental full scale plant in the UK". Presented at IAWPR workshop on Biological Phosphorus Removal. Paris, September 1984.

9. Cooper, P.F. et al (1977). Recent advances in sewage effluent denitrification: Part 1. Wat. Pollut. Control, 76, 3, 287-300.

Discussion on Papers 7 and 8

DR H. O. PHELPS, Water and Sewerage Authority, Trinidad and Tobago
The question concerns the values of coefficients used in the design of stabilization ponds, aeration lagoons and other treatment possibilities. There is considerable variation in recommendations made in the literature. Therefore the experience of Dr Huiswaard would be, in this regard, particularly welcome.

We are led to believe that temperature is the principal variable in adjusting coefficients from one temperature to another but there is gathering evidence that variations in the flora and fauna found in ponds and lagoons have significant effects on the biological reactions inherent in treatment processes.

DR P. J. HUISWAARD
The first-order rate constant k applied for designing facultative ponds usually varies between 0.25 per day and 0.40 per day at 20 $^\circ$C. The flora and fauna in ponds affect the biological treatment significantly by the following mechanisms

(a) oxygen production by algae
(b) adsorption of organic matter and microorganisms on to water plants
(c) nutrient uptake by water plants
(d) algae consumption by fishes.

The presence and the growth of different types of water plants are also greatly influenced by the temperature of the water and the air and may therefore play a role in the choice of the constant k. The choice of k is still empirical and varies greatly from country to country, or region to region.

Much research should still be made to obtain a universal method for determining the value of k.

Common formulae for calculating k are

$$k_T = 1.2 \times 1.085^{T-35} \quad (1)$$

$$k_T = 0.26 \times 1.085^{T-20} \quad (2)$$

Formula (1) results in values of k that are about 35% higher than those from formula (2), so formula (1) results in 35% smaller pond areas than those from formula (2). The choice of which formula to use will depend on local circumstances and experience.

PROFESSOR T. VIRARAGHAVAN, University of Regina, Canada
In the case of facultative aerated lagoons, usually mixing levels should be greater than 6 HP/10^6 US gal (to maintain 50 mg/l ss in suspension). This generally is enough to give dispersion of oxygen and not complete mixing. Complete mix aerated lagoons require higher mixing levels and settling of lagoon effluent. Further reference may be made to Professor Eckenfelder's work in this area.

DR P. J. HUISWAARD
The power for mixing of 6 HP/10^6 US gal, equivalent to 1.2 W/m^3, may be sufficient for aerated facultative ponds.

The power requirement will also depend on the strength of the type of waste water fed to the pond.

DR T. G. LETON, University of Port Harcourt, Nigeria
What is the Author's experience with regard to the problems and convenience of operation of anaerobic ponds?

DR P. J. HUISWAARD
An anaerobic pond is a very good system for the pretreatment of relatively strong organic waste water. The system may be applied to a wide range of waste water types, e.g. waste water from food/agricultural industries, paper industries and leather industries.

Seeding with adapted sludge in a proper manner is very important, just as the control of pH and the prevention of shock loadings are. Bad odours are an eternal problem with anaerobic ponds. Therefore the location of these ponds has to be selected with great care.

MR B. LUNOE, Hifab International AS, Norway
Concerning the use of ponds in rural development, here is an example from a school project in Zambia, where the programme called for water-borne sewerage. There were two requirements

(a) a sufficient and continuous water supply
(b) functional operation and maintenance.

In this case, both conditions failed completely resulting in collapse of the system.

PAPERS 7 AND 8: DISCUSSION

The conclusion is that it is not sufficient for engineers to apply only their theoretical knowledge, but it is important to ensure that the local conditions and what may realistically be obtained from the local system are understood.

DR P. J. HUISWAARD
We completely agree with this useful information.

MR B. N. THYAGA RAJA, Bangalore Water Supply and Sewage Board, India
What is the percentage of energy produced from the sewage gas? Can methane and carbon dioxide gases separated from the sewage gas be bottled and used for industrial applications?

DR P. J. HUISWAARD
An amount of 1 kg COD removed gives 0.3 Nm^3 of biogas, and 1 Nm^3 of biogas is equivalent to 1.2 kWh of electricity.

Biogas usually consists of about 75% of methane gas (CH_4) and 25% carbon dioxide gas (CO_2). CO_2 and CH_4 can be separated by conveying the biogas through a solution of hydroxide, but this separation is not really necessary before using the biogas as a source of energy in the treatment plant. CH_4 gas is not easily bottled because of the very high pressure needed.

MR M. A. ALDER, Wessex Water Authority, UK
Biogas can quite simply be cleaned and can have advantages for process efficiency and storage. However, cleaning processes have an energy cost and are never 100% efficient and it may be found that cleaning is too costly.

The electrical energy from biogas varies but as a guide 20-25% from small generation units with a maximum of 35% in larger units is typical.

Could the Authors of Paper 8 comment on

(a) the dissolved oxygen setting for the control of activated sludge processes
(b) the overall running cost of instrumentation, control and automation as applied to sewage treatment, in particular the staffing quality and the life span of the equipment with associated replacement costs?

DR L. R. J. VAN VUUREN, National Institute for Water Research, South Africa
Currently there are about 150 ORBAL activated sludge process plants in operation, developed in South Africa about 20 years ago, ranging from 1000 to 70 000 population equivalents served.

The process is robust, simple to operate and requires little maintenance.

Its special features can be summarized as follows

(a) low capital cost: when using a standard design with local materials labour cost savings of more than 50% can be realized.
(b) simplicity of mechanical equipment with only one moving part and little maintenance
(c) ability to deal with shock loadings
(d) flexibility to incorporate biological denitrification

The ORBAL plant is finding increasing use in developing areas because of its robustness of performance.

MR E. W. WARNER, Water and Sewerage Authority, Trinidad and Tobago
I am interested in the package sewage treatment plant. I doubt whether the inflow of sewage into the plant could be stopped to incorporate anoxic zones in it to upgrade it. Could Mr Rachwal indicate some of the problems that he encountered in having to keep the plant operating while he carried out the modification works and how he solved these problems?

MR M. A. HARTLEY, Simon-Hartley Ltd, UK
Considerable work has been done by the Authors of Paper 8 in the study of oxidation ditch systems. Recently in-ditch clarifiers have been introduced. Does Mr Rachwal have any information concerning the effectiveness of such clarifiers and their cost advantages, if any?

MR C. E. S. RAO, Hindustan Construction Co. Ltd, India
Aeration cages in oxidation ditches have been affected by corrosion in tropical situations.

Owing to the high costs of energy and the availability of adequate power, it is uneconomical to invest in the activated sludge process. Further the maintenance and operation of such large plants requires high skill, which is not easily available in developing countries.

The energy charges are proportional to the size of the demand and for larger plants the rates are proportionally higher. Hence the activated sludge process tends to be prohibitive in costs particularly in evaluation and capital cost.

Hydrogen sulphide is highly corrosive in sewage gas in tropical countries, causing damage to digestors and gas holders. This is a major problem and costly protective measures have to be taken.

Sewage gas in being used for domestic and industrial use by the Bombay Municipal Corporation on a limited scale.

MR D. B. COOK, World Bank, USA
A system has been developed for the treatment of sewage from small towns in Israel. A combination of engineers and farmers working together had produced effluents that were suitable for root irrigation from a treatment system consisting of anaerobic ponds and relatively deep (6-10 m) lagoons, the lagoons being basically storage reservoirs (Paper 7, ref. 2).

I would also like to comment on the sewage treatment options available to large cities in developing countries and the energy needs of treatment. In the exhibition there are pictures depicting the sewage rehabilitation currently under way in Cairo. It is understood that the sewage treatment proposals for the East Bank sewers include activated sludge treatment – after extensive pumping. According to the figures quoted in Mr Thomas' Paper the costs of energy may amount to US $8-10 million per annum, less any saving which may result from gas generation. These costs can be considerable in low income societies and I hope that the authorities in Egypt have discounted the operation and maintenance costs of running the new system when forecasting future tariff charges.

With respect to the use of stabilization ponds for sewage treatment for larger cities consider Melbourne and the sewage from a proportion of the city (approximately 1.75 million people). The sewage here is treated at Warabee Farm in stabilization ponds by grass and land filtration. The area of the farm is greater than 3000 ha but thousands of sheep and cattle pass through each year. Izmir, Turkey (population 1.5 million), which has a climate similar to Melbourne's, is considering using ponds. Land in Izmir may be available in the airport 'funnel' approach at prices around US $2 per metre.

DR P. J. HUISWAARD
We agree with your comments which suggest that waste stabilization ponds should be applied whenever possible in developing countries. Unfortunately their application is, in many situations, restricted by the shortage and/or high price of land.

MR J. GOAD, Binnie & Partners, UK
In response to Mr Cook's comments on the Cairo waste water project, one of the primary objectives of the East Bank project is to eliminate sewage flooding from the streets of Cairo. This is being achieved by the construction of a tunnel system which will result in the release of capacity in the existing collector system for further expansion of the secondary sewerage system, using local materials and skills, and the elimination of the majority of existing pumping stations which incur heavy maintenance and operation costs. Particular conditions in Cairo which affected the selection of the solution to Cairo's sewerage and sewage treatment problems were

(a) the high density of development
(b) the relative abundance of the water supply and major increases on the treatment and distribution of water
(c) the high value of land in and around the Nile valley and the delta: the government of Egypt is anxious to preserve agricultural land and to extend it by desert reclamation using sewage effluents.

The cost of running the activated sludge plant was considered at the time of master planning and so also were many other cost factors including land values and pumping costs when arriving at an optimum solution.

DR L. R. J. VAN VUUREN, National Institute for Water Research
With regard to the Clariflow Blue Circle process, is phosphate removed by the chemical dosing and sludge blanket clarification?

MR G. A. THOMAS and MR A. J. RACHWAL, Thames Water, UK
A lime-based chemical is used and about 90% phosphate removal is achieved.

MR L. O. WILD, Laurie Wild Consultants, UK
Could the Authors comment in relation to the energy part of their Paper: have there been any advances in the use of heat pumps at Thames Water?

DR H. O. PHELPS, Water and Sewerage Authority, Trinidad and Tobago
In the design of aeration lagoons, the requirement for power is invariably determined by considerations of mixing rather than oxygen transfer. There are conflicting recommendations in this regard. The experience of Dr Huiswaard would therefore be valuable.

DR P. J. HUISWAARD
The design of completely mixed aerated ponds is usually based on the quantity of oxygen that has to be supplied for the desired biochemical oxygen deficiency removal efficiency.

DR T. VIRARAGHAVAN, University of Regina, Canada
Both in Professor Mara's approach and in the recent World Bank report by Mr Arthur, the coefficient of bacterial decay k_b is taken to be the same for each pond (anaerobic, facultative, maturation pond) in the pond system. This may not be theoretically valid. This also applies in the case of the reaction rate coefficient k in the design of pond systems ($L_e/L_o = 1/(1+kt)$). However, the assumptions made generally give final values not far removed from the actual values encountered. I would like to comment on the tannery waste treatment proposed in Chittagong, Bangladesh. A similar approach (screening plus primary clarifier plus anaerobic lagoon plus aerated lagoon plus polishing lagoon) was developed in a demonstration tannery waste plant in Tamil Nadu, India (refs 1 and 2).

It will be advantageous to draw from the experience of a system operating in the Indian subcontinent in this regard.

REFERENCES
1. Pollution from tannery industries in India. Environment India Publications.
2. VIRARAGHAVAN T. and DAMODARA RAO T. Treatment of tannery wastewater – a case study. Industrial Waste Conference, Purdue University, West Lafayette, May 1986.

C. F. SKELLETT, MSc, CChem, MRSC,
MIWES, MIWPC, Wessex Water Authority, UK

TN17. Energy saving in the activated sludge process

The activated sludge process requires significant amounts of electrical energy, generally in the order of 0.5 kWh/kg of biochemical oxygen demand removed. Considerable effort has been made by manufacturers to improve the oxygenation efficiency of their equipment, although some claims of improved efficiency need to be treated with care. In particular it is important to remember that results obtained in clean water trials are frequently not repeatable when treating sewage. The design of a new plant can incorporate features to obtain maximum energy efficiency, but the majority of activated sludge plants were constructed at a time when energy was relatively cheap; consequently it is important that these plants are examined to determine whether or not improvement can be made. The small and medium-sized activated sludge plants are considered in particular and simple techniques are described that have been employed within Wessex Water Authority to improve the energy efficiency of existing plants.

2. The relative usage of electrical energy in the activated sludge process for a typical treatment plant is

aeration	85%
returned activated sludge pumping	12%
surplus activated sludge pumping	2%
settlement tank scrapers	~1%
control equipment, e.g. actuated weirs	<0.5%

AERATION

3. Aeration is by far the major use of electricity and is often carried out continuously. The purpose of aeration is to maintain sufficient dissolved oxygen (DO) to support aerobic microbiological activity in the mixed liquor; over-aeration can lead to poor mixed liquor settleability and wastes energy. Traditionally, only large treatment works have been equipped with DO control systems, e.g. for the control of air flow on diffused air plants or variable depths of immersion for surface aerators. The latter, in practice, often leads to the plant operating for periods at inefficient depths in terms of oxygen input per unit of power. The majority of small or medium-sized plants were constructed with, at best, manual control systems. For these plants the use of DO sensors and simple programmable logic controllers (PLCs) can significantly reduce power costs.

4. An aeration plant at Chippenham, serving a population equivalent of 20 000, has been equipped with a simple system, produced in-house, in which DO is controlled to a set point of 5% by intermittent aerator operation. A timer override operates mixing cycles with the aerators. This prevents excessive settlement during extended periods when aeration is not required. The system has reduced aerator operation by 20%, saving around £5000 p.a. Further indirect savings have been made at this site by using spare PLC capacity to control the operation of other equipment used for functions that are not time critical. By carrying out these operations while aerators are not running, maximum demand can be minimized. As a result of this approach the load factor at this site is very good - averaging 62%; this is reflected in the low electricity cost of less than 4p/kWh. The overall cost of the control equipment for this site was £2880.

5. At smaller sites where the use of a PLC may not be justified, DO control using simple timers can be effective. At Bowerhill sewage treatment works, serving 7200 population equivalent, a 23% reduction in aerator operation has been achieved, saving over £1600 p.a. for an investment of £2500.

6. A new small plant with a design population equivalent of 700 is operating with significant over-capacity which gave serious problems due to over-aeration. The use of timer control coupled with a variable speed motor fitted to one aerator has stopped the over-aeration and reduced energy usage by 50%, saving around £500 p.a.

7. Occasionally DO control is suitable, particularly if mixed liquor suspended solids are high. At a site with this problem and with substantial over-capacity, control of the aerators has been effected by the use of an ultrasonic sensor to detect the arrival of an inflow to the plant (all inflow is pumped). The system, which cost around £1000, is expected to save about £3000 p.a. - approximately 50% of the total energy usage of the plant.

SLUDGE PUMPING

8. Returned sludge pumps operate for long periods, sometimes continuously. Screw pumps

are frequently used to prevent damage to the biological flocs and these produce a continuously variable flow rate which at a fixed speed is dependent on sump level. The pumps are normally designed to handle peak flows, with motors sized accordingly. At lower rates of return the motor may be significantly underloaded and consequently operating at relatively low efficiencies. The use of thyristor-based equipment, which optimizes underloaded motor running to match the required work, can make small savings - around 5%. This makes the system uneconomic as a retro-fit, but for a new plant when used in place of a conventional starter the additional cost is small, making it a viable option. Additionally soft start/stop facilities are included reducing transmission and bearing wear. Motors also generally run cooler, thus extending their life.

9. Surplus sludge pumping is generally an intermittent operation with limited scope for significant energy saving.

SETTLEMENT TANK SCRAPERS

10. Scraper bridges are usually driven continuously by low-powered motors, costing around £100 p.a. to operate. The use of cheap, flexible timers makes it possible to use intermittent scraper operation. However, care must be taken to ensure that denitrification does not occur in the settlement tanks.

CONTROL EQUIPMENT

11. Actuated weirs are often fitted to control the discharge from an aeration tank. Setting these at a fixed level makes minor savings in the cost of valve operation, but more significantly the aerators can be maintained at their most efficient level, i.e. maximum oxygen input per kilowatt hour. Running time can then be minimized to produce energy savings.

SUMMARY

12. A number of simple cost-effective methods are available which can significantly reduce energy consumption at small and medium-sized activated sludge plants. The major savings are achieved by ensuring that aeration is at the minimum level required to maintain the microbiological activity within the plant. The precise techniques applied will depend on a detailed examination of the plant operation.

W. G. DAVIES and D. E. SMITH, Blue Circle Industries plc, Newcastle-under-Lyme, UK

TN18. The use of an upward flow sludge blanket process for sewage treatment

The use of the upward flow sludge blanket process for sewage treatment stems from earlier experimental work by Portsmouth Polytechnic, supported by the Southern Water Authority. It has been developed fully by Blue Circle Industries plc, and is now marketed as the Clariflow process.

2. The Clariflow process is a single stage sewage treatment, which produces a quality of final effluent very suitable for direct discharge to estuary or sea. The removal of suspended solids and biochemical oxygen demand (BOD) is enhanced significantly in this process over normal primary sedimentation by the formation of the suspended sludge floc blanket. Large-scale bacterial removal is also achieved because of the strongly alkaline nature of the coagulant, thus enabling the bacterial level in the receiving sea water to be kept below the standards set by the EEC directive for bathing beaches.

PROCESS OPERATION

3. An inorganic coagulant is added to raw screened sewage in order flocculate the suspended and colloidal matter. In passing through a coagulation tank the sewage achieves full flocculation as the flocs are allowed to develop in size. The flocculated sewage then flows into the upward flow tank where a suspended sludge floc blanket is created. The floc blanket is maintained in position by adjusting the velocity of the upward stream to counteract the gravitational forces on the flocculated particles. The blanket level is monitored constantly and various desludging strategies are put into operation dependent on the position and rate of rise of the blanket. The floc blanket acts as a fluidized filter medium, entrapping additional small flocs and fine suspended matter to clarify the effluent further. The very heavy flocs and gross suspended matter settle to the bottom of the tank. The final effluent is discharged from the top of the tank. The sludge is removed both from the base of the unit, as a thick sludge, and from the blanket, as a low solids sludge (approximately 0.5-2% solids). The low solids sludge readily settles to higher solids concentrations (up to 15%) and the combined sludge can be easily filtered to a relatively dry, inoffensive, cake for disposal.

4. The whole process is computer controlled based on the quality of the final effluent. The quality of the effluent in terms of turbidity, pH and conductivity is measured and recorded at regular intervals and this information, together with other data, such as flow rate and pH of the incoming sewage, is recorded in the computer for determining control strategy.

5. In order to reduce installation costs and make the process flexible in application, a modular design concept has been adopted. Each module, which is essentially the upward flow tank, has been designed to be capable of dealing with the sewage from 3000-10 000 people. In this manner, the process is suitable for townships in the population range of 3000 to greater than 10 000. In addition, as the populations of the townships increase, further modules can be added to deal with the increasing sewage level.

OPERATIONAL EXPERIENCE AND RESULTS

6. A full-scale prototype plant was erected in 1984 and ran for 12 months at Sandown, Isle of Wight. Throughout the operational period, the specially developed dedicated computerized control system operated very effectively and allowed the plant to run largely unattended.

7. The results obtained from the prototype were better than anticipated from earlier work. Consequently, a completely new treatment plant using the upward flow process has now been built at Sandown, for treatment of all the town's sewage prior to discharge to the sea.

8. The results can be summarized as follows.

(a) At least 90% of suspended matter has been removed from sewage.
(b) There has been a greater than 99% removal of bacteria from the sewage effluent.
(c) There has been a substantial reduction in BOD and removal of other pollutants such as phosphates.
(d) Odour problems associated with sewage works have been removed.

9. The blanket process is extremely good at removing suspended solid matter. Overall, the removal of solids appears to be largely independent of flow rate through the plant and not dependent on the solids content of the incoming sewage, but, as expected, it is dependent on the nature of the sewage. However, this latter effect can be corrected by increasing the pH. The suspended solids content in the effluent is consistently low and

mostly below 50 mg/l; this is extremely good in comparison with traditional primary settlement, which achieves about 200 mg/l suspended solids content of the final effluent. Typically, the influent sewage in the summer had a solids content of approximately 600 mg/l and in the winter 300 mg/l.

10. Reduction in BOD has been found to parallel solids removal and generally a 70% reduction in BOD is noted. The Clariflow process will not, however, remove soluble BOD.

11. Extensive investigations have been carried out on the effectiveness of the process in removing bacteria, as this is the key parameter requiring improvement to meet the EEC directive. Summer levels of bacteria were about 2×10^8 bacteria/100 ml which is about ten times higher than those recorded during the winter. The investigation demonstrated conclusively that the effect of alkalinity, together with that of the blanket filtration, could remove in excess of 99% of the incoming bacteria. In general terms it is considered that the effluent bacterial level should be reduced over a high proportion of operational time to approximately 10^4 bacteria/ml. This concentration of bacteria on dilution into the sea would reduce to below the EEC requirement of 10^4 organisms/100 ml.

FUTURE DEVELOPMENT

12. Present developments are concerned primarily with the use of the process in conjunction with conventional secondary treatments, so as to enable it to be used for inland applications.

B. S. PIPER, MA, MSc, MICE, Institute of Hydrology, UK

9 Water resources for irrigation

Increased area of irrigated agriculture - essential for meeting future world demand for food - cannot be achieved without an increase in the supply of water. This paper discusses the growing demands for irrigation water and its availability on a global scale. Techniques for the assessment of the available water resources are given, followed by some comments on less conventional sources of irrigation water. Recent developments in data collection are outlined and the paper concludes with some remarks on water resource planning.

INTRODUCTION

1. An irrigation system is just one of the components that make up the agricultural production system. Irrigation may be technologically complex, such as a drip system that supplies precisely metered quantities of water, fertilizer and pesticide to a plant's roots, or very simple: for example a vegetable crop watered by bucket from a nearby pond or river. Nevertheless, the objective of irrigation is the same in both cases, namely to reduce or eliminate water stress upon the crop in an attempt to provide the right conditions for optimal growth and so improve the economic return.

2. The factors that control whether or not a given irrigation scheme is successful are very complex and their interrelationships are still poorly understood. There is a set of economic constraints, as well as social factors which may include amongst others: local farming practices, class or tribal conflicts, maintenance of the water delivery system and so on. There are physical factors which in principle can be measured objectively; these include the availability of water and local climate and soil characteristics. Provided the land is suitable for agricultural production and that water of the required quality can be reliably supplied, then irrigation should in theory be beneficial. Consequently, accurate determination of available water resources is a prerequisite of successful irrigation development.

3. This paper discusses those techniques of water resource assessment which are appropriate for the countries where irrigation is most urgently required. Paradoxically, those regions where irrigation has the greatest potential for increasing the security and volume of food production (frequently areas of rapidly rising populations) are often those regions where detailed knowledge of local hydrology is lacking. Consequently, many of the most modern methods in hydrology are inappropriate and hence are not discussed here.

4. After a brief outline of the increasing needs for water for irrigation and its availability, some of the appropriate methods of water resource assessment are described. Some developments in the collection and interpretation of hydrological and meteorological data - which must underpin all efforts at scientific irrigation through reliable estimation of available water resources - are then outlined. The paper concludes with some remarks about the processes of water resource planning and allocation.

NEEDS FOR IRRIGATION WATER AND ITS AVAILABILITY

5. At the risk of duplicating material included in other papers, it is useful to outline the role which irrigation plays in current worldwide food production and the extent to which this may be expected to increase in the future.

6. Participants at a recent workshop (ref. 1) argued that a timely and well-managed supply of water and its effective use on the land are among the most crucial requirements if newer forms of high yielding agriculture are to be successful. Irrigation, where properly executed, can transform the production potential of agriculture and probably represents the only prospect for food security for a large proportion of the inhabitants of many developing countries.

7. It is also maintained that recent successes in keeping food production ahead of population growth stem largely from irrigation, and half of the increases in grain production since 1950 can be attributed to the joint use of fertilizer, improved seeds and irrigation.

8. FAO reports that despite projected increases in agricultural production, a large number of people in developing countries will still be below the critical food intake level by the year 2000, assuming FAO's normative GDP growth rate of 7.7%, which itself assumes improved performance over present conditions

(ref. 2). Irrigation has been identified as one of the essential ingredients in boosting food production, the others being high-yielding crop varieties, fertilizers and pesticides. To maintain output at or near its theoretical maximum, a high level of these inputs is required. This in turn demands great reliability in the water resource if high investment in other inputs is to be justified. So without irrigation, the potential benefits of scientific agriculture would be lost (ref. 1).

9. Whilst the evident potential for improving world food production through irrigation must be recognised, the temptation to see irrigation as a universal panacea should be resisted. In the developing world, over 80% of the arable land is under rainfed agriculture. It could be argued that improving world agriculture depends upon solving the problems that influence the reliability of rainfed agriculture.

10. For the future, it has been suggested that irrigation could provide 50% of the increase in crop production needed between 1980 and 2000 (ref. 3) either by the improvement of existing irrigation schemes or the continued development of higher yielding crop varieties or by bringing new land under irrigation. The International Food Policy Research Institute (IFPRI) estimated that over the next decade extension of the land under irrigation would account for about 60% of the food production increases it projected for all developing countries (ref. 4). The IFPRI is not alone in arguing that substantial benefits would also accrue from improvements to existing schemes. However, it seems unlikely that the amount of irrigated land can be substantially increased without a dramatic increase in the supply of water for agriculture.

11. As a first step in assessing the worldwide availability of water, it is useful to consider Stanhill's (ref. 5) comparison of the annual fluxes of rainfall and evaporation over the land surfaces of the globe. The difference between the two fluxes represents the water that is potentially available for exploitation. Fig. 1, drawn using rainfall data from Baumgartner and Reichel and their estimates of evaporation (ref. 6), shows that the maximum availability of water is around latitude 50°S. However, the land area of this region is very small and has a sparse population.

12. Another important feature of the graph is the reduction of rainfall in the mid-latitudes, which are the regions of maximum atmospheric demand for water; consequently the potential supply is also low. Moreover, the inadequacy of the potential supply is exacerbated by the high spatial and temporal variability of rainfall over these regions. Nevertheless, it is here that many of the world's major centres of population are found and where population growth rates are high. As Pereira has noted (ref. 7) the proportion of the world's population living in the areas of least reliable water supply is increasing.

13. As far as water resources are concerned, the total area of irrigated land cannot be greatly increased except by more intense and efficient use of hitherto plentiful supplies, or the exploration, evaluation and development of more marginal and less reliable sources of supply. Such sources might include desalinated saline or brackish water or treated sewage effluent. These possibilities are considered briefly below. However, to achieve the higher yields associated with improved seed varieties, the controlled and timely application of water is even more important than for traditional methods and seed varieties.

SCHEME DESIGN AND ASSESSMENT TECHNIQUES

14. Several points implied in the above discussion have a direct impact on the input required from the water resource engineer or hydrologist. Firstly, what sort of development is being considered? Does it involve only improvement in the management of an existing scheme, the water supply for which already exists in the form of storage reservoir or wellfield? Or is a completely new source of supply to be exploited for irrigation? The scale of the project is also important. For a large development, the costs of detailed hydrological analysis might only be a very small proportion of the total cost. On the other hand, the budgets typically available for small schemes seldom allow for any specialist hydrological analysis at all.

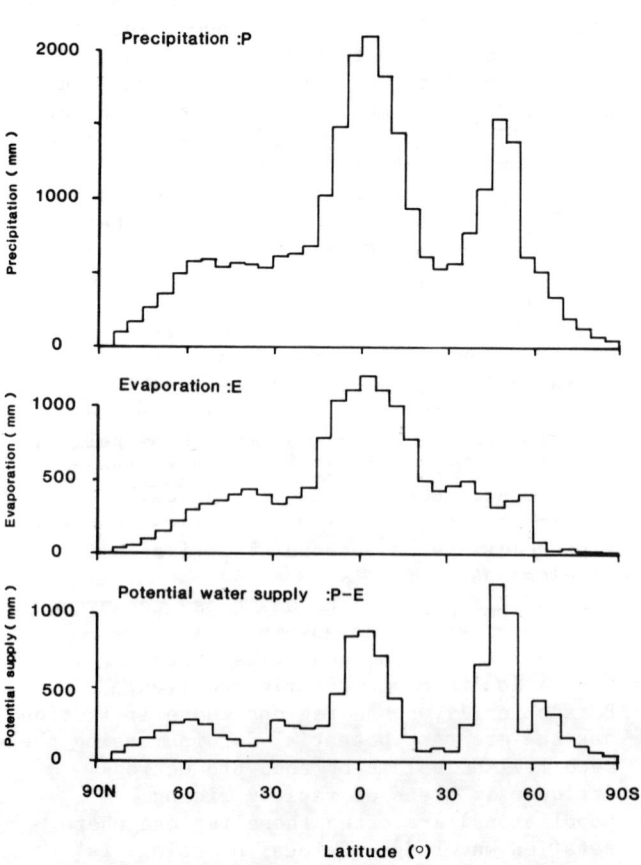

Fig. 1. Precipitation and evaporation over the land surfaces of the globe

15. Whether new development or rehabilitation is being considered, three aspects are common to all irrigation developments, namely the assessment, management and protection of the water resources. Agronomists working with the scheme designers will provide details of the water demand of the irrigated crops and field and transmission losses. The scheme designer then needs to know from the hydrologist the quantity of water available and its seasonal distribution, the reliability of the supply and whether or not the source is renewable.

Surface water resources

16. Typically, many large scale schemes rely on some form of gravity diversion or pumped abstraction from a flowing water course, which itself may be regulated by a storage reservoir upstream. In these cases long flow records at or near the dam site may be available and will usually be sufficient for design and operation. The long term mean flow and its annual and season variability at the site of a irrigation scheme downstream can then be determined from the reservoir's operation policy. In cases where flow records are too short for hydrological design purposes, they may be extended by some form of hydrological model that uses a longer period of rainfall records as input.

17. On smaller schemes, where financial resources are limited, it may not be possible to carry out detailed hydrological analysis. Yet the reliability of the water resource is perhaps even more important in such cases, so that some form of regional analysis may be appropriate.

18. Rainfall records are more readily available than streamflow records, they cover a wider area and are available for longer periods. Consequently rainfall is generally used as the basis for simple regional water balances. These may show up anomalies in individual records, as well as allowing estimation of the mean annual runoff for any ungauged catchment within the region.

19. Empirical statistical techniques can be used to relate flow indices from the observed records to measurable physical characteristics of the respective catchments. For example low flow analysis can be used to identify the limits of reliable irrigation. Typically low flow sequences from a large number of flow records are analysed by extreme value techniques to give critical cumulative curves corresponding to a given return period. Such curves, in which flow is expressed as a proportion of the long-term mean annual runoff, are generalised for different climatological regions. Then, by using the mean annual runoff at any site within the region (estimated from observed records or catchment characteristics), the family of curves can be used to estimate the cumulative low flows at that site for any duration of up to several years, and for a specified return period.

20. This type of technique, adapted from methods originally developed for the United Kingdom (refs 8 & 9) has been successfully used in various parts of the world. Examples are low flow studies in Malawi and Sri Lanka and flood studies for Indonesia, Thailand and Malawi; a world wide database of flood information has been compiled and has been used to calculate regional flood frequency curves for over 50 countries (ref. 10).

Groundwater resources

21. The estimation of available groundwater resources presents a completely different range of problems as intensive and expensive exploration is often required. Groundwater reserves are measured indirectly through a combination of test drilling, pumping tests, long-term monitoring of water table fluctuations and more recently, various remote sensing techniques. Remote sensing cannot be used extensively for quantitative resource mapping unless ground truth measurements are also carried out.

22. Experience suggests that one of the most productive lines for ground water exploration derives farily simply from analysis of the traditional settlements of an area. The physiographic setting of the settlements, the nature of their water supplies, the size and permanence of the settlement, are primary sources of data regarding the occurrence of groundwater and its long-term reliability within the climatic variation of a region. Local exploration for groundwater develops almost invariably from analysis of traditional sources yet most systematic regional studies rarely attempt to make use of this information.

23. Although surface water and groundwater have been discussed separately, it is the interaction between the two that becomes crucial in regions where water resources are scarce. In recent years the potential for using surface and groundwater in conjunctive use schemes has become increasingly important, as this can increase the total water availability to meet short duration peak water requirements that can occur at predictable points in the cropping calendar.

24. It would be false to assume that separate assessment of surface and groundwater resources is desirable just because the scientific disciplines, the type of data and the methods for analysis are themselves different. Rainfall and its variation within the climatic, topographic and geological setting promote both runoff and underground recharge. The assessment of the renewable water resources of whatever area may well derive from a similar understanding and interpretation of common meteorological data. Indeed it is essential that the two be considered together so that a bias of explanation via either geological or meteorological factors is avoided. For example, minor ephemeral runoff processes

outside the major water courses of an area make a very significant contribution to groundwater recharge in major arid and semi-arid regions. If groundwater and surface water resources are assessed separately, the importance of this contribution may be overlooked.

Other sources of irrigation water

25. Finally, other less conventional sources of water should be mentioned although to date they have been rarely used for large-scale irrigation. A wide range of desalinisation plant now exists for making sea water fit for domestic and agricultural use; but desalinisation of sea water is generally only economically viable when the water produced can demand a high price, or the energy required commands a negligible price. In practice, desalinisation has been restricted to providing essential water supplies for domestic needs at sites where no other sources of water are available.

26. However, brackish water of moderate salinity (8 dSm^{-1}) is now being used without desalinisation to irrigate salt-tolerant crops such as barley or cotton (ref. 11). Unfortunately this usually requires a high-technology drip irrigation system with its associated costs both in terms of equipment and precise management and operation. Special care is needed to prevent the accumulation of salts in the soil and to avoid damage to the physical structure of the soil itself. Hence its potential may be limited to high value cash crops and not to the large volume staple crops so desperately needed in many parts of the world. Treated sewage or industrial effluent contains significant amounts of nitrogen and phosphorous and its use for drip irrigation of cotton (ref. 12) can almost eliminate the need for chemical fertilization. However, even if there were no objections on health grounds to using tertiary effluent for irrigating food crops, the costs of the additional treatment would be extremely high, so the economic limitations that apply to desalinisation are yet again a constraint.

27. Guidelines for cloud seeding to augment precipitation have been published recently (ref. 13); these cover a variety of aspects of weather modification including social, legal, environmental and economic issues as well as scientific aspects of the subject. There is at least one well-documented case where an operational cloud seeding programme made a significant contribution to irrigated agriculture (ref. 14). Unfortunately, the number of agricultural areas in which precipitation enhancement through cloud seeding is both likely to succeed and where it would also benefit rain-fed crops is very small (ref. 15). A rainfall enhancement programme can only succeed when favourable meteorological conditions exist; reliable data in real-time on cloud formation and development are therefore required.

SMALL-SCALE IRRIGATION

28. Small-scale irrigation is not a new idea, but over the past 30 years the development of large-scale schemes has seemed a more effective way of improving agricultural production. Interest in small-scale irrigation has grown as dissatisfaction with the performance of large-scale schemes grows, and as fewer sites for large-scale development remain. Underhill (ref. 16) argues that both large formal and small informal schemes are needed, though to date little work on the implementation of small-scale schemes has been published.

29. Common to the development of schemes of all sizes is the need to assess the physical resources of climate, water and land. Underhill (ref. 16) draws attention to the problems and solutions is assessing these resources that are particularly relevant to small-scale irrigation. He notes that the concept of Rapid Rural Reconnaissance (RRR) described by Chambers (ref. 17) may be a cost-effective way of collecting the necessary information. Stern (ref. 18) provides much of the basic technical information needed for developing irrigated agriculture on a small-scale.

30. In the context of water resource assessment, the use of highly trained experts and sophisticated methods may not be justified for individual schemes. However they would be appropriate for hydrological studies carried out on a national or regional scale - as discussed in paragraphs 19 & 20 - using all available hydrological and meteorological data. Simple design manuals would be produced which would be invaluable for the rapid assessment of water resource potential of many different locations.

DATA COLLECTION AND ANALYSIS

31. The general outline of the various methods above has a constant theme: data. Without relevant hydrological or meteorological data, it is virtually impossible to make a satisfactory assessment of available water resources. Aspects of data collection, processing, storage and analysis are now discussed, followed by a summary of some of the recently developed technologies that could have a major impact on the routine management of data collection networks.

32. Apart from improved methods of recording and transmitting data - possibly in real-time as far as long term data collection for water resources assessments are concerned - the basic principles of measuring rainfall, streamflow and groundwater levels have in general remained unchanged over the years. Certainly, radar techniques and remote sensing are useful for short term forecasting and operational decision-making, but it is still the depth of rain on the ground and the water level in a river or tubewell that need to be measured.

33. One of the most significant advances has been the development of solid state loggers to record data. These can be left for several months before the device is interrogated to retrieve the readings. Alternatively, data can be transmitted from a data collection platform (DCP) to a central receiving station for rapid processing. Nevertheless, field sites still need to be visited regularly to verify the correct functioning of the measuring device.

34. These methods of data collection greatly facilitate the task of processing as the data area already in digital form for immediate computer processing. However the temptation to assume that because the data are in digital form they are necessarily correct should be strongly resisted. Routine quality control remains an essential task; this can be greatly facilitated by interactive software packages designed for microcomputers with graphical displays of the time series of data. The same packages can also produce summary tables for yearbooks or other publications; data are thus more readily available for the design and operation of irrigation or other water resource development schemes.

35. In addition to allowing improved access to data - again it must be stressed that this does not necessarily imply anything about their inherent quality - computers should allow information to be interpreted for timely operational or other management decisions.

36. To conclude the discussion of data collection, a note of caution must be sounded. The scarcity of hydrological data will remain a serious constraint on the accurate assessment of water resources and on the reliability of scheme design. Unfortunately there is no real indication that the importance of maintaining adequate hydrological and meteorological networks is fully recognised, particularly in countries where financial constraints may be severe. Modern methods of data collection, transmission, routine processing and storage can certainly ease some of the problems, but there appears to be a danger that these may be regarded as a substitute for improving, if not maintaining, the basic standards of data collection.

WATER RESOURCE PLANNING

37. Many of the best sites for large water resource developments have already been used, and there are increasing objections on enivronmental and social grounds to the exploitation of the sites that remain. On existing schemes there needs to be more emphasis on improving performance using existing water resources. Thus any action to build new irrigation systems or to extend or improve the performance of existing schemes requires basic data on resource availability.

38. Consequently in those areas where continued water resource development for irrigation is being considered it is necessary to continue investment into the collection and processing of the relevant data. As is the case whenever data are collected, the allocation of resources to acquire and process hydrological information must be selective.

39. Priorities should be focussed on potential projects that respond to urgent needs and appear technically and economically feasible of implementation. To provide a framework in which to assign priorities for data collection it is necessary to establish an overall strategy for coordinated development of regional rural water resources; this should identify weaknesses and gaps in the existing data base.

40. One possible approach could be based on the three-phase planning process to reassess and prepare national policies and project plans as proposed by WHO in advance of the 1981-1990 Water Supply and Sanitation Decade. The three phases were: Rapid Assessment Reports; National Plans and Policy Review; Project Plans. A broad assessment of water resources would be carried out using similar procedures (ref. 1), and full account would be taken of all the competing demands for scarce water. The emphasis of water resource assessment would then shift away from project specific analysis, where a single development is considered in isolation, towards a more coordinated, regional approach.

REFERENCES
1. CARRUTHERS, I.D. (ed.), 1983. Aid for the Development of Irrigation. Report on the Development Assistance Committee Workshop on "Irrigation Assistance", Paris, Sept 1982.
2. FAO, 1979. Agriculture : Towards 2000 FAO, Rome.
3. MATHER, T.H., 1983. Water in the Service of Food Production. World Water '83, Thomas Telford Ltd., London.
4. ORAM, P.A., ALIBARUHO, J., ROY, S. 1979. Investment and Input requirements for accelerating food production in low income countries by 1990. Research Report No 10. International Food Policy Research Institute, Washington D.C.
5. STANHILL, G., 1982. The World Water Problem. Proc. 10th Int. Conf. on the Unity of Science (Seoul). ICF Press, New York, Vol.1, 453-489.
6. BAUMGARTNER, A., REICHEL, E. 1975. The World Water Balance. Elsevier, Amsterdam.
7. PEREIRA, C., 1985. Irrigation for Future World Food Production. ICID Bulletin, 34(1), 1-11.
8. NATURAL ENVIRONMENT RESEARCH COUNCIL, 1975. Flood Studies Report, Volume 1, Institute of Hydrology, Wallingford.
9. INSTITUTE OF HYDROLOGY, 1980. Low flow studies report. Institute of Hydrology, Wallingford.
10. INSTITUTE OF HYDROLOGY, 1984. Research Report, 1981-84, 53-56.
11. MASS, E.V., HOFFMAN, G.J. 1977. Crop Salt Tolerance - current assessment. J.Irrig. Drainage Div. ASCE, 103 (IR2), 115-134.

12. ORON, G., BEN-ASHER, J., DE MALACH, Y., 1982. Effluent in trickle irrigation of cotton in arid zones. J.Irrig. Drainage Div. ASCE, 108(IR2), 115-126.
13. Guidelines for cloud seeding to augment precipitation. J. Irrig. Drainage Div. ASCE, 1983, 109(1), 111-182.
14. GAGIN, A., NEUMANN, J., 1981. The second Israeli randomised cloud seeding experiment: evaluation of the results. J. Appl. Met., 20, 1301-1311.
15. STANHILL, G. 1985. The Water Resources for Agriculture. Phil. Trans. R. Soc. Lond. B, 310, 161-173.
16. UNDERHILL, H.W., 1984. Small-scale Irrigation in Africa in the context of rural development. FAO, Rome.
17. CHAMBERS, R. 1981. Rapid Rural Appraisal: Rationale and repertoire. Public Admin. and Develop., 1, 95-106.
18. STERN, P.H., 1979. Small Scale Irrigation. Intermediate Technology Publications Ltd/IIIC, London.

G. LE MOIGNE, Irrigation Adviser, World Bank, Washington, USA

10 Trends in irrigation

Since 1950 irrigation in the world has gone through a period of rapid expansion of about 2 per cent per year to reach some 271 hectares gross at the present time. This rate of expansion is slowing down slightly and giving way to rehabilitation and modernisation so that existing systems can derive fuller benefits from modern agricultural technology. Furthermore in many areas of water shortage, modernisation is a pre-requisite to expansion because only by water economy on existing areas can new land be served. Management of irrigation and water management in particular has become a subject of great importance and many changes of attitude are taking place with a new emphasis on farmer participation in the decision making process. There is a need for vast new drainage systems to be built in the old irrigated areas. This represents one of the more challenging engineering tasks of the immediate future. Of the new technologies it is expected that those to penetrate the developing countries more rapidly will include water control systems; low pressure sprinklers and drip irrigation. The contribution from irrigation to total agricultural production will continue to increase, particularly in Asia.

INTRODUCTION

1. This paper addresses the broader trends that occur in the irrigation sector as a whole. The more specific trends are to be dealt with by the speakers in Sessions 2B and 3B tomorrow.

2. Although the author has been engaged in irrigation as a consultant for many years the perspective presented here is influenced by his experience as firstly an agricultural Division Chief and currently as Irrigation Adviser for the World Bank. Hence the paper will focus on developing countries and will deal with trends that occur not only in the technology of irrigation but also in related matters such as management, operation, maintenance, institutional factors and other issues that are of direct concern to financing agencies.

GROWTH IN IRRIGATION

3. Since 1950 the irrigated areas in almost all parts of the world have expanded about threefold. Table 1 gives the gross irrigated areas by continents and indicates a growth rate of a little over 3 per cent per annum. In recent years there has been an expansion of gross area at a rate of almost 5 million hectares a year of which India alone has accounted for about 1½ million hectares a year.

TABLE 1

GROSS IRRIGATED AREAS BY CONTINENTS

(million hectares)

	1950	1960	1970	1986
Europe (incl. part USSR)	8	12	20	29
Asia (incl. part USSR)	66	100	132	184
Africa	4	5	9	13
North America	12	17	29	34
South America	3	5	6	9
Australia & Pacific	1	1	2	2
Totals	94	140	198	271

Source: Phil. Trans.R.Soc London A316 355-368 (1986)

4. It is not expected that this global expansion will continue at the same rate in the future for several reasons. Firstly, limitations in the amount of land and water are beginning to become severe constraints; secondly, with the high potential yields of new crop varieties, the rehabilitation and modernisation of existing irrigated areas brings better economic returns than expansion. Thirdly, there are clear signs of over-production of certain irrigated crops - notably rice - which occupies about one third of the irrigated areas of Asia. Rice production will however need to grow at about 2% per year for domestic production to match the growth in consumption and there will also be a demand for a more varied diet that will in turn require modernisation of the irrigation networks to meet the more exacting demands on the

timeliness and reliability of water supply.

5. In short, the current trend is to give priority to the better use of the existing irrigated land either for traditional crops or for diversification into new cropping patterns. In selected areas and in certain countries there will however remain a considerable rate of expansion. Notable examples where the emphasis is still on expansion are Algeria, Brazil and Nigeria - the last two being countries that are embarking on major irrigation for the first time. In India the policies are likely to be framed within a balanced investment in both expansion and improvement. India still has problems of population expansion and although the potential for yield improvement is large, as every farmer knows, the push towards greater total production is always more assured with expansion of cropped area than with enhancement of yield.

6. The development of large irrigation schemes will require progress in solving international riparian issues that influence water use in some of the worlds' major rivers such as the Ganges and most of the African rivers (with the notable exception of the Senegal river).

7. To summarize the trends in the expansion of irrigated areas in the world over the next decade, it is judged that it will proceed at a slower rate than hitherto, perhaps between 3 and 4 million hectares a year. This represents a growth rate of about 1½ per cent per annum compared with the achievement of 2 per cent in the recent past. The growth in irrigated crop production, however, should be well above the 1½ per cent level, perhaps more than double, resulting from the current efforts being made to modernise and improve the existing systems and to raise the quality of management. An unfortunate exception to this last generalisation is Africa to which further reference will be made.

REHABILITATION AND MODERNISATION

8. The Green Revolution made available to farmers, both small and large, some new crop varieties that have dramatically changed the whole pattern of production in the world today. It has placed a premium on high-yielding land with good rainfall or irrigation. As a demonstration of the disparities in production caused by the new crop varieties it is interesting to note that 30 years ago the rain-grown cereal yields in the poor conditions of the African Savannah were about one third of those in areas of advanced agriculture. Today the proportion is about one twentieth.

9. The objective in India and much of Asia must be to bring yields much nearer those of countries with more advanced crop husbandry. There are signs of progress with irrigated crop production but rainland cropping in the poor climates of much of the developing world remains disappointing in its response to modern agricultural inputs. The high potential yields of new crop varieties favours all measures that improve the quality of farm inputs including water supply, drainage, fertilizers, pest control, pesticides and above all good crop husbandry. It is not therefore surprising to find that today all countries with extensive irrigation are placing emphasis on rehabilitation and modernisation of existing irrigation projects. Usually there is a need for a package of actions extending from the physical improvement of canals and structures to the re-organisation of the institutions that operate them.

10. Morocco for example has adopted a highly successful strategy that covers a wide range of actions including land consolidation, major changes to water conveyance systems (including the conversion from canal to piped reticulation) and important institutional changes involving the promotion of co-operatives and Water User Associations. Development of these associations however is rather recent in the modern irrigation sector as the Moroccan authorities first wanted to promote industrial crops and train their farmers in irrigation and crop husbandry.

11. In India, the first important move towards modernisation and remodelling was in 1970 with the launching of the Command Area Development (CAD) programme. The aim was to introduce a more integrated approach to water management below the outlet that serves the agricultural units (usually about 150 hectares with about 30 cultivators). While the CAD programme has fulfilled many of its objectives it is hindered by lack of progress in the modernisation and management of the main canal system. This is perhaps not surprising because the old systems have now become an integral and fixed part of the local geography. Their modernisation involves considerable physical works which are difficult to execute owing to the need to keep services open. An example of the magnitude of these physical works is found in the remodelling of the Eastern Yamuna canal in Uttar Pradesh, India. It is the oldest canal command in the country and serves about 200,000 hectares. The physical works include the remodelling or rebuilding of 63 bridges, 73 escapes, 103 major drains and 14 syphons together with extensive canal improvements. Similar programmes are in hand for 4 other canal commands in Uttar Pradesh covering 1 million hectares. In India there will be an ever-increasing emphasis on rehabilitation and modernisation but with more attention to the integration of the agriculture and the irrigation engineering - this might be described as a close link between the CAD programme and the physical reconstruction programme of the type adopted for the Eastern Yumuna.

12. The nature of rehabilitation and modernisation work varies widely from country to country but has the common objective of better water management leading to a more equitable water supply to the cultivators and greater efficiency in water use.

WATER MANAGEMENT

13. In recent years water management has become a subject of primary importance in many countries and has been given emphasis by the international agencies. The International

Commission on Irrigation and Drainage gave water management special focus in its working groups and conferences. A support group of the Consultative Group for International Agricultural Research (CGIAR) created the International Irrigation Management Institute (IIMI) in Sri Lanka in 1984. The World Bank is supporting the enhancement of water management standards in many countries.

14. In India, a Water Management Division was set up in 1984 in the Ministry of Water Resources and initiated a number of projects. There is currently a proposal to proceed with a $150 million Natural Water Management Project which will benefit a number of Indian states with initial emphasis on the south of the country.

15. The fundamental objectives of any programme of water management are to maximise production through the more timely delivery of the right quantities of water to the farmer. Such endeavours are not new but they have attracted greater emphasis by the conditions of today - particularly by the limitations on total available water, the ever increasing pressure to raise cropping intensities and the more exacting water requirements of the new high-yielding crop varieties.

16. Certain clear trends are emerging from the studies of water management carried out over the last few years. These may be summarised as follows:-

a) More responsibility is being given to the cultivators - particularly for the operation and maintenance of the irrigation systems at the tertiary level - by the creation of "water user associations" with executive power.

b) There is a recognition that in order to ensure the good functioning of water user associations, the water delivery to the heads of the tertiary canal systems must be assured by good management in the main and secondary canals. Thus there is now more of a total system approach in contrast to the earlier farm level approach.

c) Although much can be done to improve management standards of existing systems, the full demands of modern intensive agriculture cannot be met without some major modifications to the physical works including the provision of better control structures. Furthermore, the economic return on canal lining in regions of permeable soil has been raised by the high opportunity cost of water and by the dis-benefits caused by water-logging and salinity.

d) More attention is being given to the application of the tools provided by modern electronics to the problems of water indenting and routing.

e) As groundwater becomes a more important contributor to total irrigation water supplies so more attention is being given to the conjunctive use of ground and surface resources.

f) Management training is now recognised as an urgent need in many countries. Furthermore the training should aim to provide stronger links between the engineering and agricultural aspects of management.

17. The endeavours that are now being made to raise water management standards and make them better capable of meeting the demands of modern agriculture are only in the very early stages. It is expected that water management programmes will intensify over the next 10-20 years. As land and water resources approach their limits, management resources must be further exploited in order to sustain growth rates in agricultural production.

WATER ECONOMY

18. Closely linked with Water Management is the topic of Water Economy. The layman is always a little dismayed to hear that engineers have not attained higher levels of efficiency for the works they create whether they be irrigation systems or internal combustion engines. Higher efficiencies come about largely as the result of economic pressures and other incentives. In irrigation there are now many reasons to raise irrigation efficiencies. Not only are the unexploited water resources limited at the headworks but also with intensification of cropping it is physically difficult and expensive to raise canal conveyance capacities to bring additional resources onto the fields. Typical overall irrigation efficiencies in Asia are about 30 per cent. Looked at from a water "loss" point of view this results from about 35 per cent loss in the conveyance system, 25 per cent in the tertiary distribution system and 40 per cent in the application to the field. Some of these losses are of course recovered in various ways. Logically, the trend is to reduce first those losses which give the best return on the costs involved and invariably that means distribution losses. This strategy is evident in the watercourse remodelling programmes in India referred to above.

19. Conveyance losses are now being tackled in many countries as part of remodelling and modernisation programmes - they take the form of the physical works described above in the example of the Eastern Yamuna canal. In typical surface water irrigation systems in Morocco the combined conveyance and distribution efficiency is about 75 per cent compared with about 50 per cent in Asia. At farm level, the application efficiencies are more difficult to raise particularly with small-holder agriculture. Australia has substantially improved application efficiencies by providing incentives to farmers to carry out laser controlled levelling of the fields and at the same time announcing crop water requirements over the local radio service. Whereas land levelling is encouraged in some developing countries the attitudes to water application are rarely governed by water economy because water is rarely paid for by volume.

20. In the absence of a volumetric method of charging the only way to achieve the correct water application is by a careful procedure of indents based on cropped areas and climatic conditions. That principle was adopted for

water management in the Gezira scheme in the Sudan. By contrast, in Asia, indents have been based mainly on achieving a constant flow at the watercourse head.

21. In North Africa quasi-volumetric methods for water charging have been successfully introduced. Water billing is based on the time for which measured flows take place through adjustable modules.

22. As the farmers assume more responsibility for water management it is expected that there will be trends towards wider use of volumetric billing and water charges - payment being effective through water user associations. As that happens a different attitude towards water economy will take place at farm level.

23. A new and interesting aspect of water economy measures arises in the need to provide better domestic and industrial supplies. In exchange for additional resources, domestic supply organisations are showing a readiness to pay for water saving works in the irrigation sector, such as canal lining. Recent examples in World Bank financed projects can be found in India and Indonesia.

DRAINAGE

24. Drainage of agricultural land falls broadly into two categories; surface drainage to dispose of runoff and prevent prolonged inundation of crop lands and sub-surface drainage to prevent water-logging and salinity.

25. In the early part of this century drainage was often omitted from irrigation projects on the consideration that the need for it was not immediate. For the most part, that policy was justified but now the stage has been reached in both arid and humid areas when there is an enormous demand for drainage. In recent years it has received special emphasis as the need for crop diversification arises in so many of the traditional rice lands. Furthermore, in some of the rice lands there are large areas that are subject to frequent flooding to such an extent that it is not feasible to make full use of modern farm inputs without major drainage, as is the case for example in Bangladesh.

26. Drainage of agricultural land has hitherto remained largely a luxury afforded only by developed countries. Now there is an urgent need to implement drainage projects on an extensive scale in developing countries. For example the Government of China's objectives for the year 2000 calls for installing drainage facilities on 17 million ha to prevent waterlogging and eliminating salinization problems. The main issue is however one of cost. Even with the technological advances that have taken place in horizontal drainage the costs are often about $2000 per hectare - if surface drainage and flood protection is also required the total cost can approach that of the irrigation system itself!

27. As with irrigation distribution, the plastic pipe has made an important contribution to drainage work and taken together with the trenchless pipe laying machines we can continue to see important advances.

SPRINKLER IRRIGATION

28. Overhead irrigation had scarcely begun to make in-roads in developing countries before it was severely hit by the rise in energy costs. Despite the introduction of low pressure equipment to offset energy costs, both mobile and fixed, very little sprinkler equipment is yet found in the developing world. (With the notable exception of Saudi Arabia where over 400 000 ha are irrigated by centre-pivot systems). However there are signs that as the opportunity cost of water rises and at the same time it becomes necessary to irrigate land with high permeability and undulating topography, as in much of Africa and South America, sprinkler irrigation will be more widely used.

DRIP IRRIGATION

29. Drip irrigation has many advantages in situations where water is scarce, soils are permeable, topography uneven and a high degree of automation is required. For these and other reasons it has received much attention in the USA, Israel, Jordan, China, Brazil, Mauritius and Taiwan. In Hawaii dual chamber sub-surface drip irrigation is used very successfully for sugar cane; in Brazil surface drip lines are used to irrigate row crops such as melons and passion fruit. In Mauritius and Taiwan the focus is on sub-surface drip for sugar cane. In China 13,000 hectares are now under drip irrigation.

30. The success of drip irrigation depends largely on the successful manufacture of low cost plastic drip lines which are resistant to weathering and to insect attack. Progress has so far been good and drip is a method of irrigation that should expand rapidly in the future.

SUB-SAHARAN AFRICA

31. In the light of recent events it is appropriate to include a specific paragraph on irrigation in sub-Saharan Africa.

32. This region of the world is not well endowed with good soils or resources. By comparison with Asia or even North Africa the irrigation potential, as measured by today's yardsticks, is relatively small and the cost of development is high. On the other hand the need for irrigation is very compelling particularly in the arid zones. The sub-Saharan regions of Africa have no long tradition in irrigation hence farmers find it difficult to adjust from extensive rain-land cropping to intensive irrigation.

33. Despite these constraints there are encouraging signs that sub-Saharan Africa is beginning to master the art of irrigation. A great help has been the development of rice varieties that respond well in the arid zones and produce yields of about 4 tons of paddy per hectare. Other crops such as irrigated bananas, sugar cane and tomatoes are now produced successfully in many African countries.

34. Although the total potential for irrigation is modest in sub-Saharan Africa when compared with Asia or the Middle East, so too are the total demands for food production. Thus the irrigated areas in the basins of the Niger, Senegal, Zambesi, Juba, Schebelli and

several other rivers that flow through arid zones are capable of being developed to fill foreseeable deficits in food production. The problems are more financial, administrative and social than technical ones. In the light of the present distressing decline in the agricultural growth rate in Africa there will be renewed efforts to alleviate the situation including active programmes in irrigation including small scale irrigation based on development of groundwater for multipurpose use (human, livestock and crop production).

THE WAY AHEAD

35. In summary:-

a) The rate of expansion in irrigation in the world will be a little slower than hitherto, perhaps falling towards about 1½ per cent per annum from previous levels of about 2 per cent. It will require progress in solving some of the major international riparian issues that influence water use.

b) There will be a continued emphasis on making better use of the irrigation schemes that exist, through rehabilitation, modernisation, improved management and enhanced institutional support.

c) Drainage will become a major activity in all countries with old irrigation systems.

d) New technologies in irrigation will penetrate developing countries more rapidly, such as control systems for water indenting and management, low pressure sprinklers and drip irrigation.

e) The contribution from irrigation to total agricultural production will continue to increase particularly as Asian countries improve both irrigation systems and management to take greater advantage of the high potential yields from improved crop varieties.

Discussion on Papers 9 and 10

DR P. BOLTON, Hydraulics Research Ltd, UK
In the developing world, over 80% of the arable land is under rain-fed agriculture. Irrigation makes an important contribution to furthering agricultural output but is sometimes a fairly expensive option requiring particular circumstances for success. Do engineers have a role in ensuring that the most advantageous combination of rain-fed and irrigated production is achieved?

The emphasis in Paper 9 on the reliability of the water source is crucial. Hydrologists are having to study this more closely. If water is the limiting resource in a project, is it right to seek high reliability of supply for a relatively small land area, or would it be preferable to design schemes with additional areas which could be irrigated in years when the supply is adequate? Taking this a stage further: should the livelihoods of farmers be safeguarded solely by the provision of adequate water storage to provide a supply at a given risk, or should the inter-annual fluctuations be countered by the storage of food in abundant seasons for less abundant seasons? Behind such questions lie some issues which are clearly hydrological, but others which go beyond the engineer's normal domain.

Two topics mentioned in Paper 9 are becoming increasingly important: the potential for water savings through the improved management of existing schemes and the need to understand the particular hydrological problems arising in small irrigation developments.

Paper 9 makes a plea for greater efforts to be made in data collection. This is indeed needed but it is unfair to put the responsibilities wholly on developing countries whose hydrologists must compete for scarce financial resources. Moreover, although high technology systems may be appropriate in some remote regions, to be accepted they must allow local personnel to attain full control over the system and the data obtained. The development of simple low cost data collection networks designed to optimize the information obtainable using the resources available poses a continuing challenge to hydrologists.

MR M. SNELL, Sir M. MacDonald & Partners, UK
I welcome the emphasis in Paper 9 on regional methods in hydrology, especially because they present the results of relatively sophisticated analysis in a way that is simple to use and thus applicable to small schemes. However, I would dispute the emphasis on high reliability and optimal plant growth. When water rather than land is scarce the best balance may involve lower reliability, with water more thinly spread (hence higher application efficiencies), especially if food is stored from year to year.

Can the Author give details of the cost of the water harvesting techniques described in the Paper?

MR B. S. PIPER
Costs of rainfall harvesting techniques vary according to the technique used and the cost of local labour. Costs of microcatchment construction in the Negev of around US $5-20 have been quoted.

Where chemical treatment or covering of the soil to increase run-off is being considered, the costs are likely to be much higher.

MR W. R. RANGELEY, Sir Alexander Gibb & Partners, UK
The following important trends in world irrigation are taking place.

(a) In Asia there is a slowing down on expansion in favour of rehabilitation and remodelling.
(b) In Asia there is a need to diversify crops on account of the overproduction of rice. Changes are needed in water management and the installation of effective drainage.
(c) In Brazil and Nigeria there is need to develop large-scale irrigation for the first time.
(d) In sub-Saharan Africa there is a need to enhance food production as protection against famine.

The adoption of appropriate technology needs to be done carefully. In the past investments in food production irrigation schemes have usually been concentrated on the more favourable combinations of land and water - very little effort went into the marginal areas.

MR C. R. C. JONES, Sir M. MacDonald & Partners, UK
I am concerned that during this International Drinking Water Supply and Sanitation Decade improvements in design and construction techniques are not disseminated easily enough.

PAPERS 9 AND 10: DISCUSSION

For example, in the field of groundwater development significant economies have been achieved by the careful choice of materials and design of boreholes based on sound hydrogeological principles. Unfortunately, there are many instances where different organizations within a country and neighbouring countries retain older and more expensive methods despite the adoption of improved techniques in other organizations.

Can the Authors comment on this problem?

MR B. S. PIPER
In reply to Mr Jones, in my Paper I did not argue that patterns of traditional settlements and sources of supply should be the only way to evaluate regional groundwater resources (see paragraph 22). Rather it is one source of information that can be extremely useful, yet is too often rarely used in practical studies. Used in conjunction with modern geophysical techniques there are many potential benefits.

PROFESSOR M. J. HAMLIN, University of Birmingham, UK
The concept of high reliability for a relatively small area against a lower level of reliability (or a greater risk) for a larger area has been mentioned.

I believe that because of the stochastic nature of rainfall input the area to be planted in a particular year also needs to be treated as a stochastic variable. Under these conditions three major problems confront the planners of an irrigation scheme at a particular location with a given hydrological regime.

(a) Given that the whole area will not be irrigated each year, the balance being given over to dry land crops, what area should be brought under irrigation?
(b) What rules can be devised to determine, at a given level of reliability, the cropped area to be brought under irrigation at the start of a particular growing session to maximize, for the chosen level of reliability the long-term average yield of the project?
(c) How should the water be scheduled to maximize the yield from the area actually planted?

The answer to each of these questions is derived from the interaction of physical, social, economic and hydrological factors.

I believe that the water resources engineer has much to contribute, particularly on the hydrological inputs, the derivation of control rules and the scheduling of water allocations.

DR Z. J. SVEHLIK, Institute of Irrigation Studies, University of Southampton, UK
It seems to me that the quantitative assessment of water resources for irrigation presents two basic problems: the reliability of the assessment and the reliability of an irrigation system.

The importance of a proper, reliable assessment is stressed in Paper 9. Any error made there will be reflected either in the lowering of the yield or in the waste of water. But how far can the approaches used in practice lead to a reliable assessment? Certainly, there are weaknesses. In my opinion there is a great weakness in evaluating the components of the water balance equation (irrigation water requirements) and the water source separately.

The problem of the reliability of the system is very complex. Considering the variability of irrigation water requirements and water resources, a certain risk of failure in water delivery is usually accepted. In an irrigation system, the failure may occur for different reasons, e.g. there is not enough water in the source or the water demand exceeds the design capacity of the irrigation network, or - in extreme conditions - a combination of both.

Each of these cases will require a different approach. A still different approach is required if water flow is controlled by reservoir(s). It should be stressed that any assessment of water resources for irrigation must be based on the reliable assessment of irrigation water requirements.

MR B. S. PIPER
It is the responsibility of the hydrologist to ensure that the assessment of water resources availability is as reliable as possible given the data and information that are available. I agree that the components of the water balance should be analysed together. If the water balance is not plausible then it must be assumed that the data and/or the analyses are deficient in some respect.

G. R. THORPE, PhD, BSc, MICE, Howard
Humphreys & Partners, UK, and
B. PANTANAVIBUL, BSc, Royal Irrigation
Department, Thailand

TN19. The Sukhothai groundwater development in Thailand

The Sukhothai groundwater development project which is currently under construction by the Royal Irrigation Department in Thailand has introduced a number of new concepts into the kingdom's continuing programme of irrigation works. The development is described briefly and areas of innovation and their impact on the agricultural community are discussed.

BACKGROUND

2. Traditionally Thailand's lowland agriculture was the single, annual cultivation of rice watered by rainfall and the spreading of the seasonal flood water of the principal rivers. Increased reliability of water supply and some second cropping has been facilitated over the past two decades by the construction of major storage dams. Notwithstanding such developments, however, water rather than land remains the principal constraint to future agricultural intensification. Consequently utilization of groundwater would increase the overall development potential but, prior to the commencement of the Sukhothai project, groundwater exploitation was limited to wells for public and industrial supply. There was only a small irrigation component from domestic wells.

3. The project area is within the valley of the River Yom, 480 km north of Bangkok. The topography exhibits low relief with flat clayey flood plain areas suited to rice growing, interspersed with old and contemporary levee areas having lighter soils on which the traditional upland crops are maize, cotton, sugar cane and soya beans. The aquifers are the sands and gravels within a 50-250 m thick alluvial sequence which overlies hard basement rocks. Recharge is primarily by vertical infiltration.

4. A comprehensive exploration and hydrogeological monitoring programme undertaken throughout the investigation and implementation phases has provided sufficient data for the calibration of a mathematical model of the aquifer, covering 1675 km^2 and incorporating all the regional groundwater recharge and abstractions. The model has been used to help in the siting of boreholes in the later stages of the development and is now available to assist in the operational management of the wellfield.

PROJECT FEATURES

5. Each well serves a discrete irrigation unit, but individual command areas join up, to avoid leaving unirrigated farms in an otherwise irrigated area. This criterion, combined with knowledge of the aquifer from exploratory drilling and testing, has determined a well spacing of 800 m. Each well yields 55 l/s and supplies a gross command area of 64 ha (58 ha net). Production wells are drilled 100-120 m deep using the reverse circulation technique. They are cased in GRP with a 356 mm pump casing to 40 m below ground and 254 mm selected GRP casing and screen below. A graded gravel pack is used.

6. The well pumps are water-lubricated, electric motor driven, lineshaft, multistage, mixed flow units which discharge through control valves and measuring equipment into a uPVC pipe distribution system. The uPVC pipe is buried 1 m below ground level and forms a dendritic system, reducing from 254 mm in diameter at the pumping station to 150 mm in diameter at the farm outlet. Individual farms can be up to 16 ha in area with an average holding of about 4 ha in rice areas and rather less where upland crops are grown. Each farmer is served with at least one valved outlet, sited at the highest point of his land, which allows him to gravitate water in earth canals to any point on his holding.

PROJECT OPERATION

7. In both upland and lowland areas it is possible to increase cropping from 100% to 200%, provided the farmers are willing and able to meet the challenges of working with the new system. It is estimated to take five years to build up to full cropping, given adequate practical help and guidance. To this end the Royal Irrigation Department provides support in three ways: the provision of a direct extension service, demonstrations on specially selected farms, and assistance with land forming. Crop varieties, crop rotations and irrigation methods are demonstrated on the projects' own farms. For upland areas, basin, straight furrow and contour furrow irrigation methods are practised so that farmers can see what potential exists for improvement on their own land.

8. At the beginning of the season a microcomputer is used to work out the water sharing schedule based on the farmers' choice of crop and area cultivated, which is vetted by the Department to ensure the practicality of the overall demand on the well. The hydrogeological model can then be used to

World Water '86. Thomas Telford Limited, London, 1987

forecast the influence of the proposed abstraction schedule as a whole on the groundwater resource so as to mininize the risk of over-exploitation.

9. Further use of the computer is in monthly operation monitoring as a routine management exercise. This gives the project manager 'touch of the button' access to the valuable parameters such as supply per unit of cultivation, pumping water levels, pump down time and energy cost per unit of water delivered. With this information, problems can be identified at an early stage and action can be taken.

PROJECT COSTS

10. The project when complete will include 104 wells serving 5700 ha (net) and will have cost $2800 per ha. The operation and maintenance cost including the provision of an agricultural extension service, energy costs and replacement will be $156 per ha per annum.

PROJECT IMPACT

11. The capital cost, which may appear high, includes all infrastructural elements such as a comprehensive network of all-weather roads, complete rural electrification and a planned land drainage system. Surface water might have provided a marginally cheaper alternative but the groundwater, being a previously untapped, replenishable resource, is valuable in an area where water availability will ultimately limit development and it should be used in conjunction with surface sources. The pipe distribution system lends itself to groundwater supplies because water can be pumped directly to distribution without sedimentation problems. The pipe system provides almost 100% efficiency of delivery at the farm turnout and also facilitates supplies to small parcels of land in topography broken by old river channels. Unlike canals, it is not dependent on gravity flow.

12. The centralization of development in the deep well project facilitates the provision of support services such as extension and credit and affords the Government a degree of control over crops and cropping patterns. This hopefully will minimize the risk of crop failure which is so discouraging to farmers in the early stages of any project.

13. The rice farmers have found it easy to adapt to irrigated cultivation and the much increased reliability of water supply has allowed them to switch to a high yielding variety of rice, doubling the yield per crop. On upland areas requiring irrigation practices new to the area, the farmers are slower to adopt the new system and in the present economic climate they are reluctant to invest in the essential land forming work. Consequently the Government is having to provide more assistance in this area than was envisaged in the days of high farm gate prices which would have been a stimulus to the farmers, encouraging both learning and investment. Notwithstanding this constraint, however, it is clear that the general wealth and commercial activity of the region has been greatly increased by the development. Moreover it has created opportunities for the expansion of local uPVC and GRP pipe manufacturing industries and broadened the scope and expertise of local drilling contractors.

G. S. PARTHSARTHY, P. N. SUTARIA,
N. M. BHATT and PROFESSOR P. M. MODI,
M.S. University of Baroda, Gujarat, India

TN20. Water resources management: its role in effective utilization of water resources in developing countries — a case study of Gujarat State, India

The distribution of water to individual fields of irrigation is a very complex facet of water management. For many reasons water management is very important to fill the gap between the potential created by the construction of big projects and irrigation water resources utilized. The estimate is put that water resources may not last beyond AD 2000 if the present style of its indiscriminate usage is continued.

2. The International Development Agency and World Bank funding agencies are also stressing the importance of the efficiency of water resource usage. For a very big project - the Narmada Project - the World Bank also emphasized the manpower requirement needed to manage efficiently the huge amount of water potential created. This has made administrators aware of the need to use water resources efficiently.

CASE STUDY

3. Gujarat State is situated in the western region of India covering 19.6×10^6 ha and has a population of 40 million. Agriculture provides 40% of the state's income. The state is predominantly urbanized with large modern manufacturing sectors, and there is an urgent need to remove constraints in growth and progress in agriculture. There are three cropping seasons: kharif (monsoon), rabi (winter) and hot-weather. Kharif crops depend mainly on rainfall. Rabi and hot-weather crops are grown with residual soil moisture or irrigation.

4. In Gujarat, by completing major, medium and minor projects tapping most of the rivers like the Tapi, Mahi, Sabarmati and Banas, good potential is created, but this is low compared with that in other states. To use the potential created efficiently, after due thought it was decided to use a rotational water distribution method to supply the right quantity of water at the right time, ensuring equitable distribution as well.

Rotational water distribution system of irrigation

5. In this system a predetermined schedule is enforced. A weekly schedule gives a predetermined time for the supply of water in a field. Farmers have also accepted this method and adjusted their farm-work schedules accordingly. Considering the basis of the cropping pattern, the water requirement of crops (month/fortnight) of 1 cusec of water is released for a command area of an outlet. Water is taken by farmers as per the schedule in that area. Suitable allowances for channel losses, length of channel and so on are considered. As the system is developed, the allowance for transportation time of the water is also considered. The weekly time schedules are maintained throughout the season. They are

Table 1. Typical group and subgroup schedule on Wagasi Subminor for rabi in 1979-80

Sr No	Group	Outlet	Sub-group	Area of irrigation: ha	Adjusted time schedule From	Adjusted time schedule To	Total hours allotted	Discharge cusec
1	1	O1/L	1	3.78	Mon. 06.00	Mon. 21.45	15.75	1.06
			2	2.93	Mon. 21.45	Tue. 10.45	13	1.06
			3	3.02	Tue. 10.45	Wed. 00.45	14	1.06
			4	3.66	Wed. 00.45	Wed. 18.45	18	1.06
			5	3.14	Wed. 18.45	Thu. 11.45	17	1.06
			6	4.40	Thu.	Fri.	24	1.06

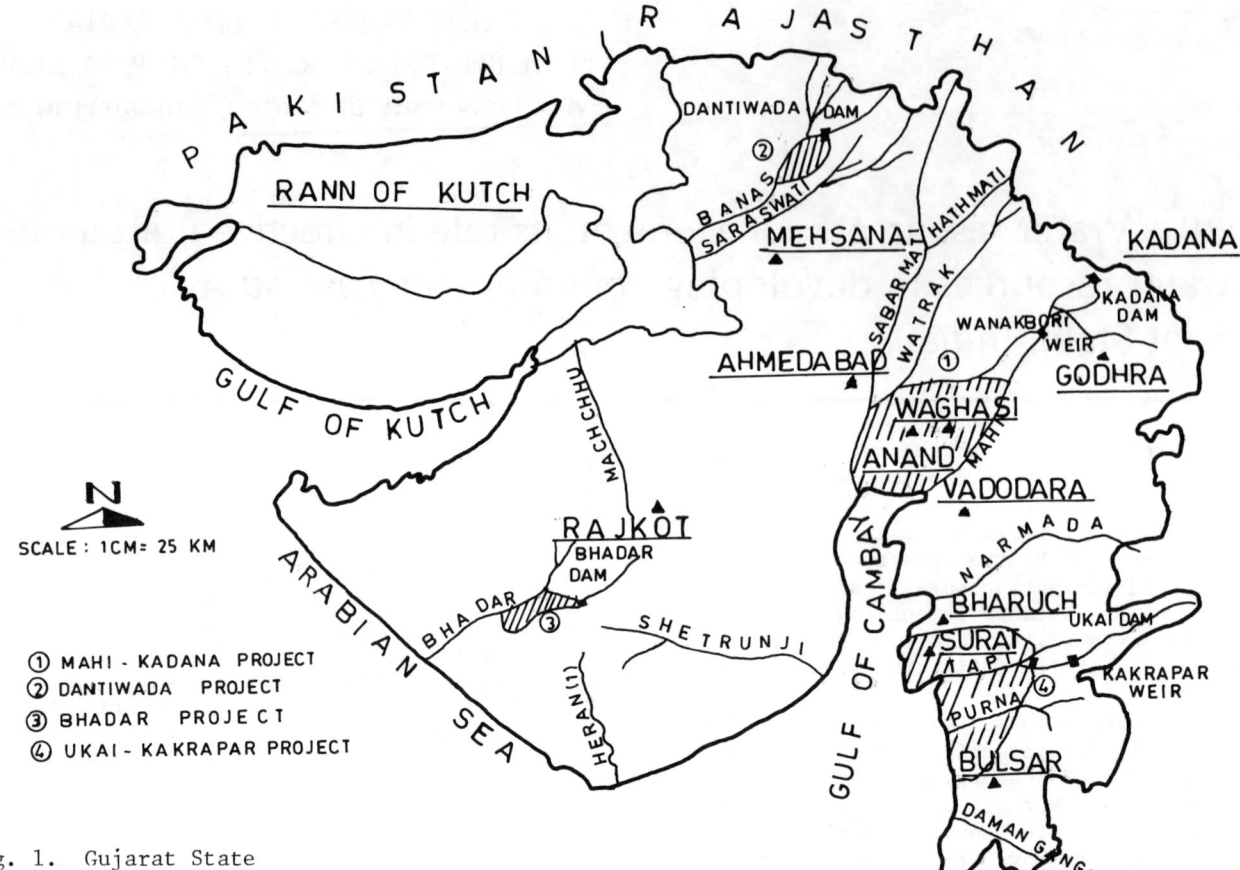

Fig. 1. Gujarat State

displayed on notice boards fixed at the outlet head.

6. A typical rotational distribution system - that on part of the Chikodhra distributory system for the rabi season in 1979-80 - is as follows.

(a) the government agronomist gives the cropping pattern in the command area as a percentage. The intensity of irrigation water requirements for each crop during the season is also given. The details are given for different crops in different months/fortnights in millimetres.

(b) From the cropping pattern, the water requirement of the crops and the crop water requirements during the month of maximum water required, the maximum water requirement for that fortnight/month is calculated. In the first part of January, for example, it is 749 m^3/ha.

(c) Assuming 65% efficiency of the outlet, the water requirement is calculated and a schedule is prepared (Table 1).

Rotational water distribution system at Dantiwada

7. The Dantiwada project is located at Dantiwada village on the River Banas in the Banaskantha district of north Gujarat; live storage is 44 430 x 10^6 m^3 and the irrigable area is 44 500 ha. Only 31 985 ha are covered by irrigation. To improve efficiency it was decided to introduce a rotational water distribution system. A 370 ha area was considered on a trial basis. The system gave good results. The World Bank also appreciated the value of the system which was then used widely in the Dantiwada project. It was found that 1 cusec of water could irrigate 17.50 ha, fulfilling the water requirements of the crops in that area.

CONCLUSIONS

8. The case study indicates clearly that the rotational water distribution system is very efficient in improving the irrigation system; it involves the judicious, equitable use of water at the right time by farmers. In addition to using available water resources properly, Gujarat state is encouraging farmers to form irrigation co-operative societies or districts to buy water in bulk from the Irrigation Department and ensures equitable distribution among its member farmers. One such society is already working well and has earned a sizeable profit since its inception. For efficient water management practice more such schemes should be developed which will help to stretch the available resources to the maximum extent.

REFERENCE

1. CHAUHAN L.J. Rotational water distribution system on canal systems of Gujarat State. Government of Gujarat, 1980.

M. A. JAEMTLID, Swedish Geological Co., Luleå, Sweden

TN21. Slingram as a water exploration method

The Slingram is a lightweight, portable, single frequency, horizontal/vertical loop electromagnetic system, developed by the Swedish Geological Co., designed for the detection of electrical conductors (e.g. orebodies and fracture zones). The Slingram is being used increasingly in water exploration in Africa. It is now in regular production by the District Development Fund in Zimbabwe, the Department of Water Affairs in Botswana and UNICEF in Nigeria.

2. The equipment, powered by rechargeable nickel-cadmium batteries, consists of a transmitter and a receiver connected by a reference cable.

PRINCIPLES OF THE SLINGRAM

3. The principles of the Slingram are shown in Fig. 1. The transmitted field (primary field) will induce a secondary field in conductors penetrated by the primary field. A resultant field, consisting of both the primary and the secondary field, will be picked up by the receiver. The resultant field is compared with the primary field in the receiver. The anomalies thereby obtained are characteristic for the shape of the conductor and its physical properties.

4. Interpretation of the shape, orientation and relative conductivity of the conductor is easily done directly in the field.

5. The depth penetration of the Slingram depends mainly on the frequency of the transmitted field, the thickness and conductivity of the overburden, and the distance between the transmitter and the receiver.

6. In areas with crystalline rocks where the overburden is 0-60 m thick and the resistivity between 10 and 1000 ohm m, the following configuration has proved to be efficient: a frequency of 3600 Hz and a 60 m coild distance.

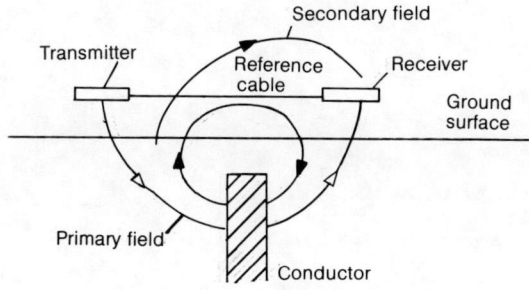

Fig. 1. Principles of the Slingram

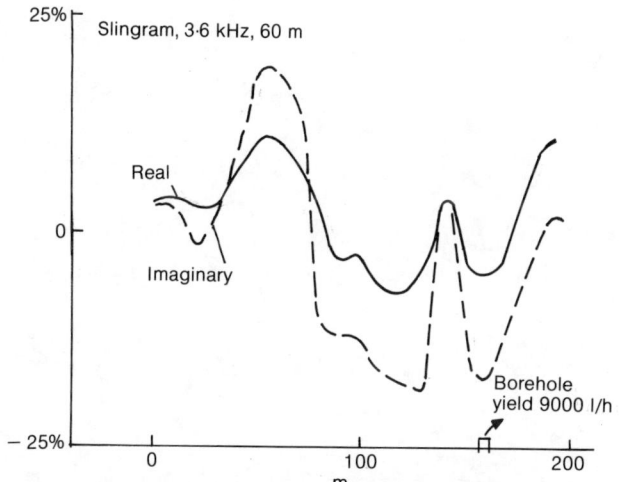

Fig. 2

FIELD PROCEDURE

7. The Slingram needs two operators: one to carry the transmitter and one to carry the receiver. Since this is an electromagnetic method, contact with the ground is not needed, which is an advantage in arid and semi-arid areas. In this way the measurements are performed rapidly. An average speed of measurements along cut-lines is 1-1.5 km/h with a point distance of 20 m.

WATER PROSPECTING AT CHIKUPO SCHOOL, ZIMBABWE

8. Chikupo School is situated in the rural areas of Zimbabwe and was one of the places in the Murewa district to be supplied with a handpump. Sites for a borehole were investigated using the Slingram, which gave the result shown in Fig. 2. The resulting well has a capacity of 9000 l/h.

CONCLUSIONS

9. The Slingram has proved to be an asset in water exploration in Africa. The District Development Fund in Zimbabwe has a success rate of 85%, the Department of Water Affairs in Botswana from 55% to 80% and UNICEF in Nigeria increased their success rate to 90%.

10. The Slingram alone will not solve all the problems in water exploration. The input and choice of geophysical methods must be correlated to the objectives and the economy of the project and the actual geological conditions.

C. R. C. JONES, BA, MSc, MICE, FGS, and
S. BEESON, BA, PhD, FGS, Sir M.
MacDonald & Partners, Cambridge, UK

TN22. A combined geophysical method for borehole siting, Kano State, Nigeria

Hydrogeological conditions in the Kano Basement Complex are similar to those described for other parts of Africa (ref. 1). Rock types include various granites, metamorphics and migmatites with rapid lateral changes in lithology. Groundwater occurs in useful quantity only in weathered zones or fissures, and is normally hard to locate. In Kano static water levels vary from 1 m to 42 m below ground and specific electrical conductance (EC) ranges from 0.02 mS/cm to 1.0 mS/cm.

2. The problem was that over 500 rural communities were to be provided with handpumped boreholes drawing water from the basement complex. An efficient and rapid siting method was essential. Resistivity methods alone had been found to be of doubtful value without good borehole calibration, given the interpretation problems that lateral variation, salinity and degree of weathering could produce. Similar problems exist with electromagnetic methods alone. Previous workers have described each method for specific investigations (refs 2 and 3), but in the rural water supply context, selection criteria and interpretation are poorly documented.

3. Calibration drilling was out of the question and it was decided to use a combination of the two methods, specifically an electromagnetic traverse (EMT) in conjunction with vertical electric sounding (VES). The main reason for this was that EMTs were rapid and each community obviously had a preferred site not related to hydrogeology, whose general area could be quickly examined.

4. The survey comprised measurement of EC and static water level (SWL) in any dug wells near the preferred area. Traverses were then carried out along any suitable track with a Geonics EM-34 low frequency electromagnetic instrument. This comprised a receiver and transmitter: each a shoulder bag of electronics plus a 1.1 m dia. coil, connected by cable to give standard separations of 10 m, 20 m or 40 m. The receiver is scaled to show both coil separation and apparent conductivity. The reading is apparent because it is a weighted mean of real conductivity distribution (inverse resistivity) in the subsurface. Weighting coefficients systematically vary with depth and coil separation and orientation (ref. 4). The EM-34 is an extremely useful traversing tool because

(a) traversing is rapid - up to 1 km/h

(b) up to six different readings per station can be taken to interpret conductivity distribution with depth
(c) no physical ground contact is necessary and results can be obtained where electrical contact is impossible, as over bare rock
(d) traverse lines can be sinuous, as along a path in standing crops.

5. The traverse used was a series of vertical dipole V readings with 10 m or 20 m station intervals and 20 m or 40 m coil separations, decided on by a critical water table depth of 15 m. In areas of interest where V readings exceeded a minimum requirement (15 mS/m for EC < 0.14 mS/cm or 12-14 mS/m for 0.14 < EC < 1.0 mS/cm) the horizontal dipole reading was taken. A feature of interest occurs wherever the V reading exceeds the H, since the nature of the weighting functions are such that this condition obtains only when the conductor is at a depth exceeding 39% of the coil separation, thus avoiding some of the uncertainty inherent in VES methods. At such a feature, station interval was closed to 5 m or 2.5 m in the traverse and orthogonal directions. If the measured width was more than 5 m, the origin of the maximum was accepted for VES checking.

6. The VES was performed with an ABEM Terrameter in the Schlumberger electrode configuration (refs 3 and 5), to sufficient AB/2 to define the rising limb on log-log paper - about 100 m in Kano. The data was then corrected for MN/2 shift (ref. 3) and preliminary interpretation made using type curves and analysis of total longitudinal conductance. This was later checked by computer or programmable calculator (ref. 5) to give the required degree of accuracy. The site was accepted for drilling, provided it had a pre-basement layer extending at least 10 m below the water table and the layer resistivity

Table 1

Groundwater EC: mS/cm	Pre-basement resistivity: ohm m
0.03-0.1	350-50
0.1-0.4	100-25
0.4-1.0	80-10

World Water '86. Thomas Telford Limited, London, 1987

lay in the range shown in Table 1.

7. Using the methodology and selection criteria outlined, results were obtained with an equipment package of capital cost about £12 500 and an output rate of 20-30 sites per month. Of the 429 boreholes drilled, 65 (15%) were abandoned (they failed to achieve the requirement of 10 1/min during a three-hour pump test). The average top of the screen was 28.3 m BGL, the average bottom of the casing 39.0 m BGL and the average SWL was 15.1 m BGL. The average specific capacity was 0.175 1/s per metre and the median was 0.081 1/s per metre.

8. Although these success rates and yields may seem small by international standards, they represent a considerable advance on some previous projects in Africa which have had failure rates of 25-50% based on the same criterion and siting wells using photointerpretation and resistivity methods alone. The combination of the rapid EMT method with the slower and depth-accurate VES method has led to the development of a useful tool in the circumstances of a thin, variable depth productive layer. It may have application elsewhere.

9. Another useful achievement has been that boreholes have been constructed with productive zones (screens), usually far enough below SWL (12.8 m average difference) to allow for adequate drawdown and seasonal variation. Average screen lengths of 9.2 m and total casing depths of 39.0 m have also been minimized and costs are low in comparison with previous work. Further, boreholes have generally been sited in the areas of deepest weathering close to settlements, so tapping the zones of greatest storage where low drawdowns minimize users' handpump efforts. Many boreholes could sustain the use of powered pumps and this suggests that the siting technique should receive more widespread consideration in situations similar to the Nigerian Basement Complex.

ACKNOWLEDGEMENTS

10. The Authors acknowledge the co-operation and kind permission of Kano State Agricultural and Rural Development Authority to present the paper.

REFERENCES

1. CLARK L. Groundwater abstraction from basement complex areas of Africa. Quarterly Journal of Engineering Geology, 1985, Vol. 18, No. 1.
2. DIRKS F. et al. Electromagnetic profiling in the investigation of small scale groundwater flow systems. UNESCO/IAH/IAHS, 1983.
3. ZOHDY A.A.R. et al. Application of surface geophysics to groundwater investigations. USGS, 1974.
4. McNEILL J.D. EM34/3 survey interpretation techniques. Geonics Technical Note TN-8, 1980.
5. PARASNIS D.S. Principles of applied geophysics. Methuen, 1982.

W. R. RANGELEY, OBE, BSc, FICE, and
R. BARNSLEY, BSc, MICE, Sir Alexander Gibb
& Partners, UK

11 New developments in surface irrigation

Surface irrigation is the predominant method used in the world covering almost 250 million ha out of a total irrigated area of 275 million ha. In the last 40 years there has been a rapid expansion of surface irrigation but based mainly on concepts and designs that were developed in the early part of the century. With the dramatic advances in agricultural technology that were precipitated by the distribution of new crop varieties in the late 1960's a new set of demands has been made on the existing irrigation systems. These demands bring to light many constraints in both the physical and institutional structures of irrigation systems. The paper reviews the responses that are taking place to these demands in this intensive form of agricultural production where water, land and human resources are so intricately interwoven.

INTRODUCTION

1. Surface irrigation, which includes all forms of irrigation based on application of water to the field by surface flow, is the predominant method used in the world. In the vast irrigated lands of Asia surface methods are used almost exclusively. Even in the United States, where overhead and drip methods have been developed to an advanced state, two thirds of the total area of 26 million hectares remain irrigated by surface methods.

SURFACE IRRIGATION METHODS

2. Fig. 1. shows the main types of surface irrigation. The dominant form is "basin" followed by "furrow" irrigation. Contour check and borderstrip methods are used in limited areas for high value crops, forage and pastures. In addition controlled or partially controlled flooding is still used in some parts of the world.

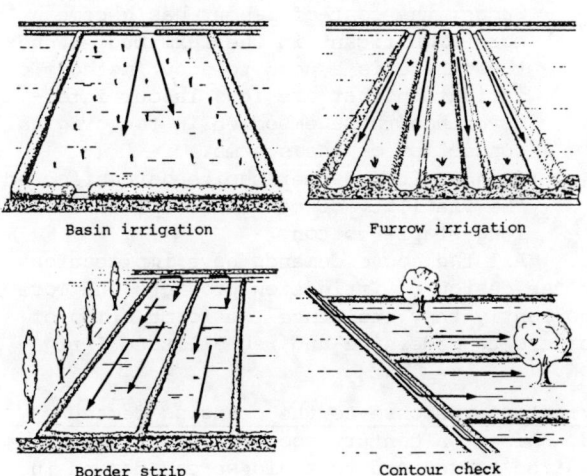

Fig. 1. Surface irrigation methods

Table 1. Distribution of surface and other irrigation methods (million ha)

Region	Surface	Other	Total
Europe	20	10	30
Asia	182	4	186
Africa	12	1	13
North America	23	12	35
South America	8	1	9
Australia & Pacific	1	1	2
	246	29	275

3. There are no reliable statistics on the proportions of the various methods used. Of the total irrigated area of the world today, which amounts to about 275 million hectares, about 90% or 246 million hectares are irrigated by surface methods.

4. Table 1. gives a very rough estimate of the distribution by surface and other types of irrigation in the world.

5. The dominant type of surface irrigation is "basin". This is used widely in Asia, not only for paddy rice, which occupies about one third of the total irrigated cropped area, but also for many other crops. It is a simple method of water application well suited to the very small fields found in many parts of Asia.

6. The furrow method of irrigation demands good management and well graded land. In its conventional form with a steady flow along a sloping furrow it is a method largely confined to developing countries. In many parts of Asia a form of irrigation is used with very short furrows in level fields with the result that it

is something between furrow and basin irrigation.

7. Contour strip and border check methods are not widely used outside developed countries where they are adopted for tree crops, forage and pasture.

8. Although it is expected that micro-irrigation, including drip and micro sprinklers, will expand and replace surface irrigation in certain situations, the rate of expansion will be slow. In fact the rate of expansion of surface irrigation, which amounts to about 4 million hectares a year, is likely to be much faster in absolute terms, than the expansion of other methods. Thus surface irrigation will remain dominant in the foreseeable future and hence becomes an important point of focus in modernization programmes.

MODERN REQUIREMENTS

The changing demands

9. There are several factors that are currently provoking new thoughts in the technology of surface irrigation. These include:

- better water management to meet the needs of modern agricultural technology;
- growing problems of operation and maintenance in developing countries;
- scarcity of further irrigable land and water resources;
- the general rise in energy costs that have taken place over the last 15 years;
- need to diversify crop production to respond more effectively to market demands.

10. Although the total area of surface irrigation has expanded almost three times since 1950, most of the new systems have been based on technologies that were developed in the first quarter of this century. The emphasis in the last 40 years has been on quantity rather than quality and there has been a deep rooted trend to stay with the known technologies and their associated institutional and social frameworks. However, it must be acknowledged that this approach has paid certain dividends as is clear from the very satisfactory rates of growth in agriculture now evident in many parts of Asia. At the same time there may have been some lack of foresight because, for example, most of the surface irrigation projects designed for rice cropping now need to be remodelled for mixed cropping in the face of an overproduction of rice. It may be argued that the design of schemes built in the last 20 years should have offered greater flexibility for adjustments to cropping patterns but until the advent of the so-called Green Revolution in the late 1960's no one was able to foresee the dramatic increases in crop yields that were feasible with the new cultivars and the more exacting demands that new crop varieties and mechanisation would impose on irrigation systems.

11. The factors listed above are imposing new demands on the irrigation system and its associated infrastructure and institutional framework and are broadly described as follows:

(a) The introduction of high yielding crop varieties coupled with mechanisation and modern agricultural inputs creates a demand for:

- better total reliability of water supply;
- more timely delivery of water to the cultivator;
- higher canal capacities because more rapid methods of field preparation give rise to higher peak irrigation demands.

(b) The growing scarcity of new sources of water means that further intensification or expansion of irrigation calls for higher efficiency of water use in existing areas.

(c) In some regions, such as Pakistan for example, the balance of availability between land and water has changed. Whereas in former times low intensity cropping was logical, now land is the limiting factor and higher cropping intensities are called for. This stresses the physical works and management institutions.

(d) There is a call for improvement in the physical structures for water delivery at farm level. In order to provide cultivators with better and more direct access to the canalization system.

(e) As stronger farming communities evolve in the older irrigation systems so there is a greater readiness, and sometimes appeal, to take over more responsibility for the operation and maintenance of the lower end of the canalisation system. This in turn has implications for the type of physical structures to be provided in the system.

(f) Changes in terms of trade in agriculture have given rise to a new set of priorities. Firstly high yielding land now carries a much higher premium than in the past - mainly because input costs are largely area related. Secondly because in some countries, such as the Sudan, the cost of labour has become more significant in the farm budget and thus there is a need to adopt methods of irrigation that are less labour intensive than those embodied in the original conception of the scheme.

(g) Changes in consumer choice have affected the viability of certain crops such as long staple cotton.

12. All the above demands have implications for the design of irrigation projects and more importantly they influence the formulation of projects to modernize and rehabilitate irrigation.

Irrigators response to the changing demands

13. Almost a century ago irrigators were required to respond to a widespread demand in the Middle-East and Asia to convert the old non-perennial inundation systems with their annual crops into controlled systems with

perennial cropping. It was a period of research and development. Major advances were made in the design of barrages on alluvial foundations with associated research in moving bed hydraulics. New forms of control gates were developed to meet specific needs such as minimizing head-loss, ease of operation and durability. There was a proliferation of designs for field outlet modules particularly in India where ingenious devices were invented, all with the general objective to minimize the effect of changes in water level on the resulting outflows. A dominant factor in all these developments was the influence of silty water and the disposal of the silt. The basic aim was to transport silt through the system onto the land and so avoid sedimentation of the canalization system - it was the problem that stimulated the work of Kennedy, Lacey and others and gave rise to the regime flow theories.

14. Today, we are once again at a point in the history of irrigation when major changes are required in order to respond to the new set of demands described above. How is the irrigation engineer responding to this situation and what are the resulting new developments?

15. Broadly, the new developments may be discussed under the following headings:
- Conservation works
- Diversion works
- Conveyance systems
- Water distribution and application at farm level
- Management, operation and maintenance

Conservation works

16. It is not intended to discuss here developments in the design of dams and storage works such as the advances in building on alluvial foundations and in the high head hydraulics of spillways. That would be a topic on its own. It is however appropriate to refer to trends in application of storage works to irrigation developments.

17. Up to the end of the first quarter of this century very few storage dams had been built expressly for irrigation. In the Indian sub-continent there was no real progress in the construction of storage dams until much later when the first major projects of Bhakra in India and Mangla in Pakistan came into service in the mid 1960's. The third quarter of this century was a period of great activity in dam construction in most parts of the world. This activity slowed down with the recession of the mid 1970's and the resulting poor terms of trade in agriculture. However, the increase in oil prices in the early and late 1970's stimulated greater interest in hydro-electric power and gave a new emphasis to dual purpose storage dam projects. Examples in India, Pakistan, Egypt and the Sudan are shown in Table 2. No new major dams have been brought into service in those countries in the 1980's but there are several under construction, such as the Sadar Sarovar dam in India with 9,500 million m3 storage and the Upper Wainganga dam also in India with 50,000 million m3.

18. The recent fall in oil price levels to those of the mid 1970's will, if sustained

Table 2. Some major irrigation storage dams with hydro-electric power facilities in India, Pakistan, Egypt and Sudan

Dam	Location	Date	Height (m)	Storage (Mm3)
Hirakud	India	1956	59	8 000
Tungabadra	India	1957	49	3 750
Gungi Sagar	India	1960	64	7 750
Bhakra	India	1963	226	9 500
Ukai	India	1972	81	8 500
Nagarjunasagar	India	1974	125	11 500
Pong	India	1974	60	8 500
Balimela	India	1977	75	3 500
Mangla	Pakistan	1967	116	7 250
Tarbela	Pakistan	1975	143	14 000
Aswan High	Egypt	1970	111	170 000
Roseires	Sudan	1966	45	3 500

lead to a serious reduction in the hydro-electric benefits to be derived from dual purpose projects. This in turn will place a bigger premium on the achievement of higher efficiencies in irrigation. There may however be some slowing down of new dam projects.

Diversion works

19. There have been improvements in the design of barrages over the last 50 years but the basic principles have not changed significantly. A comparison of some of the physical characteristics of the Sukkur and Chasma barrages shows some of these improvements and how they have developed.

20. The Sukkur barrage was constructed on the Lower Indus in 1932 whilst the Chasma barrage was completed in 1971 some 620 km upstream. Both barrages are designed for a flood of 27,000 m3/s but Chasma is 356 m narrower between abutments (1084 m compared to 1440 m). This was achieved partly by reduction in pier widths from 3 m to 2.1 m and by elimination of intermediate abutment sections but mainly by higher intensity of discharge through the openings at Chasma. The openings are 18.3 m wider at both barrages but Chasma has only 52 compared to 66 at Sukkur and so the discharge intensity has been increased by approximately 27%.

21. Three factors which perhaps allowed higher discharge intensities to be adopted at Chasma are:
(a) The gates at Chasma are approximately 50% higher than at Sukkur (most of the gates at Chasma are 8.2 m high whereas most of the gates at Sukkur are 5.6 m high). The type of gate also changed. Chasma has radial gates whereas the older Sukkur barrage relies on vertical lift gates with counter weights.
(b) The foundation at Chasma was densified by vibrofloatation to increase allowable bearing pressures.
(c) Greater confidence in energy dissipation predictions, though both barrages relied

on physical hydraulic models to aid design.

22. Lastly, Chasma was constructed in reinforced concrete in a period of 4 years whereas Sukkur was constructed of locally quarried masonry in the much longer period of 7 years.

23. Turning to the subject of irrigation offtakes, there has been more emphasis on reducing the amount of silt diverted into irrigation canals to reduce maintenance costs but more importantly to provide the reliability of supply demanded by modern agriculture. For example, the Major Eastern Kosi Canal in India, with a capacity of 300 m^3/s, used to be closed for desilting for about 20 per cent of the time but since constructing a silt excluder at the point of diversion, closures have been completely avoided.

24. Silt excluders take various forms according to the nature and size of the rivers and the characteristics of the silt load. For major rivers curved training walls and undersluice tunnels are frequently used. For small streams, drop inlet structures, built in the bed of the stream and vortex tubes are used.

Conveyance systems

25. Conveyance systems for surface irrigation are predominantly open channel systems rather than closed pressure conduits. Open channels are more suitable for the flat gradients generally found in the irrigable areas of the world. For pump schemes the smaller head losses achieved with open channels make them less energy demanding than closed conduits. However, open channels are not easily controlled and with the modern agricultural demands described above there is a new focus on ways of achieving better water control - through a combination of new physical works and improved management.

26. In recent years there has been a trend towards large and long conveyance canals. For example, the Kara-Kum canal, built in the USSR in 1975, has a capacity of 285 m^3/s and a length of approximately 1,500 km; in the USA the California Aqueduct, completed in 1974, has a length of 1329 km of which some 1065 km is canal; in Pakistan, the Indus scheme, completed in 1971, includes eight link canals of up to 736 m^3/s capacity with a combined length of 650 km and in India the recently constructed Rajasthan canal has a maximum capacity of 524 m^3/s and a total length of 685 km.

27. The two factors that have stimulated these major canal projects are firstly the need to transfer water over long distances from areas of plenty to areas of overutilization, and secondly the development of large and efficient excavating machinery. Fig. 2. shows a bucket wheel excavator engaged on the construction of the Chasma-Jhelum link canal in Pakistan.

28. Canal lining is a field where engineers have sought to improve the performance and at the same time reduce the costs of lining. There are numerous types and variations but the three more important in modern practice are concrete, clay and buried plastic membrane.

Fig. 2. Bucket Wheel Excavator (3823 m^3/hour)

Traditional brick linings are still used where labour is plentiful such as in Pakistan and India. Concrete linings remain popular because of their low frictional resistance, tolerance of high velocity flow, resistance to weed growth, robustness and simplicity. They are usually installed to reduce leakage, either to conserve water or to prevent waterlogging and drainage problems, but their greater carrying capacity is a valuable side benefit wherever gradients are adequate to take advantage of the higher water velocities possible with hard surface linings.

29. Clay lining, rather like the traditional puddled clay linings used extensively in the old navigation canals in the UK but now placed by modern earth moving equipment, has several advantages. It uses naturally occuring material which if found nearby can provide very economical lining. If damaged by seepage, uplift, or maintenance operations, it is easily repaired and therefore well suited for use in developing countries.

30. The use of plastic membrane lining for canals has developed apace with the wide range of plastics now available. Probably the most successful to date is a thin film of PVC or Polythene buried under approximately 0.4 m of soil and gravel to provide protection from mechanical damage, ultraviolet light, uplift pressures and scour. Current practice in the USA is to use 0.5 mm thick plasticised PVC sheeting whereas British engineers tend to favour polythene sheeting because it is cheaper and does not lose plasticity with age.

31. In most open channel conveyance systems water levels have to be controlled at intervals along the canal to maintain water levels for irrigation offtakes and to prevent scour due to shallow, high velocity flow. In large parts of Asia systems of what may be described as "up-stream" control have been used for decades. By this system the water supply is released by head-regulators and cross-regulators and delivered to the secondary or tertiary canal system (laterals or farm outlets) on pre-arranged schedules. Up-stream control is the only feasible system where supplies at the headworks are unpredictable as in the case of run-of-river diversions if equitable distribu-

tion is required.

32. In recent years various forms of float operated automatic upstream control gates have been developed. Their use is largely confined to Mediterranean countries, the USA and Australia but recently, they are finding application in the developing countries.

33. The advent of telemetry, remote control, computer control and the desire for on-demand irrigation has fostered downstream control systems. The control gates are operated in response to changes in downstream conditions. The simplest form is to maintain a constant downstream level immediately downstream of the gate but more effective is to maintain constant level at the downstream end of the downstream reach. A comparatively recent development is to maintain a constant volume of water in the downstream reach, the volume usually being assessed from two water levels, one at each end of the downstream reach. This arrangement reduces the required height of canal banks for safe operation and so reduces the cost of the canal construction and the width of land taken by the canal. Upstream and downstream automatic control arrangements are illustrated in Figure 3.

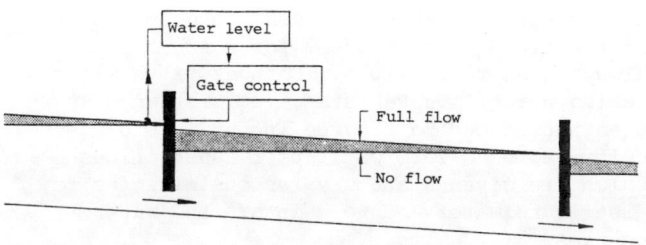
Upstream control

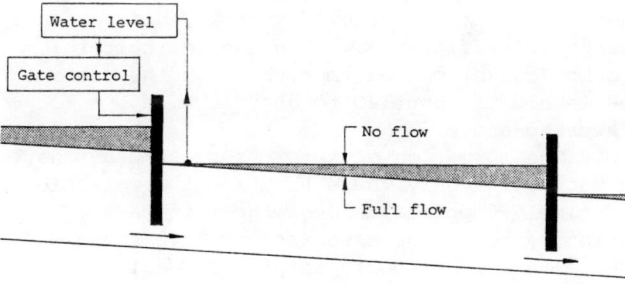
Downstream control

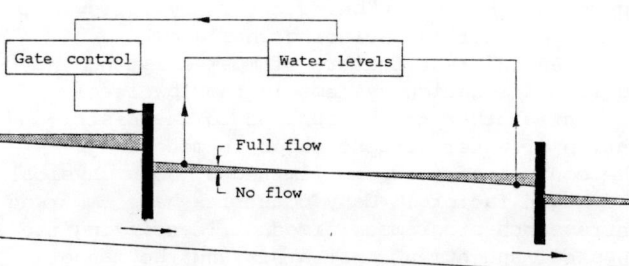
Constant downstream volume control

Fig. 3. Canal control

34. A recent development relating to control methods is the use of computer programmes to simulate the operation of a canal to test various control methods and operating procedures. System models can run tests at low cost and provide information which could otherwise only be gathered on reduced scale physical models or on the full size system, at great expense.

35. The tertiary canals, which distribute water within the field are of importance despite their smallness in size. Each hectare of irrigated land will have approximately 140 m of tertiary canal associated with it and so the total length of tertiary canals on a scheme can easily exceed that of the larger canals. They are thus significant in terms of capital cost, maintenance requirements and seepage losses.

36. Measures to overcome the problems of maintenance and leakage include:
- Reconstruction by removing weed bearing material and forming new banks with clean soil;
- lining with concrete and other materials;
- replacement of ground level channels with elevated precast concrete channels;
- replacement pipes.

37. The last method is finding application where shortages of land, water and labour are dominant considerations. The move to pipe distribution has been facilitated by the availability of low cost plastic pipe. Pipes can be laid on the surface but a recent development is the use of 50 mm to 300 mm diameter flexible hose underground. The hose is inflated prior to backfilling and the ground provides the resistance to internal pressures in service.

38. Pipe distribution systems facilitate automatic or remote control of irrigation to reduce labour requirements and to improve the uniformity of water application.

Water distribution and application at farm level

39. Until recently little attention had been given to the efficiency with which water is used in the field. Exceptions occured in regions of severe water shortage but in most parts of the world, the farmers were left to their own devices and the procedures they adopted were, and largely remain, very wasteful. In India and Pakistan where the agricultural block is a unit (chak) of about 145 ha, it has been found that the overall irrigation efficiency is as low as 42%. With good management and well controlled water application at the field it should be possible to raise the overall efficiency to about 60 - 65%. In Pakistan such an improvement in efficiency would produce a saving of 4 or 5 times the useful capacity of Tarbela reservoir after allowance for groundwater recovery and return flow. With such savings in mind Pakistan has embarked on a far reaching programme of on-farm water management. India has been pursuing a similar programme of command area development aimed at a co-ordinated approach to enhanced productivity with emphasis on the links between irrigation and agriculture. Basically these programmes have four major objectives in the

water sector:
- improve the tertiary canal system;
- level the fields;
- provide better outlets into the water courses;
- in some regions, create water-user associations, as referred to in the next section of this paper, to provide more direct management at field level.

40. The methods of water application to the field remain the traditional ones of flood or furrow.

41. Of the above improvements in water management at farm level the most effective is perhaps that of field levelling and here we are seeing dramatic changes with the introduction of laser control techniques coupled with computerised plotting of field surveys and earthwork planning. Better and more frequent levelling of fields is vitally important if the best response is to be obtained from the new improved crop varieties, particularly the short straw high yielding cereals. Even the aquatic rice plant can suffer from too much immersion!

42. In some situations, such as the Sudan, special and difficult problems have arisen with water application to the field. Traditionally, water was served into small level basins 18 metres x 18 metres but today this is too demanding in labour. Attempts have been made to introduce long and carefully graded furrows but precision and the necessary standards of management have been difficult to achieve. In the long term the authors see the answer to this and many other problems in better and highly mechanical large scale land levelling coupled with improved water control. In some countries a further pre-requisite to the improvement of field irrigation efficiency is land consolidation.

MANAGEMENT, OPERATION AND MAINTENANCE

43. In recent years there has been a growing awareness that management, operation and maintenance of surface irrigation schemes have been unduly neglected. The deficiencies in these activities now represent serious constraints on production in many parts of the world.

44. With more modern forms of irrigation, such as sprinkler and drip, these deficiencies are not so widespread for the reason that they are usually associated with more advanced agriculture as a whole.

45. The factors that have provoked the sudden if not dramatic, interest in management, operation and maintenance are:
(a) The new crop varieties are much more responsive to good husbandry than were the traditional varieties.
(b) In many developing countries management standards have fallen and maintenance has been seriously neglected.
(c) The marginal cost of expanding the traditional irrigation schemes has become so high that there is a move towards their intensification. This is more demanding of management and operating skills than was the old low intensity cropping.
(d) The deteriorating terms of trade for agriculture means that costs must be reduced to make irrigation worthwhile.

46. As part of the general move to improve water management, operation and maintenance, there is a tendency to sponsor the creation of water-user associations among the cultivators with the aim to delegate to them a greater responsibility for operation and maintenance. Such associations have long been established in developed countries whilst in most developing countries the responsibility for the running of the irrigation networks has remained in the hands of a central agency, usually the irrigation department.

47. In the USA water-user associations have become highly developed. They have autonomous boards of directors, employ managers and staff, have their own workshops and maintenance plant and co-ordinate water scheduling within the district covered by the association.

48. In the large surface irrigated areas of Asia both the creation and the successful operation of water-user associations present many difficulties compared with the situation in the USA. For decades the farmers in Pakistan and India, for example, have been accustomed to the dominant and paternal role of government. Experience has shown that, often, they have been unable to band themselves together even to undertake the elementary task of the maintenance of the field water courses for which they have always been responsible. Changes are however taking place. In Pakistan, legislation has been introduced to provide incentives to the cultivators to form water-user associations which are given statutory responsibilities to distribute water and to maintain the watercourses.

49. The aims of the current programme in Pakistan are to set up the water-user associations and renovate the water courses at the same time. It is believed that in this way the irrigation efficiency within the agricultural blocks (chaks) can be improved from the current low levels of about 40 to about 60%, as discussed above.

50. The formation of water-user associations is successful only where they receive reliable and timely supplies through the main canalization system. The associations have shown themselves to be poorly adapted to dealing with irrigular deliveries or situations of water scarcity. This underlines the importance of the irrigation network as a whole, an approach that is now being carefully followed in many countries including Thailand.

51. An inherent feature of most existing surface irrigation systems is the low level of control that can be applied to the distribution of water without major re-modelling. The control of flows is limited by both physical and human factors. Many countries have embarked on research programmes aimed at identifying the physical and human constraints and the manner in which they might be relieved. Recently, a new international institute with bases in Sri Lanka and Pakistan (International Irrigation Management Institute) has been established with the objective of assisting

countries to enhance their management standards. Although IIMI does not have an exclusive focus on surface irrigation it will inevitably find itself doing so in the light of world priorities.

52. The current emphasis on improvements in the management, operation and maintenance of irrigation systems is expected to continue for many years. The task is large indeed and the sociological constraints, if not also the physical ones, will take much time to overcome. In all irrigation systems water, land and human resources are intricately interwoven to form well established and complex patterns of production for food and fibre.

M. D. SQUIRE, NDA, NDAgrE,
BScHons, Project Manager,
Irrigation Systems, Wright Rain
Limited, UK

12 Recent developments in sprinkler and drip irrigation

Recent developments in sprinkler and drip irrigation attempt further to solve or reduce some of the problems which exist with irrigation systems. Some of these problems include unmanageable labour requirements for operation of the system, high consumption of energy, non beneficial total irrigation costs, poor water use efficiency and application uniformity, variable water quality and inadequate capacity to apply the required irrigation scheduling to the crop. This paper illustrates some of the recent developments which have occurred in the various system concepts showing the efforts which have been made to attempt to eliminate or reduce the problems.

TRADITIONAL SPRINKLER IRRIGATION SYSTEMS

1. Traditional sprinkler irrigation systems cover a wide range of application methods. They developed from the manufacturer's ability to supply lightweight portable piping and sprinklers able to apply water fairly uniformly to the crop. These systems have the advantages of being simple and light-weight, versatile, having a low capital cost and being very adaptable.

2. A typical traditional sprinkler system comprises a set of rotating sprinklers on risers connected to lightweight tubing and forming a sprinkler lateral line. The lateral line is stationary while irrigating but is moved at the end of the set, (Fig.1.& 2.).

3. Some of the problems of high labour requirements, high energy requirements, poor application uniformity and short life and wear have at least been reduced.

Fig.1. Hand move sprinkler system

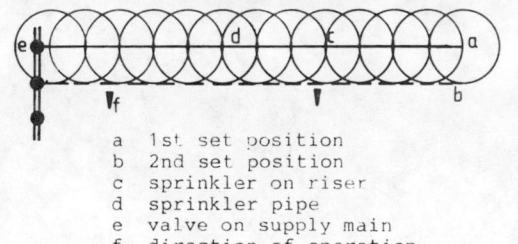

a 1st set position
b 2nd set position
c sprinkler on riser
d sprinkler pipe
e valve on supply main
f direction of operation

Fig.2. Plan view of sprinkler system

Labour

4. Traditional sprinkler systems have high labour requirements due to the need to reset the sprinklers and sprinkler lateral lines after every irrigation set.

5. Sequence valves used under each sprinkler head enable each sprinkler to irrigate in sequence along the lateral, (Fig.3.).thus reducing each lateral pipe diameter and lateral cost.

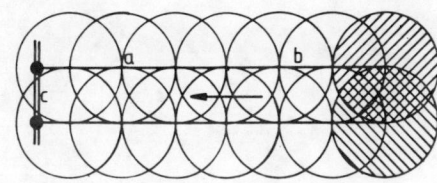

a sprinkler on riser fitted with sequence valve
b sprinkler line
c supply main

Fig.3. Sequence valve system

6. Lateral valves controlling the operation of all the sprinklers on the lateral at the same time provide further automation control, (Fig.4.).

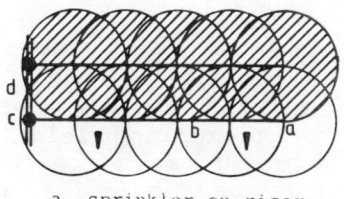

a sprinkler on riser
b sprinkler line
c lateral valve
d supply main

Fig.4. Lateral valve system

Energy

7. Energy requirements for the system have been reduced a small amount by the use of low friction loss automatic valves on automatic systems, (Fig.5.).

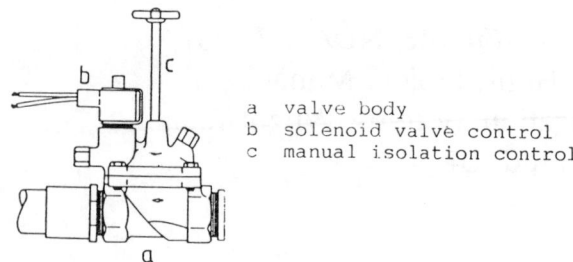

Fig.5. Automatic control valve

8. The development of special nozzle combinations and shapes which are able to operate at lower pressures but to still give an acceptable droplet size, achieve a greater saving in energy, (Fig.6.)

Fig.6. Various nozzle combinations

9. A combination of labour saving and energy saving has been achieved with the development of close spacing solid set mini sprinkler systems. These use low operating pressure small sprinkler heads working on close spacings laid out semi permanently within the crop. The development of various plastics has made these systems possible, achieving an acceptable capital cost with low energy requirements, (Fig.7.).

Fig.7. Solid set mini sprinkler system

Uniformity

10. It is essential to apply water uniformly to the crop for optimum crop yields. Sprinkler trajectory angle, rotation speed, beats per minute, spacings, pressure nozzle combinations, and wind all play an important part in affecting the application uniformity. Developments especially with computerised sprinkler design have improved the performance of sprinklers but they are still subject to the effects of the climate, (Fig.8.).

Fig.8. Factors affecting

Life/Wear

11. The use of new types of plastic materials within irrigation systems especially for the sprinkler or application heads has brought about a greater resistance to wear and a longer operating life, (Fig.9.).

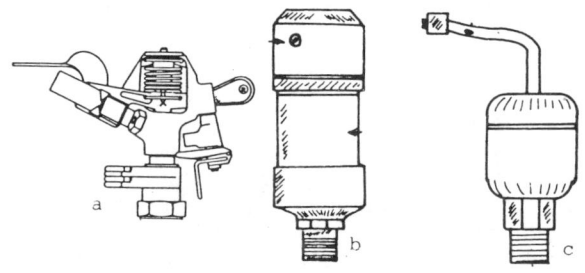

a Impact drive sprinkler
b Gear drive sprinkler
c Reaction drive sprinkler

Fig.9. Sprinkler types

SELF TRAVELLER IRRIGATION SYSTEMS

12. The self traveller irrigation system comprises a gun on a trolley, a flexible hose and a hose wind in mechanism. The self traveller has the advantages of low labour requirement, suitability for operation in any field shape, and compactability for ease of movement, (Fig.10a and Fig.10b.).

Fig.10a. Self traveller irrigation system

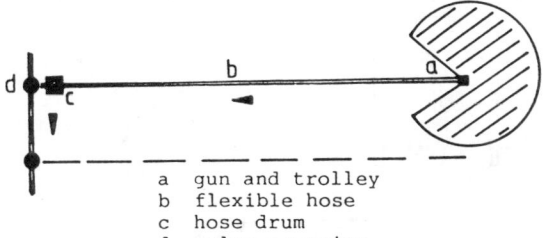

Fig.10b. Self traveller irrigation system

a gun and trolley
b flexible hose
c hose drum
d valve on mains

13. Developments to overcome the problems of application uniformity and operating pressures have produced the following.

Uniformity

14. Self traveller guns are subject to wind drift and wind distortion due to the large curtain of water which is exposed to the climatic conditions. This combined with variations with wind in speed have led to non uniform application depths over the lane area irrigated by the machine. The use of computers in setting up the application required by the system and also in controlling and keeping the wind in speed uniform from the start to finish of the irrigation set has meant higher standards in uniformity are achieved, (Fig.11.).

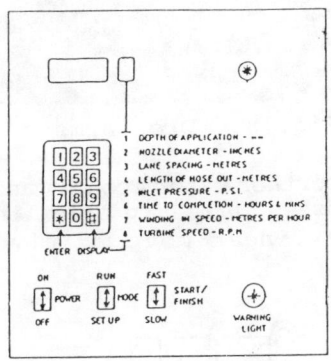

Fig.11. Control unit

Operating Pressures

15. Where guns are used it is necessary to have high pressure to throw the water the great distance required in order to irrigate as large an area of land per set as possible. Further improvements in gun performance include reduction of operating pressure with acceptable droplet size by the use of alternative nozzle types and by good design of gun drive systems and flow passages, (Fig.12.).

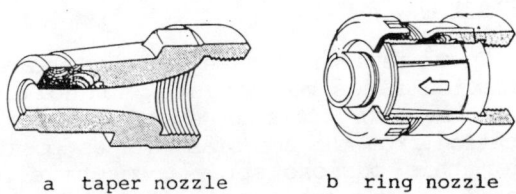

a taper nozzle   b ring nozzle

Fig.12. Nozzle assemblies

16. Booms have also played a major part in reducing the operating pressure required by the system and therefore reducing energy requirements. Booms normally comprise a series of sprinklers mounted on a lateral pipe assembly which replaces the single gun application head, (Fig.13.).

Fig.13. Irrigation boom

MOVING LATERAL IRRIGATION SYSTEMS

17. Moving lateral systems include centre pivot and linear move systems.

18. The centre pivot comprises a central water input point about which a series of wheeled towers rotate, (Fig.14.).

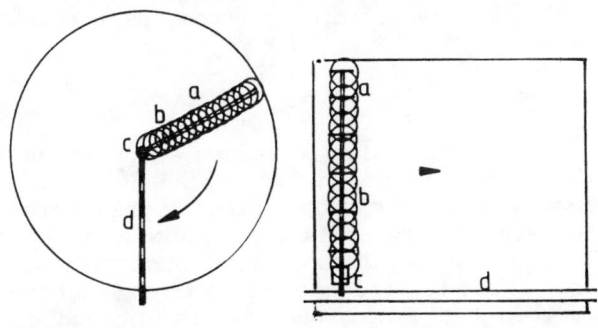

a application heads   a application heads
b towers   b towers
c pivot point   c input drive point
d supply main   d water source supply

Fig.14. Centre pivot

19. The linear move is similar to a centre pivot except that the wheeled towers travel in a straight line instead of rotating about a central point.

20. These systems have low labour requirements, low fuel consumption and low capital costs provided that investment is spread over a large unit area.

21. Further improvements have been made in energy conservation, uniformity of application, losses from the system and infiltration.

Energy

22. The development of low pressure spray heads operating at less than half the pressure of traditional sprinklers give a great energy saving, (Fig.15.).

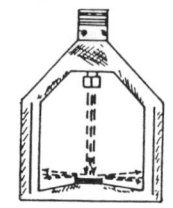

a   Full circle spray head

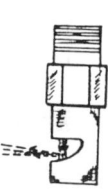

b   Part circle spray head

Fig.15.   Spray heads

23. The use of more efficient electric drive motors for the tower spans and improved gearbox drive systems both combine to give lower power requirements, (Fig.16.).

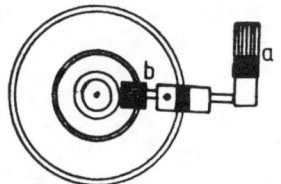

a   motor drive unit
b   planetary gear box

Fig.16.   Drive system to wheels

Uniformity

24. The use of computer models to calculate out the application specifications for the irrigation heads on the pivot ensures a high coefficient of uniformity from the irrigation system.

Losses

25. In an effort to reduce evaporation and wind drift losses new spray heads have been developed. They use the rotating principle but operate at low pressures, producing larger droplet sizes. The mounting of spray heads on drop tubes to irrigate just above the crop or between the crop rows has also proved to be advantageous in reducing losses, (Fig.17.).

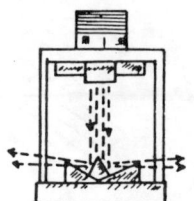

a   Internal rotating   b   External rotating

Fig.17.   Spray heads

26. Drop arms fitted with very low operating pressure splash plates used in conjunction with basins on linear move systems has been developed in areas of especially high winds and soils with low infiltration rates, (Fig.18.).

Fig.18.   Drop arms

27. For further water economy and reduction of spray losses, trailing drip lines are run behind the linear move or pivot system, (Fig.19.).

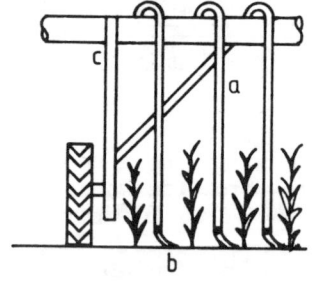

a   drop tube
b   drip line
c   tower unit

Fig.19.   Trailing drip lines

Infiltration

28. Spray jets which have high instantaneous application rates may cause runoff and ponding problems. It is necessary to spread the application from the heads as much as possible and for this purpose spray booms have been developed. These booms are placed at an angle to the direction of travel and may have several spray jets fitted to each boom, (Fig.20.).

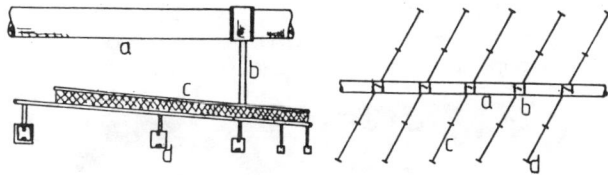

a tower span pipe
b drop tube
c spray boom
d spray nozzles

Fig.20.   Spray booms

Fertigation

29. Because of the high uniformity of application from the pivot and linear move systems it is possible to apply both fertiliser and chemicals through the irrigation system. This reduces total crop production costs by avoiding the need to apply fertilisers and chemicals in the traditional methods, (Fig.21.).

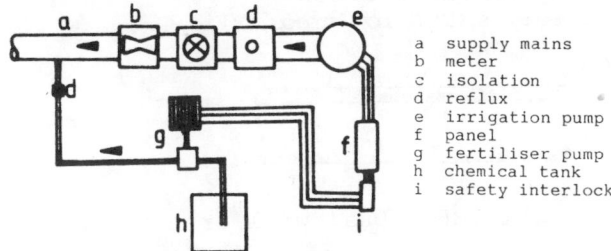

a supply mains
b meter
c isolation
d reflux
e irrigation pump
f panel
g fertiliser pump
h chemical tank
i safety interlock

Fig.21.   Fertigation equipment

Supply Hose

30. A partial limitation of both centre pivot and linear move systems is that they may be restricted to irrigating one land area. To try and improve the mobility of these systems, equipment is being developed where the supply hose can be wound in and out on a drum fixed to the main chassis of the irrigation machine, (Fig.22.).

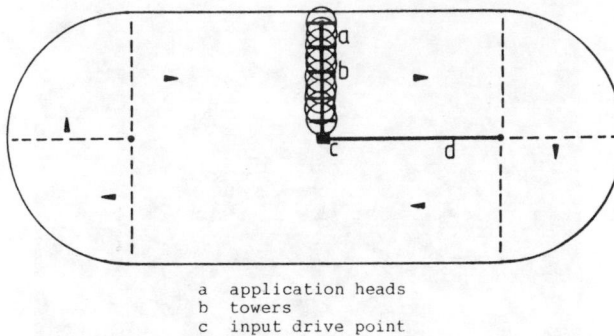

a application heads
b towers
c input drive point
d supply hose

Fig.22.   Hose wind linear move

LOCALISED IRRIGATION SYSTEMS

31. Localised irrigation systems include drip and micro jet irrigation for the irrigation of the crop root zone and not irrigation of the total soil area, (Fig.23.).

Fig.23.   Drip wetted soil area

32. The in field equipment comprises small bore polyethylene lateral pipe often laid down each crop row. Emitters or microjets are positioned along the lateral so that they irrigate the root zone of each plant, (Fig.24.).

Fig.24.   Drip irrigation layout

33. These irrigation systems have the advantages of low energy requirements, water economy and low labour.

Automation

34. If the irrigation equipment is laid out for the duration of the crop growing season it is possible to fully automate the equipment reducing the labour requirement to a minimum, (Fig.25.).

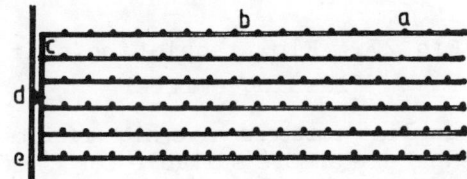

a emitters
b drip lateral
c header pipe
d control valve
e supply main

Fig.25.   Drip equipment layout

Mobile Drip Lines.

35. Taking advantage of the improved characteristics of both piping and fittings mobile drip lines have been developed. The drip lines are moved from set position to position during the irrigation cycle. The ability to reset the lateral lines results in a reduction of the capital cost of the equipment as each drip line irrigates a number of set positions, (Fig.26.).

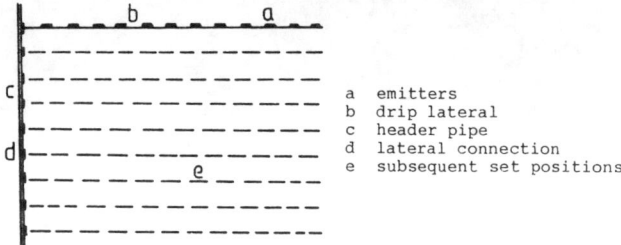

a emitters
b drip lateral
c header pipe
d lateral connection
e subsequent set positions

Fig.26. Mobile drip system

Emitters/Discharge Orifices

36. The requirements for emitters to operate over a range of pressures giving the same flow rate, to be resistant to clogging, to be self flushing and to be damage resistant, have led to the development of several types of emitters.

37. On line emitters have the ability to be flow regulating, self flushing and pressure compensating, (Fig.27.).

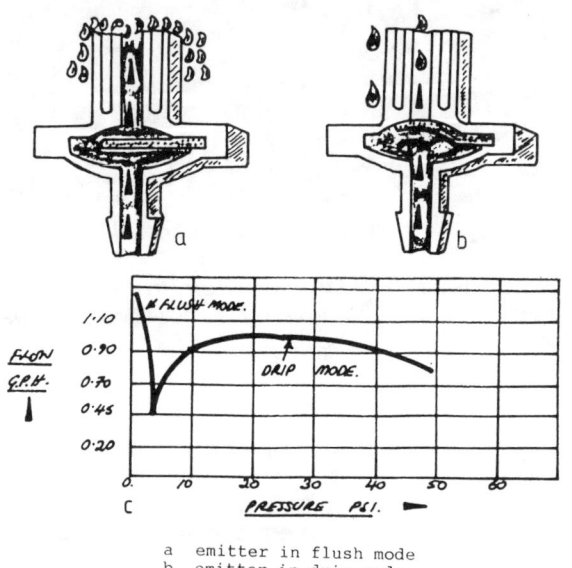

a emitter in flush mode
b emitter in drip mode
c performance graph

Fig.27. Flow regulating/self flushing emitter

38. In line emitters have high resistance to damage and are a more suitable for the mobile drip systems, (Fig.28.).

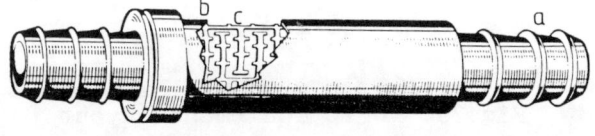

a connection to lateral pipe
b cut away section
c flow paths

Fig.28. In line emitter

39. Multiple outlet emitters are an economical solution where a large root zone of the crop has to be irrigated, (Fig.29.).

Fig.29. Multi outlet emitter

40. Drip tubing comprising a dual wall is an economical solution for the irrigation of closely spaced row crops, (Fig.30.).

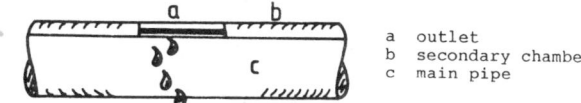

a outlet
b secondary chamber
c main pipe

Fig.30. Dual wall drip tubing

Micro Jets and Mini Sprinklers

41. These types of application heads are very suitable for crops growing on wide spacings and having large root zones which need irrigating. One micro jet or a series of micro jets is able to irrigate a larger root zone of the crop and can be a more economical solution than drippers in certain applications. Again these have been introduced and made available due to the developments in the plastics industry, (Fig.31.).

Fig.31. Microjets irrigating trees

AUTOMATION

42. Automation in irrigation systems gives a range of benefits, these may include: Reduction of people required for system operation, accurate application depths applied to the crop, greater control of water volume applied, automatic recording of applications, direct control relationship between crop water requirements and the climatic conditions within the field, (Fig.32.).

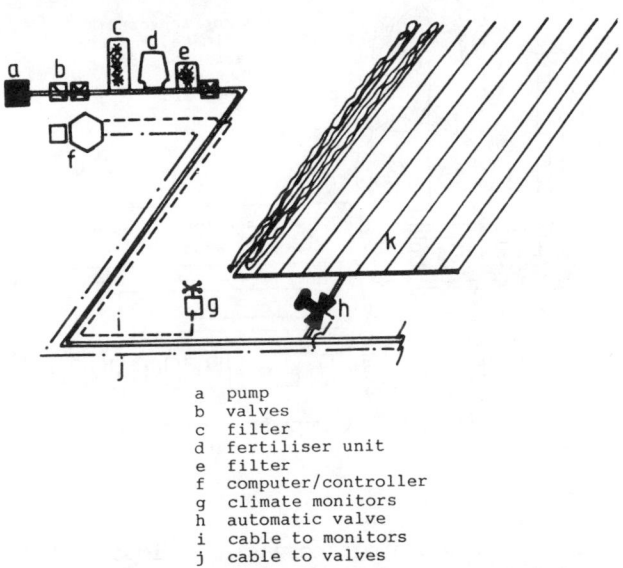

```
a  pump
b  valves
c  filter
d  fertiliser unit
e  filter
f  computer/controller
g  climate monitors
h  automatic valve
i  cable to monitors
j  cable to valves
```

Fig.32. Automation of irrigation systems

Controller/Computers

43. Developments within the elctronics industry have meant that it is possible at an economical price to introduce automatic control for the irrigation system. The controllers may be very simple units operating on a time clock basis or they may be very complex computer systems utilising and assessing a wide range of data, (Fig.33.).

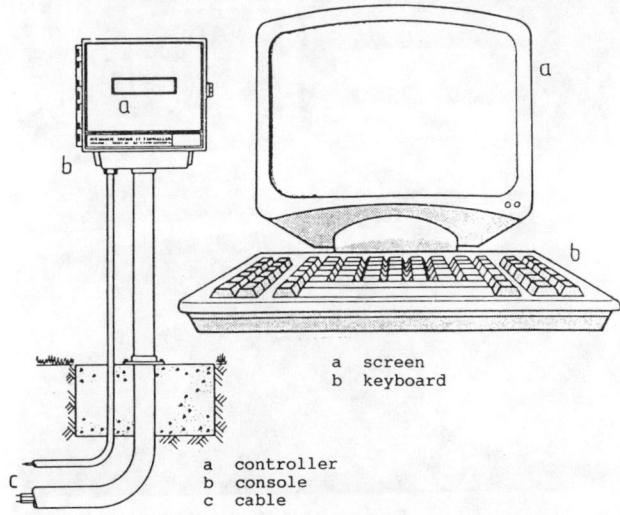

```
a  screen
b  keyboard

a  controller
b  console
c  cable
```

Fig.33. Controller and computer

Fertigation

44. Where irrigation systems have a high uniformity of application it may be economically attractive to apply the fertilizer and other chemicals through the irrigation system. This method of application produces great savings in labour eliminating some of the traditional application practices, (Fig.34.).

Fig.34. Fertiliser unit for irrigation system

Filtration

45. Water quality, mechanical, chemical or bacteriological is important especially in the localised irrigation systems which utilise small discharge orifices. It is therefore necessary to have adequate filtration. To ensure that the filtration is carried out correctly it is possible to fit automatic flushing circuits to the filters and these can be controlled by time clocks or by pressure differential, (Fig.35.).

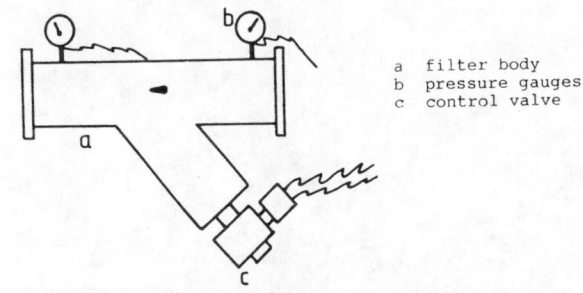

```
a  filter body
b  pressure gauges
c  control valve
```

Fig.35. Automatic filter

Valves

46. Valves in the field which the operate the irrigation may function either on a volume basis or on a time basis. These are linked back to the central control system to apply the correct amount of water to the crop, (Fig.36.).

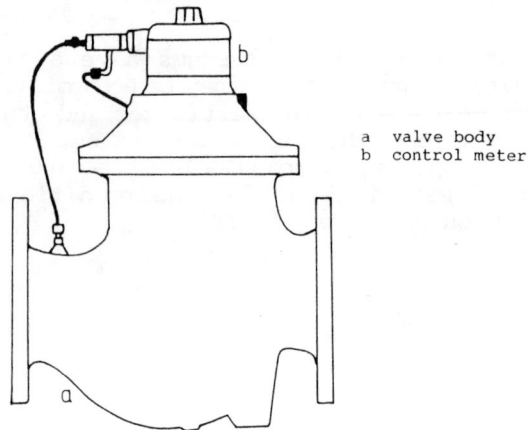

Fig.36. Volumetric automatic valve

a valve body
b control meter

Infield Sensors

47. Infield sensors can be used to determine the soil moisture conditions, and the climate conditions. They can also be linked into equipment such as lysimeters which indicate actual soil conditions and evaporation pans which indicate theoretical water requirements, (Fig.37).

Fig.37. Monitoring equipment

Scheduling

48. It is important to be able to apply the correct amount of water to the crop in the required number of days per cycle. Automatic equipment makes this much easier to do especially where computers are being used to control and evaluate all aspects of the crop situation and water requirements, (Fig.38).

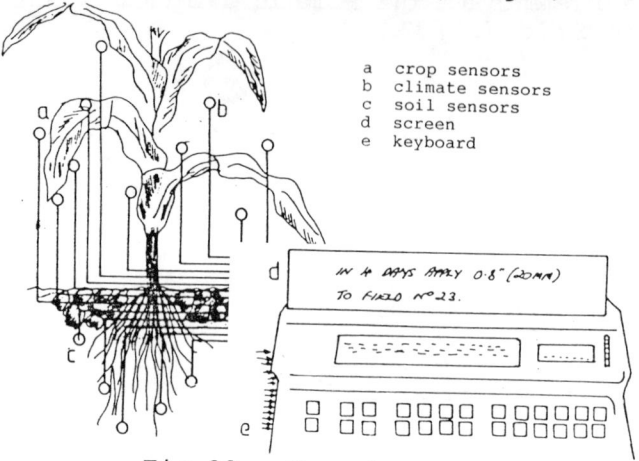

a crop sensors
b climate sensors
c soil sensors
d screen
e keyboard

Fig.38. Computer system

ECONOMICS

49. It is becoming increasingly important to ensure that the irrigation system specified is the most economic to do the job required. It has been difficult in the past to assess the true economic situation for irrigation systems when wishing to economically evaluate them. This applies both to the different irrigation concepts and for the designs within the specific concepts, (Fig.39).

Fig.39. Computer for economic analysis work

50. The use of computers has made this economic evaluation available in a manner which can be used and applied to all projects and feasibility studies, (Fig.40).

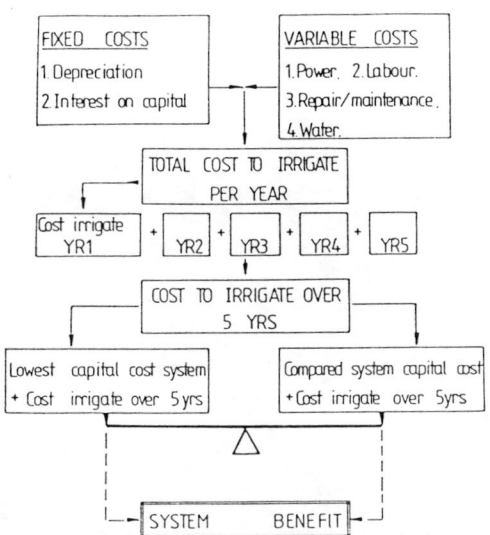

Fig.40. Economic analysis flow chart

Factors to Consider

51. Many factors need to be considered for the economic analysis of the irrigation system including both fixed cost and variable cost elements. Factors to consider include the depth of water applied to the crop, power requirement for the irrigation system, the cost of power, labour costs, interest on capital and inflation for both labour and power, (Fig.41).

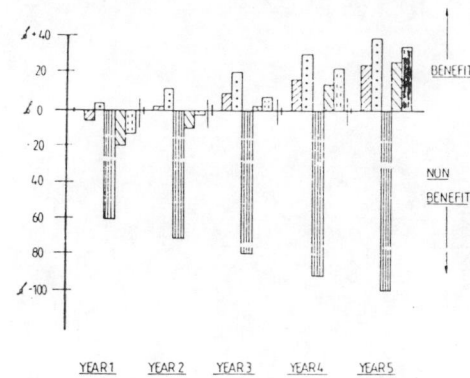

Fig.41. Analysis graph alternative concepts

Specific Sections of Design

52. During the design of the alternative irrigation systems it is possible to consider the effect of varying capital costs of the equipment and the effect that these have on the variable cost element of the irrigation system. This is applicable to all sections of the design including pump units, supply mains and the infield irrigation equipment, (Fig.42).

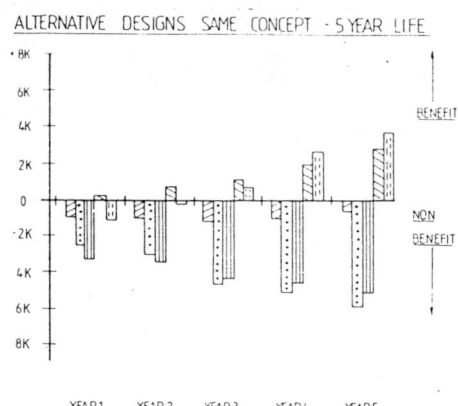
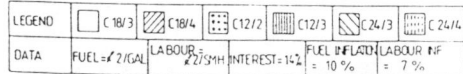

Fig.42. Analysis graph alternative designs same concept

53. Great benefits have been obtained by utilising these computer programmes and they make much clearer the required information needed for economic evaluation of the designs. (Fig.43).

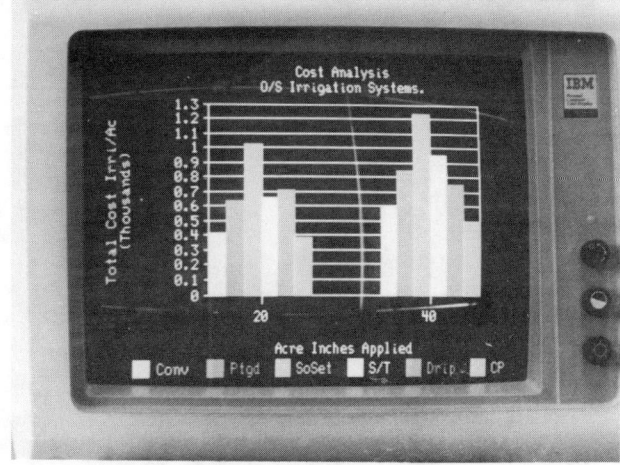

Fig.43. Screen display

SUMMING UP

54. This paper has attempted to highlight some of the recent developments in sprinkler and drip irrigation but in the time it is not possible to discuss all the work that has been carried out.

Discussion on Papers 11 and 12

DR R. MANIFOLD, Pilkington Reinforcements Ltd, UK

Mr Le Moigne and Mr Rangeley have both referred to the lining of canals as a means of improving irrigation systems. In Mr Rangeley's Paper the significance of lining the long lengths of tertiary canals in irrigation schemes is discussed.

Several Indian states are looking at glass-fibre-reinforced cement (GRC) watercourses and structures under a UNDP/FAO project which examines new lining materials.

GRC - basically a cement mortar reinforced with CemFIL alkali-resistant glass fibres - has been available for a long time now, mainly as a lightweight façade or architectural component, as in civil engineering applications, such as permanent formwork. It combines the compressive properties of the matrix with the tensile and bend properties of the glass fibre to give a strong, impact-resistant, lightweight material used typically in sections 6-25 mm thick.

It was realized that in addition to these properties it also had excellent water contact performance - low permeability, good erosion and corrosion resistance, and low surface roughness (Manning's n value was recently measured as 0.012 at Southampton University). Use has already been seen in drainage applications, particularly in the UK, and it was felt that GRC could have major potential in irrigation for liners, channels, canalets and structures.

This has been borne out by discoveries of its use in Swaziland and Southern Africa for just such applications. Although many advantages can be pointed to, it is invariably GRC's light weight and thus its handleability which provides major cost benefits. For example, in Southern Africa 2 km of semicircular watercourse liner were installed in only one week by a team of three unskilled labourers and the farm manager. Also, recent trials in Egypt have shown a tenfold increase in installation rate over existing site-cast concrete channels.

In India, under the UNDP/FAO project five states have shown an interest in CemFIL GRC and substantial trials are to be carried out in Maharashtra and Rajisthan. Following a recent FAO study tour to the UK by senior irrigation personnel from four of these states, recommendations are being placed to purchase additional production facilities in India to service this vast market further.

With interest rising in many other countries there may soon be another material to add to those mentioned in the Paper.

MR V. R. ROWLAND, Pauling PLC, UK

The Paper referred to silt exclusion. Silt excluders designed to separate the lower and more heavily silt-laden part of the flow are included in several structures earlier than Chasma, and conservation dams have included low level desilting sluices aimed at scouring the deposits of silt formed during retention during the appropriate part of the cycle. Would the Author comment on the success of these installations and any reports of this?

Evaporation control is not mentioned. Has there been any significant development in this area and in particular on the possibilities of liquid or boom-suspended membranes to reduce evaporation?

The factors stressed as important during the earlier discussion, quality of management, water efficiency and capital investment requirement, are more readily applied over larger rather than smaller areas and there are examples of resettlement schemes involving major capital investment and the use of valuable land reserves which are subsequently farmed inefficiently and eventually support only the farmers and their families on a subsistence level contributing virtually nothing to the national product. Does the Author see a likelihood of a move to larger plantation developments on either a private or a government-controlled basis?

MR A. JONKMAN, ACROSS, Kenya

New irrigation works or intensified irrigation are mainly made for world market production. The world market for rice, cotton and sugar is diminishing. Which new crops are of interest for these new developments in irrigation?

MR R. BAHAR, World Health Organisation, Switzerland

What are the new developments in the control of aquatic vegetation in canals using engineering methods apart from the use of herbicides and biological control methods?

Aquatic weeds are an important element for food and shelter of vector and intermediate hosts of diseases (mosquitos, malaria vector and snail intermediate hosts of schistosomiasis).

World Water '86. Thomas Telford Limited, London, 1987

PAPERS 11 AND 12: DISCUSSION

MR J. HENNESSY, Sir Alexander Gibb & Partners, UK
In the UK there is a thriving industry in manufacturing canal maintenance equipment such as weed-cutting apparatus (boat or land mounted). A possible contact for information is Mr D. Noble, Secretary, The Association of Drainage Authorities.

MR C. M. BOLT, Hexagon Development Services, UK
I should like to make the following points.

(a) Canal weeds are inhibited by muddy or silty water.
(b) Water use efficiency and management are assisted by land shaping on farms and watercourse lining and a reduction in the watercourse hydraulic slope, thus increasing the irrigated area.
(c) Land shaping is sometimes better done by local methods, although there will be plenty of scope for laser-guided systems.

MR A. J. WOODFORD, Sir Alexander Gibb & Partners, UK
In his Paper, Mr Rangeley has referred principally to the case of surface sources for irrigation. Could he elaborate on the influence of groundwater development on methods of surface irrigation?

MR E. A. BURLEIGH, Atkins Land and Water Management, UK
The recent improvements in sprinkler and drip irrigation have been described and no doubt the efficiency of water use of the various systems has improved. Could the Author indicate in what areas better efficiencies of water use have been made and what efficiencies of water use can be expected for drip and sprinkler irrigation?

MR M. D. SQUIRE
Better efficiencies of water use have been achieved in most of the various irrigation concepts which are now used. Normally the efficiency of water use relates to the application head which is being used for the specific concept type. Developments and improvements in nozzles for sprinklers and guns were illustrated in the Paper and these would fall within this category.

Efficiencies of water use for drip irrigation and for sprinkler irrigation will vary according to the climate and the application head being used.

Typical losses incurred by drip irrigation will be between 5% and 10% only while typical losses for sprinkler irrigation can range from 10% to about 30%.

MR J. STEINER, Rudolf Bauer GMBH, Austria
The alternatives to big guns on hard-base travellers are, for example

(a) two smaller guns at lower pressure
(b) a boom with sprinklers
(c) a boom with low pressure spray nozzles.

There is a different intensity of precipitation with these systems.

MR M. KEEN, Arab World Agribusiness, UK
The subsurface net system described by Mr Steiner is an Austrian invention called Agronet, whose makers feel that in Saudi Arabia more work and tests had to be undertaken before they were ready to launch the product commercially.

Can Mr Squire comment on the extent to which flushing is required to leach out salts where soil salinity is building up in semi-arid or arid lands, and could he forecast the extent to which money will have to be spent on underdrainage pipes to convey the leached water away?

MR M. D. SQUIRE
The leaching out of salts from the soil where salinity is building up is achieved by flushing water through the soil and carrying the salts away to drainage. The leaching requirement for any specific soil, water quality and crop will depend entirely on the site conditions in which the crop is being grown. There are different tolerances of crops to salinity and different effects of salinity on soil types.

The extent to which money will have to be spent on underdrainage pipes to convey the leached water away is a question which cannot easily be answered as it relates very much to the area in which the crop is being grown and all the relevant criteria.

MR B. S. PIPER, Institute of Hydrology, Thailand
There have been many comments on the advantages of land levelling on irrigation schemes using sophisticated techniques such as laser levelling.

What are the disadvantages of this procedure, particularly in areas where the soils are poor and the topsoil layer is thin?

The technique is called 'land consolidation' in Thailand; in the north-east where the soil structure and fertility are poor its success has been limited.

MR M. SNELL, Sir M. MacDonald & Partners, UK
The disadvantages of land levelling are

(a) the difficulty of achieving the desired planes because of swelling in cut areas and settlement in fill areas
(b) the limited plot size because of this and also the limited soil depth.

With regard to techniques, in one case in Iraq a modern laser-controlled system was superseded by numerous tractors with box scrapers, controlled by surveyors and pegs: this gave better results and lower costs.

MR W. C. HULSBOS, Euroconsult, The Netherlands
On the mechanical clearing of canals to eradicate the aquatic weeds especially designed equipment is available, but the use of this equipment in many irrigation systems is severely hampered by the lack or non-existence of maintenance roads, especially along the minor canals. For an efficient use a redesign might in many cases be required. It is stressed that for newly designed systems maintenance roads along the minor canals are being taken into account.

MR G. NEVILLE, Sir William Halcrow & Partners, UK
With regard to the low evaporation losses achieved by certain types of sprinkler nozzles, how significant is evaporation in terms of overall efficiency and what range of losses is achieved by the various sprinkler nozzle types?

MR M. D. SQUIRE
Evaporation losses as part of the overall efficiency of the system can be very low for systems such as drip irrigation and can be much higher with systems which use either sprinklers or guns. Evaporation losses from drip irrigation are virtually nil up to 1-2%, while sprinkler or gun irrigation systems may approach 20-25%.

The various sprinkler nozzle types that are used have a marginal effect on the evaporation losses but many of the evaporation losses relate to the nozzle diameter and the pressure at which the nozzle is operating giving a resultant droplet size. The size of the droplet greatly affects the evaporation loss that occurs.

DR T. W. TANTON, University of Southampton, UK
Most irrigationalists would agree that the standard of maintenance of most irrigation projects is declining, but it is interesting to note that at the same time the educational standards of irrigation staff within most countries is improving and it might be expected that the level of competence to operate and manage projects would also improve.

Unfortunately maintenance can be evaluated in economic terms but must be paid for in money. It is this lack of an adequate budget which in most countries leads to inadequate maintenance. Although misappropriation of funds may account for some cases of poor maintenance the main cause is the inability of farmers to finance it. In most countries increases in population and/or political pressure has decreased farm sizes to a level where farmers cannot live and pay for essential maintenance. The small plots place the farmers in a poverty trap where they cannot pay for essential inputs which might give them sufficient yield to pay for improved maintenance to maintain high yields. The result is that for the foreseeable future many projects will be unable to finance essential maintenance.

DR Z. SVEHLIK, University of Southampton, UK
The recent development in sprinkler and microirrigation was influenced, as Mr Squire explained, by the requirement to reduce the labour and energy inputs, to increase the water use efficiency and to lower the cost. Considering the various types of equipment described in the Paper and by applying these criteria, there should be an idea of the best sprinkler system now available. It is interesting that, for example, the self-traveller system which does not perform well according to the criteria (high energy input, high cost, relatively limited type of water of application) is popular among farmers, in some countries. Could Mr Squire elaborate on the popularity of the various types of sprinkler equipment?

MR M. D. SQUIRE
The popularity of the various types of sprinkler equipment depends not only on the personal choice of the client but also on the capital cost of the equipment and the ability to operate the equipment in the field.

In many areas of the world the traditional hand-moved sprinkler line equipment is still the major system that is used owing to its low capital cost. Where labour is a problem developments to semi-automate or to automate fully these hand-moved sprinkler lines have occurred thus reducing the labour requirement. Only a small percentage of the total irrigated land area uses semi- or fully automatic systems.

The popularity of other types of irrigation equipment such as drip irrigation, centre pivot irrigation, etc. varies greatly from country to country and is highly dependent on the money available, equipment preferences and suitability for the specific country.

MR M. KEEN, Arab World Agribusiness, UK
Having seen laser-operated flood levelling systems in California and North Africa recently, and being able to compare them with similar fields not subjected to the effect of a laser beam, I can confirm that there is a great improvement in water efficiency, and in the evenness of growth of the crop, where water flows steadily and without obstruction over the land from the intake end. Does Mr Rangeley see a successful future for the regular laser levelling system for field work?

I have had the opportunity to test the water distribution efficiency of the reel-in spraygun irrigation machine on my own farm in East Anglia where it is used to spread slurry. From the highly uneven distribution of the slurry deposition, which is visible when the land is dry enough for inspection, it is apparent that even a moderate wind can disturb the spread pattern significantly.

J. STEINER, Rudolf Bauer GmbH, Austria

TN23. New developments in water distribution with hard hose travellers

For many years hard hose travellers have been irrigating cultivated land without requiring much labour. In the past, all hard hosed travellers were equipped with wide-range rainers. In the course of the development of these irrigators the application of bigger and more efficient machines became possible requiring increasingly larger nozzle diameters at the wide-range rainers. However, this development had disadvantages.

(a) To distribute the water sufficiently (optimum break-up of the water jet) with larger nozzle diameters the nozzle pressure had to be increased, entailing higher energy inputs.
(b) Large nozzle diameters led to large droplets hitting the ground and crops with higher energy. As a result the soils were puddled and young and sensitive plants were damaged.

These disadvantages can be overcome by using two smaller wide-range rainers per irrigator which are equipped with smaller nozzles, or by using a boom-type trolley with small sprinklers or low-pressure spray nozzles (Figs 1 and 2).

HARD HOSE TRAVELLERS WITH TWO WIDE-RANGE RAINERS
2. Instead of one wide-range rainer with a large nozzle (e.g. 36 mm dia.) two wide-range rainers with smaller nozzles (e.g. 26 mm dia.) spaced 2 m apart are mounted on the sprinkler sledge. The total amount of water is thus divided into two halves and can be applied at a lower nozzle pressure.

Table 1. Nozzle comparison

Single rainer	Double rainer
1 x 32 mm dia.	Approx. 2 x 22 mm dia.
1 x 34 mm dia.	Approx. 2 x 24 mm dia.
1 x 36 mm dia.	Approx. 2 x 26 mm dia.
1 x 38 mm dia.	Approx. 2 x 28 mm dia.

3. When using double rainers the maximum strip width is reduced by 10-15% because the spray ranges are smaller at lower nozzle diameters and nozzle pressures.
4. Both rainers are mounted at precisely calculated sector adjustments so that the water is applied evenly over the entire strip width. The pressure required at the sprinkler nozzles can be cut by up to 1.0 bar when the nozzle diameters are appropriate. Despite this fact a better water distribution and rain quality is achieved than with a single rainer. The nozzle comparison given in Table 1 shows that when irrigating with double rainers considerably smaller (i.e. more efficient) nozzles can be used.

BOOM-TYPE TROLLEY WITH SMALL SPRINKLERS OR LOW-PRESSURE SPRAY NOZZLES
5. Using a boom-type trolley, which can be equipped with low-pressure spray nozzles or small sprinklers, has further major benefits

Fig. 1. Hard hose travellers with two wide-range rainers

Fig. 2. Boom-type trolley with small sprinklers or low-pressure spray nozzles

regarding precipitation quality and efficiency. These boom-type trolleys are engineered to be set up, dismantled and moved to the next working position by one person only, with hardly any time and labour involvement. They are partly available with freely swinging booms and adjustable track widths fitting any terrain and ensuring trouble-free irrigation on sloping ground.

6. When fitted with sprinklers boom-type trolleys of up to 40 m in width are currently used. With an additional spray range of the sprinklers obtained at both ends the wetted strip widths cover up to 70 m. When fitted with low-pressure spray nozzles trolleys up to 50 m in width can be used. With an additional spray range of the system with spray nozzles the wetted strip widths cover up to 58 m.

7. Connection pressure and intensity of application are different for boom-type trolleys with sprinklers and with spray nozzles.

Sprinkler equipment

8. Unlike wide-range rainers, which require nozzle pressures of up to 5.0 bar for adequate rain quality depending on the nozzle diameter, small sprinklers on boom-type trolleys with nozzle sizes of 5-10 mm can already be operated with a nozzle pressure from 2.5 bar. The small nozzle diameters enable a remarkably finer jet break-up as well as smaller droplets with less energy, leading to high rain quality and water which is sprinkled gently on the soil and plants.

9. The intensity of application ranges from about 10 mm/h to 30 mm/h, being substantially higher than the intensity with wide-range rainers.

10. Boom-type trolleys with small sprinklers are suitable for irrigating medium to heavy soils.

11. Sprinkler spacings on the booms as well as nozzle diameters can generally be adjusted to the desired application capacities; to achieve a more uniform application over the entire irrigation strip width the sector operation of the sprinklers has been computerized.

Low-pressure spray nozzle equipment

12. Boom-type trolleys with low-pressure spray nozzles are the most energy-saving method of irrigation with hard hose travellers. Dependent on the nozzle sizes these spray nozzles can be used from 1.7-2.0 bar nozzle pressure.

13. Rain quality and uniform water distribution comply with the values of boom-type trolleys equipped with sprinklers. Only the intensity of application with low spray range of the spray nozzles depending on the sledge retraction speeds of up to 150 mm/h is considerably higher than the intensity of the sledges equipped with sprinklers. Therefore, the boom-type trolley with spray nozzles is designed mainly for light and medium soils.

14. However, it has to be emphasized that the infiltration rate (mm/min) of the soils is remarkably higher during the first 10-20 min of the irrigation process than after a longer irrigation time. Duration of the high intake rate of the soils when irrigating with spray nozzles (e.g. at a retraction speed of 50 m/h) is only 1-11 minutes, thus lying within the period where soils show high infiltration rate characteristics.

W. C. HULSBOS, Euroconsult, Arnhem, The Netherlands

13 Recent developments in horizontal drainage

The early drainage systems of agricultural land consisted of open ditches. In the 19th century the first sub-surface drains were introduced. A scientific basis for drainage design was provided only about 50 years ago. Throughout the world probably about 200 million ha still requires some form of drainage. It is estimated that more than 100 million ha need pipe drainage. In the last 30 years rapid developments have occurred in design procedures, drainage materials and construction methods, e.g. use of larger and faster drainage machines, and replacement of clay and concrete pipes with flexible PVC or polyethylene pipes. Finally, cost aspects and expected future developments are discussed briefly in this paper.

INTRODUCTION

1. Groundwater drainage or groundwater depth control in agricultural lands is needed in low-lying areas and most irrigated areas where groundwater tables are shallow. Groundwater depth control improves the soil-water-air relationship in the root zone to ensure optimal plant growth. This control is not simply prevention of groundwater table becoming too high but is management of the fluctuations in the groundwater table so that crops do not suffer from excess or in special cases shortage of water.

2. Groundwater depth can be controlled through pumped wells, but this means of groundwater control known as vertical drainage can only be used if the aquifer is rather pervious and deep. Groundwater depth can also be controlled by a series of parallel drains. This system, which is referred to as groundwater drainage, is sometimes combined with a surface drainage system which may consist of a system of parallel shallow ditches. A surface drainage system discharges excess surface water but does not control groundwater depth.

3. Horizontal drainage may comprise open drains or buried pipes with openings for groundwater to enter, or a combination of open drains and buried pipes. Groundwater drainage by means of buried pipes is also referred to as sub-surface drainage, closed drainage, or tile drainage. In this paper the term pipe drainage is used.

HISTORY OF HORIZONTAL DRAINAGE

3. As recent developments in horizontal drainage can only be understood in the context of the past, these developments are briefly reviewed mainly in relation to the Netherlands. Since the 10th century land has been reclaimed by constructing systems of drainage channels and ditches. In the 11th century, dikes were constructed to protect the land from flooding from the sea and rivers. Windmills, which are typical of western Holland, have been used since the 14th century to pump water from low-lying polders into rivers and canals. As a result of increasing population and demands on agricultural production, drainage was introduced to improve in-field hydrological conditions. That is, a system of drainage was introduced to collect excess surface water and to discharge groundwater through the soil into a disposal system of open ditches or buried collector pipes.

4. Originally farmers used brush wood, peat or organic porous materials to facilitate the flow of water through the soil to the ditches. Later, they used roof tiles of baked clay (from which the term tile drainage originates), pipes of the same material and, in some cases, concrete pipes. Now perforated corrugated PVC pipes are used widely because of the relatively low price and ease of installation.

5. Until comparatively recently, design of pipe drainage systems was based on practical experience. In the mid-19th century, Henri Darcy, a French physicist, developed the basic equation for groundwater flow but is was not until Hooghoudt (ref. 1), a Dutch drainage engineer, formulated an equation to calculate drain spacings that this empirical technology evolved into a science. Since then, many drain spacing equations for both steady and non-steady state flow have been developed for a wide range of conditions.

6. Revolutionary development has taken place in drainage technology since the Second World War, especially in methods of installation and materials used. Originally pipe drains were installed by hand, but increasing labour costs have led to the development of pipe laying machinery. Most

World Water '86. Thomas Telford Limited, London, 1986

of these machines also excavate the trench in which the pipe is laid. The latest development is trenchless pipe laying, but as yet this is not suitable for drain depths exceeding about 1.50 m, because the tractive force is excessive beyond this depth. Another aspect is that smearing of the soil around the drain can be expected, thus reducing the drainage flow into the pipe.

7. Originally, pipe drainage was installed in the humid regions where high water-tables resulted in insufficient aeration of the root zone, thus leading to asphyxiation of plant roots. More recently, groundwater depth control was also introduced in arid zones, where evaporation exceeds rainfall and where natural drainage is inadequate. Pipe drainage is used to leach excess salts originating from groundwater and irrigation water which accumulate in the root zone. In arid zones buried collector pipes are often preferred because the deep tile drains which are used to control the capillary rise of salt require deep collectors, and open collector ditches would lead to considerable loss of cultivable land.

AREAS DRAINED

8. Pipe drainage is found mainly in temperate and arid zones, and to a very limited extent only in humid tropical areas where drainage is limited to surface drainage of excess rainfall. Information on the world's drained areas and area still in need of drainage is limited. And, as usually no difference is made between surface drainage and groundwater control, the available information does not allow an accurate estimate of the areas which would require pipe drain systems. The data in the following paragraphs do not pretend to indicate more than only an order of magnitude.

Temperate zones

9. Pipe drainage was first introduced in temperate regions, especially in low-lying areas. In Table 1 an indication is given of areas under drainage and areas requiring drainage. Similar figures for pipe drainage only or even for groundwater control are not available. In countries such as the Netherlands, most areas requiring sub-surface drainage, with the exception of the new polders, have been drained, and drainage works are limited to renewal of poorly functioning systems. In countries in southern and eastern Europe, relatively large areas still need to be provided with this type of drainage.

Table 1 - Area under drainage and area requiring drainage (million ha) Source: Refs. 2-3

	Land area	Agricultural area	Drainage Existing	Drainage Required
Europe (excluding USSR)	473	230	38	41
Asia (excluding USSR)	2677	1020	32	38
Africa	2965	1010	2	5
North and Central America	2140	618	68	67
South America	1754	546	8	11
Australia	843	517	1	2
USSR	2227	606	12	55
Total	13079	4547	161	219

10. An additional difficulty in estimating the areas requiring drainage is the development in cropping systems and growth in economic value of the agricultural production factors which will increase the demand for drainage. Some areas may even be considered for re-drainage with a more intensive system because of higher demands. On the basis of the information of Zonn and Nosenko (ref. 2,3) it is estimated that of the agricultural land in the temperate zone of Europe, including USSR and North America:

- 25-50 million ha of the area already drained is still in need of pipe drainage
- 50-100 million of the area requiring drainage needs pipe drainage.

Thus it is estimated that a total of 75-150 million ha still has to be provided with a pipe drainage system.

Arid zones

11. The first pipe drainage systems in an arid region were probably installed in the USA, early this century. In other arid zone areas pipe drainage has only been introduced in the last 10 to 20 years. In Egypt, it was introduced on a large scale in the early 1960s and to date more than half the agricultural area requiring groundwater depth control has been drained. In other countries in the Middle East and North Africa pipe drainage has also been introduced on a large scale. In other areas such as South America and Asia, pipe drainage has recently been introduced.

12. Also, only limited statistical data are available on the total area in need of pipe drainage but it is evident that only a small part of this total area has been provided with a drainage system. A further difficulty in estimating the requirement is that areas underlain with a very permeable and deep aquifer can also be drained by installing tubewells. Where the effluent is fresh and can be used for irrigation, tubewells have an additional advantage over horizontal drains. It is estimated that 25 to 50 million ha require sub-surface drainage in this zone of which about 10-20 % already has a system.

RECENT DEVELOPMENTS

Drainage criteria

13. Depth of drainage required to control the water-table in humid areas is usually 1 to 1.5 m, which is less than that to prevent salinization caused by capillary rise in dry periods without irrigation. In arid areas, the minimum depth required for lateral drains ranges from about 1.5 m in fine and coarse soil to 1.8 m in medium-textured soil (ref. 4), and greater in areas of seepage. In practice, where soil and other conditions permit there is a growing tendency to install drains at depths greater than the required minimum. The minimum depth in the 160 000 ha Abu Ghraib drainage scheme in Iraq is 2.2 m. The same depth criterion has been applied for the design of drainage system in pilot areas in India and for new projects in Pakistan, but for projects in the USA even

greater depths have to be considered. A minimum depth of 2.2 m for lateral drains implies that the collector pipes in composite drainage systems may be 3 m deep and more.

14. The advantages of drain installation at depths greater than the minimum required are:
- extra safety margin to prevent salinization
- wider spacing permissible if drains cut into more permeable sub-soil strata
- wider spacing possible because of the greater hydraulic head generated in the flow toward the drains, and
- drains function more efficiently as a result of the more regular and continuous discharge.

On the other hand, drain depth may be limited by soil conditions; by the capacity of the drain construction equipment; and by difficulties of deep drainage in water-logged, unstable soil.

15. The discharge capacity required is a major criterion in the design of a drainage system. It determines the drain spacing and the discharge capacity of pipes and ditches. The discharge required is determined from soil-water balance studies for different periods of the year, taking into account irrigation, rain, evapotranspiration, seepage, and the water required for leaching of salts and salinity control. The horizontal flow component, either inflow (seepage) from or outflow (natural drainage) to neighbouring areas, is perhaps the most difficult parameter in the soil-water balance to assess. However, in areas with highly permeable aquifers, such as in the Indus basin, the horizontal flow component influences the areal variation of the drainable surplus and hence assessment of the design discharge for various parts of the drainage area. Therefore, in practice two-dimensional and three-dimensional finite-element groundwater flow computer models are being used increasingly to determinate the drainage discharge for various parts of large projects.

16. Spacing of lateral drains determines the performance and to a certain extent the drainage costs per unit of area. The required spacing is related to the discharge capacity required and to the hydraulic properties of the aquifer in which the drains are located. Analytical and numerical calculation procedures are now available to determine the permissible spacing. This approach using groundwater flow analyses is based on the early steady-state drainage equation of Hooghoudt (ref. 1). Subsequently, analytical steady-state and non-steady-state flow equations have been developed and refined to simulate field conditions which may be very complex (refs. 5-9).

17. The wide availability of computers and programmable calculators has made it possible to carry out calculations which were considered previously to be too difficult and too costly. By means of analytical and numerical calculation models and by computer processing account can be taken of soil stratification, anisotropic hydraulic permeability, and steady or non-steady flow regimes to determine optimal spacing of the drains.

Drainage designs

18. The singular system (A in Figure 1) is the traditional method for drainage of flat land in humid areas. Initially, it also served as a model for the drainage required for reclamation of saline areas in arid and semi-arid regions. The system consists of pipe drains which flow directly into an open collector ditch. The length of the pipe drain is usually not more than 200-300 m, in particular to allow for cleaning with flushing apparatus. The main advantages of this system are: simple lay-out and construction; visibility of performance at outflow pipes; and easy maintenance. However, the large length of open drains required implies loss of valuable agricultural land and high construction costs (see Table 2), need for regular maintenance, and additional costs for crossing irrigation systems and roads.

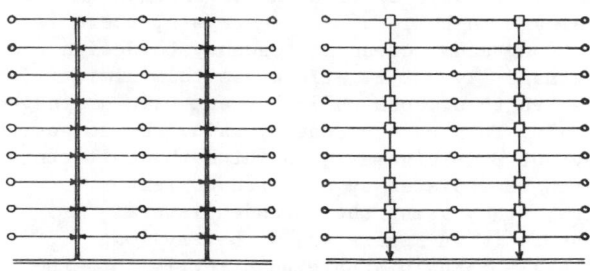

A singular system laterals of 250 m.

B composite system laterals of 250 m.

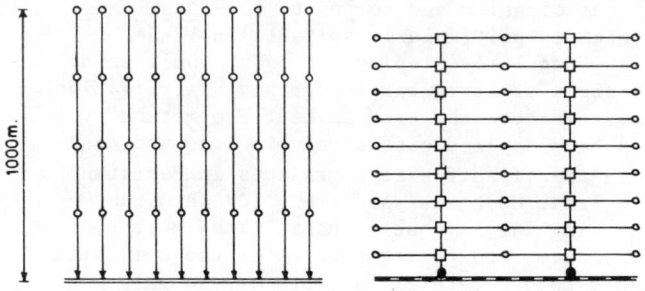

C extended singular system laterals of 1000 m.

D pumped composite system

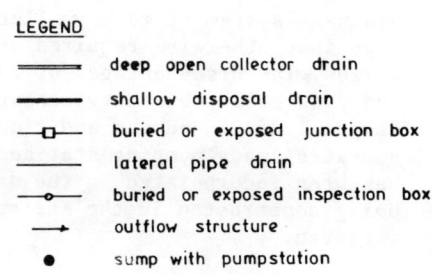

LEGEND
- deep open collector drain
- shallow disposal drain
- buried or exposed junction box
- lateral pipe drain
- buried or exposed inspection box
- outflow structure
- sump with pumpstation

Fig. 1. Pipe drainage lay-out alternatives

Table 2 - Excavation volume for and land loss due to drain ditches*

Depth (m)	Excavation (m³/ha)	Land loss (%)
1.6	108	2.7
2.1	174	3.5
2.6	254	4.3
3.1	350	5.2

* Drains with bed width 1.0 m, side slopes of gradient 3:2 a 2 m berm, a maximum spoil height of 2 m, and spaced at 500 m intervals.

These disadvantages are more important for the deep drainage required for salinity control in arid areas than for shallow drainage in humid areas. Therefore, there is a strong and increasing tendency to design drainage lay-out in arid areas to minimize the length of open collector and main drainage canals. This can be done by using the composite lay-out B in Figure 1, in which the collector ditch is replaced by a buried collector pipe, or by using the extended lateral lay-out (C in Figure 1), with very long lateral drains discharging directly into the main system.

19. The composite system has been used extensively in Egypt. However, this has a complicated structure of many lateral collector joints, to check performance of the system, and also maintenance of buried junction boxes is difficult. The extended lateral lay-out (C) overcomes most of the disadvantages of the singular and composite lay-outs. This system is particularly suitable in large flat, regularly sloping areas but application in areas of irregular topography or irrigation and road networks are limited. The method has been applied extensively in Iraq and has been specified for some projects in Pakistan and Iran. Cost analyses (ref. 10) show that the singular lay-out is usually the most expensive, and that the costs of the composite and extended lateral lay-outs are roughly similar.

20. The need for deep drainage canals in the composite and extended lay-outs can be reduced (see D in Figure 1) by connecting the discharge from the underground system to a pump to pump water into a shallow disposal drainage system of much smaller cross-section than that otherwise required of deep collector drains. The disadvantages of this method are the possibility of salty seepage from the disposal drain and the additional costs of operating many pumping stations. This system has been incorporated in the drainage network being constructed in the Khairpur project in Pakistan.

Drainage materials

21. Traditionally, drain pipes have been made of baked clay or concrete in lengths of 30 to 50 cm and diameters for lateral drains in the range 5 to 10 cm. In about 1955, plastic pipes were introduced in field drainage schemes. The first of these drains were made of smooth plastic pipes with slits as perforation. Subsequently in about 1967, lighter, cheaper, flexible corrugated piping was introduced. At present most drain piping is made of corrugated plastic, which is usually either polyvinyl chloride (PVC) or polyethylene (PE). In Egypt, until a few years ago mostly concrete pipes were used, but now plastic piping is being used extensively. Between 1958 and 1978, plastic pipes replaced clay pipes in the Netherlands. Lightness, flexibility and ease of installation give plastic pipe many advantages over clay or concrete pipe. Corrugated perforated plastic piping for lateral drains is available in the diameter range 5 to 20 cm. Use of perforated and non-perforated corrugated plastic piping for collector drains is also increasing, with diameters of up to 40 cm and even 60 cm being used. Corrugated plastic pipes can be machine-laid to a depth of 4 m, even in difficult unstable soil.

22. Generally, drain pipes are covered or surrounded by filter material to prevent their becoming silted up with soil entering through the perforations, but in stable clay soil, such as in Egypt, filter material is not used. The filter most often applied is a graded gravel envelope 8-10 cm thick, depending on the texture of the surrounding soil. Graded gravel is in many ways an excellent envelope material: it protects against silting up; is highly permeable; and provides a good bedding for and sidelong support to the pipe. For the last reason, the use of gravel or sand is necessary for large-diameter corrugated pipes. However, graded gravel, if not available locally, may be very expensive, and therefore considerable attention is being paid to the search for a good, reliable, and cheap substitute. In Europe, organic cover material, such as peat litter and coconut fibre, is used, but these materials decompose rapidly when not permanently under water, and therefore cannot be recommended for use in drainage systems in arid and semi-arid regions. Synthetic filter materials seem to be a better alternative. In the 1960s, the glass fibre tissue was introduced and about ten years later polypropeen tissue was introduced. Since then many other synthetic materials have become available. Tests on new materials (ref. 11) indicate that thin glass fibre tissues are durable but sensitive to blockage especially for chemical compounds. But some thick materials such as polypropene fibres and polystyrene pellets appear promising with a large durability and low sensitivity for blockage by soil or chemical compounds.

Drain construction

23. Now practically all drainage systems are constructed with drainage machines which install in one operation the drain pipe plus gravel or other filter material to the required depth and grade. Two types of drainage machines are available: drainage trenchers; and drainage ploughs for trenchless drain installation.

24. The first drainage machines were trenchers with a digging chain or digging wheel. Trenchers, and particularly chain trenchers, are still the main drainage machines, and are available for a wide range of installation depths, pipe diameters and soil conditions. They can be classified into two groups: machines for installing lateral drains at depths up to 2.5 m and with drain diameter up to 20 cm; and machines for installing collector drains of larger diameter and depth up to 4 m. Since the introduction of the first drainage trenchers in Europe in the 1950s, machines have been improved, modified and adapted to the latest technical developments and to user demands.

25. Heavier and more powerful machines have been developed for drain installation at great depth and of large diameter pipe in accordance with the requirements for salinity control in arid and semi-arid regions. The development of the drainage trenchers quoted as standard by one of the main manufacturers in the Netherlands is given in Table 3.

26. Trenchless drainage is a more recent technique in which a plough with a blade similar to that of a mole plough or a subsoiler is pulled through the soil to make a tunnel in which the drain pipe, introduced through or directly behind the blade, is installed. Although the principle of trenchless drainage has been known for some time, the technique could not be used satisfactorily until a suitable flexible pipe, the corrugated plastic drain-pipe, became available. Toward the end of the 1960s, many types of trenchless drain ploughs were developed and trenchless drainage became more common. The method reduces costs because of the greater production speed, but also has disadvantages, including the risk of soil compaction around the drain, less reliable depth control, and limited installation depth due to the rapid increase of the tractive power requirement as depth increases. For the last reason, trenchless drainage has not as yet been used extensively in drainage for salinity control in arid and semi-arid regions (ref. 13).

27. Drains have to be installed at the appropiate depth and gradient which can be controlled visually by sighting boards. However, most modern machines are equipped with laser controlled depth regulation, which may be either fully automatic in which the laser signal is coupled directly to the depth control mechanism, or semi-automatic in which depth is adjusted manually by the operator.

28. Depth regulation by laser does not guarantee that a pipe will be laid correctly. With trencher-type pipe installation, trenchers need to left open until supervisory staff have checked that the pipe has been installed correctly. In unstable soil, and generally in deep drainage, the trench collapses behind the machine, which makes checking the installation very difficult, if not impossible. With trenchless drainage, checking is not possible. An experimental new development is the 'black box'. This is an electronic device mounted on the drainage machine which registers accurately and continuously the exact depth and gradient at which the pipe is laid.

COST OF PIPE DRAINAGE

29. The cost of a drainage system depends on many factors. The remarks here made will be limited to some recent data on the cost of installed pipe drains per metre. To transform these to cost per ha of drained land the drain spacing should be known. Also data on the cost of the open or closed collector system and the cost of the main system which transports the drainage effluent out from the area should be available. For a cost comparison of all these factors, the complete drainage system should be analysed, which is beyond the scope of this paper.

30. The cost of installation of pipe drainage includes the cost of the pipes, filter material and installation. These costs depend on the pipe diameter, type of filter material, the depth of installation and the construction method. All these factors have been influenced considerably by recent and rapid developments in technology and this will continue to be the case in the near future. Estimated installation cost of pipe drainage in various countries are set out in Table 4. There appears to be a wide variation in cost. Drains with a diameter of

Table 3 - Review of selected types of Hollanddrain® trencher since 1955

Year of introduction	Type and model	Power (hp)	Max. depth	Max. width	Weight (t)
1955	Tyred wheel trencher	55	1.75	0.22	8
1959	Crawler trencher	120	1.80	0.22	9
1964	Semi-tracked crawler trencher	150	2.00	0.25	10
1966	Crawler trencher ESL175	160	1.75	0.30	14
1967	Crawler trencher ESL210	160	2.10	0.35	14
1972	Crawler trencher GSL250	200	2.70	0.50	19
1979	Crawler trencher BSS320	325	3.20	0.50	30
1982	Crawler trencher GSS325	300	3.25	0.40	25
1983	Crawler trencher BSS360	325	3.60	0.65	35
1983	Crawler trencher BSS400 Super	390	4.00	0.90	45
1974	Trenchless drain plough	200	1.20		
1982	Trenchless delta drain plough	275	2.20		

Source: Steenbergen, Utrecht, The Netherlands.

Table 4 - Estimated cost of installation of pipe drainage with flexible PVC piping including a gravel filter

Country	Year	Pipe diameter (mm)	Installation depth (m)	Cost (US$/m)
The Netherlands*	1985	60	1.2	0.7
Iraq	1985	65	2.2	4
		100	2.2	4.5
		200	3.0	8.5
Dominican Republic	1985	200	2.5	6.8
China	1985	80	1.6	1.7
Egypt	1983	80	1.8	1.3
		200	2.2	1.5
Peru	1982	100	1.8	2.2
Pakistan	1983	100	1.8	4
		200	2.2	11

* filter material is here coconut fibres

60-80 mm installed at a depth of 1.50 m or less cost US$ 0.7-1.7/m. Drains at a depth of 2-3 m cost US$ 4-11/m. It appears that the depth has more influence on the cost than the diameter.

31. The cost tendency in the Netherlands is that installation cost decreases slowly. In 1977 the 60 mm PVC pipes with coconut filter were installed for US$ 0.8/m. On the other hand new filter materials may cause the cost to increase: using polypropylene fibres instead of coconut means a cost increase of US$ 0.1/m in 1985.

EXPECTED FUTURE DEVELOPMENTS

32. Although it is difficult to predict future developments, some trends may be discerned on the basis of recent developments. As modern horizontal drainage is a relatively new technique and large areas of agricultural land throughout the world are in need of drainage, we are still in a period of very active product development, to use a marketing term.

33. In general, it is expected that sub-surface drainage will be preferred to open drainage. The requirements for efficient farm management will put increasing demands on the economy of the farm operations and accessibility of fields. Therefore, wherever possible open ditches will not be used in drainage systems for groundwater control, and this will also decrease maintenance requirements and cost. No new exciting developments are expected in drainage theory. With the aid of computer calculations more accurate water balances and estimates of drainable surplusses can be made. There will probably be a growing tendency to apply non-steady-state equations to establish optimal drain spacings and depths. However, practical usefulness of this more sophisticated approach will be severely hampered by the limited knowledge of the reaction of plant growth on fluctuating groundwater depth.

34. Although no radical changes in drainage designs are foreseen, it may be expected that the drainage criteria will continue to be raised as a result of the increasing value of the agricultural production, which justifies higher investments and decreases the willingness of the farmers to take risks.

35. Further developments are expected in the use of new drainage materials, especially for envelope or filter materials around the pipes, and in mechanical construction methods of pipe drains. The construction of pipe drains has always fired the imagination of inventors. Numerous trials have been carried out on new materials and construction methods during the last few decades. Although drainage practices are well established, these trials will continue in the hope of increasing efficiency in drainage construction. These developments may reduce the cost of drainage systems and also make them more reliable.

REFERENCES
1. HOOGHOUDT S.B. Bijdragen tot de kennis van enige natuurkundige grootheden van de grond, no. 7. Verslag Landbouwkundig Onderzoek, 1940, vol.46, p.p. 515-707.
2. ZONN I.S. and NOSENKO P.P. Modern level of and prospects for improvement of land reclamation in the world. ICID Bulletin, July 1982, vol. 31, no 2.
3. ZONN I.S. and NOSENKO P.P. Land Drainage in the world. ICID Bulletin, January 1976, no 2.
4. FAO Drainage design factors. FAO Irrigation and Drainage Paper 38, 1980.
5. KIRKHAM D. Seepage of steady rainfall through soil into drains. Trans. Am. Geophys. Union 1958, vol. 39. 892-908.
6. TOKSOZ, S. and KIRKHAM, D. Steady drainage of layered soils: I, Theory; II, Nomographs. J. Irr. and Drain. Div., Am. Soc. Civ. Eng. 97: IR1, Papers 7985: 1-17; 7986: 19-37
7. ERNST L.F. Grondwateroverstromingen in de verzadigde zone en hun berekening bij aanwezigheid van horizontale evenwijdige open leidingen. Verslag Landbouwkundig Onderzoek, 1962, vol. 67, no.15.
8. DUMM L.D. Validity and use of the transient flow concept in subsurface drainage. Paper presented at ASAE meeting, Memphis, TN, United States, 4-7 December 1960.
9. KRAIJENHOFF van de LEUR, D.A. A study of non-steady groundwater flow with special reference to a reservoir-coefficient. De Ingenieur, 1958, vol. 40, 87-94.
10. BOUMANS J.H. Extended singular drainage: an interesting alternative lay-out for drainage of irrigated lands. ILRI Publication 25, Proceedings of Drainage Workshop, Wageningen, 1979.
11. STUYT, L.C.P.M. De toepassing van geotextielen in ontwateringssystemen voor cultuurgronden I.C.W., 1985.
12. NAARDING, W.H. A review on international experience with trenchless versus trenching drainage machines. ICID tenth congress - Athens 1978 8.3 question 34.2
13. BOUMANS J.H. Modern drainage technology for the reclamation of saline land. Int. Symposium on the Reclamation of Salt affected Soils, Beijing 1985.

G. R. HOFFMAN, BSc, FICE, FIWES,
R. F. STONER, BSc, FICE, and J. H. PERRY,
BSc, MICE, Sir M. MacDonald and Partners,
UK

14 Mechanisms in watertable control

High watertables and transient waterlogging affect crop yields by reducing the amount of oxygen available for root respiration. Although the physiological processes involved are well understood, very little usable data are available on yield response to high watertables and transient waterlogging. Such data are essential to drainage engineers especially those involved in the design of vertical drainage systems, where depth to watertable can be varied. Guidelines are required and, to produce such guidelines, research work needs to be undertaken to establish the effects of all the governing parameters on the yield response to watertable depth relationships. The terms of reference for such research work should be prepared by drainage engineers who, in order to do this, need to be familiar with the mechanisms involved and the current status of available data.

INTRODUCTION

1. One of the advantages often quoted for vertical drainage is the inherent flexibility in depth to controlled watertable. For such an advantage to be utilised, however, guidance is necessary on yield response to watertable depth and this guidance needs to cover such governing factors as crop species, soil type, climate, watertable salinity and the effect of transient waterlogging. Waterlogging, for the purpose of this paper, implies a watertable within the root zone.

2. Current drainage reference texts are invariably orientated towards the design of horizontal drainage systems. The inflexibility of such systems together with the favourable economics for deeper, wider spaced subsurface drains (optimum drain depth is usually about 2.5 m) generally leads, in terms of watertable control, to a conservative design. Knowledge of yield response to watertable depth is not seen as a prerequisite for horizontal drainage design methodology, which probably explains why current guidance on optimum watertable depth can be vague and sometimes conflicting.

3. The most common advice to drainage engineers is to ensure that the watertable remains below the root zone which, for a cropping pattern that includes several crops, is taken to be the larger of the root zone depths. Some designers allow the watertable to rise above this level for a short period (perhaps 2 or 3 days) but generally this is not recommended. Other factors may influence the depth to watertable such as salinisation of fallow land, trafficability and crop water use from the capillary fringe.

4. The main aim of a drainage system, irrespective of whether it is vertical or horizontal, is to ensure that high watertables and waterlogging are not constraints on crop yield. To achieve this aim a knowledge of crop response to watertable depth is clearly important, especially as drainage systems cannot and do not maintain constant watertable depths. Even under normal irrigation practice watertable fluctuations occur and with heavy rainfall or over-irrigation the watertable can rise sufficiently to cause transient waterlogging.

5. Compared with horizontal drainage, vertical drainage offers two main advantages: firstly, transient waterlogging could be prevented or, at the very least, the time period reduced merely by pumping for extra hours. Secondly, it would be possible to vary the depth to the watertable to match the stage of crop growth and hence root development. This second advantage would be important where a crop obtains part of its water requirements from the watertable. In order to implement such operational procedures there are two main aspects of watertable control to be considered: firstly, the optimum depth required at different stages of growth of the crops proposed and secondly, the effect of transient levels, above the optimum, on crop yields.

6. This paper examines the present state of knowledge of a crop's response to watertable depth and transient waterlogging. The aim is to promote amongst drainage engineers a better understanding of the mechanisms involved so that they can encourage research into the essential relationships for optimum crop yields.

THE EFFECTS OF WATERLOGGING

7. Before discussing the effects of waterlogging it would be useful to define this term once again, as waterlogging and a high watertable have, for this paper, different meanings. Despite the fact that waterlogging is caused by a high watertable, waterlogging implies a watertable within the root zone as distinct from a high watertable which can be defined as one which restricts root development. The effect of high watertables will be discussed later but essentially the effect on a crop is similar to that of waterlogging.

8. The response of crops to waterlogging is complex but it is not the water itself that is harmful (provided that salinity levels are low); the critical factor is oxygen depletion and consequential anaerobic conditions in the root zone. There is considerable literature on the general effects of waterlogging on plant growth and recently Kozlowski edited a series of monographs, texts and treatises on all aspects of the subject (ref. 1). Although such texts describe in detail all the physiological processes involved, they are not directly relevant to the drainage engineer who is more interested in the response of particular crops. However, an understanding of the key parameters is important to enable an objective assessment to be made of currently available data.

9. Oxygen is required for root respiration and by soil organisms. In a well aerated, unrestricted root zone the oxygen removed from the soil by this demand from plants and organisms is readily replaced by diffusion from the atmosphere. Under waterlogged conditions the oxygen content of the flooded part of the root zone can drop rapidly, because oxygen diffuses in water about 10^4 times more slowly than in air (ref. 2). If waterlogging persists anaerobic conditions will develop in the affected part of the root zone which will ultimately kill the roots. If total waterlogging of the root zone occurs then the crop will eventually die.

10. Even with only transient waterlogging the rapid depletion of oxygen can cause appreciable yield reductions, especially if the waterlogging occurs during a critical stage of crop growth. The severity of the effect will depend on the proportion of the root zone waterlogged and the length of time that waterlogging persists.

11. Waterlogging not only reduces oxygen availability but also increases the retention of the gaseous products of root respiration, mainly comprising carbon dioxide and ethylene. Carbon dioxide rarely accumulates to toxic levels (ref. 3) but the increased ethylene concentrations can inhibit phosphorus uptake and stimulate the outgrowth of adventitious roots in some crop species (ref. 4). The role of adventitious roots remains in doubt, although most authors consider that these roots replace many of the functions of the original defunct root system (ref. 1).

12. Nitrogen supply may also be affected by waterlogging; yellowing leaves is a common phenomenon of waterlogged fields. The extent to which this is attributable to nitrogen deficiency is not clear (ref. 2). The reducing conditions associated with an anaerobic environment do however reduce nitrate to nitrous oxide or nitrogen rendering it nutritionally valueless to plants and resulting in a gaseous loss of soil nitrogen. It has been reported that applications of additional nitrogen offset some of the effects of waterlogging (ref. 5) although not all research has found a similar interaction.

13. It can be concluded therefore that the effects of waterlogging on plants are primarily due to oxygen deficiency the extent of which depends on the rate of consumption of oxygen by plants and by soil organisms. Oxygen availability is affected by the soil texture and the degree of waterlogging: the rate at which available oxygen is consumed depends on the crop species and soil temperature. Temperature is an extremely important parameter because oxygen consumption increases rapidly with temperature. Experiments carried out on the effects of transient waterlogging on wheat yields (ref. 2) showed that oxygen consumption approximately doubled for a $10^{\circ}C$ increase in mean soil temperature.

YIELD RESPONSE TO WATERTABLE DEPTH

14. Although considerable progress has been made towards fully understanding the effects of waterlogging on plant growth, quantifying, in terms of yield, the effects of both waterlogging and high watertables has proved to be more complex. Currently available data have generally been derived either from controlled experiments or field measurements. Experimental data are usually obtained from lysimeters and relate to a fixed watertable depth; field data usually assume the watertable was at a fixed depth by averaging depth measurements over the growing period. The various data are occasionally reviewed in an attempt to produce general guidelines on crop response to watertable depth (ref. 3, ref. 6). However, this task is extremely difficult as most of the data are country specific and, with field measurements, other parameters influencing crop yield are uncontrolled.

Constant watertable depth

15. Despite the wide variety of data it is possible to draw general conclusions on the effect of watertable depth on crop yield. Assuming a constant watertable depth over the whole growing period then two basic relationships are possible and these are shown in Fig.1. Curve (a) relates to an irrigated crop not relying on water from the watertable and curve (b) relates to a crop relying (after establishment) on water from the watertable. The shift in the maximum yield position and then yield

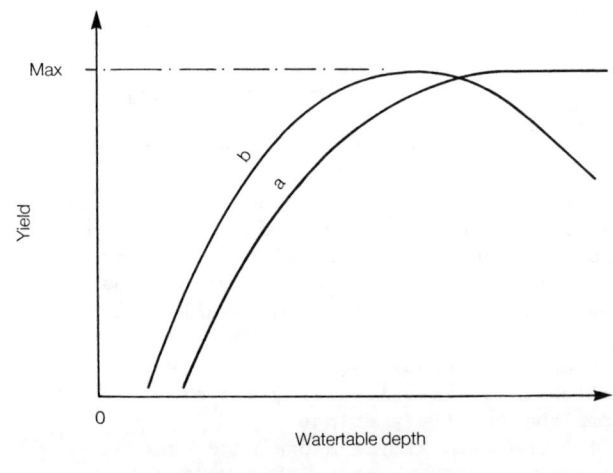

a – crop unreliant on watertable
b – crop reliant on watertable

Fig. 1. Generalised yield response to watertable depth relationships

decline with increasing watertable depth shown for curve (b) indicates that with increasing watertable depth the crop's ability to obtain sufficient water from the capillary fringe becomes a constraint on yield.

16. The position of maximum yield shown in Fig. 1 will depend on the crop species and soil type. Differing climatic conditions will not greatly influence the position of maximum yield, but the shape of the response curve will be affected. As previously discussed, soil temperature determines the rate at which available oxygen is consumed and thus increasing soil temperature increases the effect of high watertables on yield. Fig. 2 shows the effect of soil temperature on the generalised response curve.

Transient waterlogging

17. Although a comprehensive set of yield response to watertable depth relationships for a variety of crops and climatic conditions would be of great value to drainage engineers, this is only part of the problem. In practice, even with drainage, the watertable will fluctuate and transient waterlogging may occur. Transient waterlogging can have a great effect on yields especially if it occurs during a critical stage of crop growth. In fact the limited available data indicate that transient waterlogging, whereby the watertable rises into the root zone for a limited period, can have more effect on yields than a static high watertable at the transient level.

18. The concept of transient waterlogging makes the whole subject of yield response to watertable depth more complex. Generally two situations can arise, firstly a low watertable, non-limiting to root development, which rises into the root zone; in this situation the transient waterlogging alone affects yields. Secondly a high watertable, limiting root development, which rises into the root zone; in this situation transient waterlogging will cause an additional effect on yield which may be dominant if it occurs for a prolonged period of time.

19. The effect of transient waterlogging is dependent on the same parameters as the effect of high watertables. Temperature is probably the most significant parameter for a given crop species as the effect on yield will depend on the rate at which available oxygen is depleted. It should be noted that transient waterlogging does not have an immediate effect as it takes a finite time for the available oxygen within the flooded part of the root zone to be consumed. Thus the watertable could rise to ground level and immediately drop again without any effect on yield.

Salinity

20. The parameter missing in the discussion so far is the salinity of the watertable. In the international context of drainage a high, saline watertable is a common problem. Normally drainage engineers consider this problem in terms of salinisation of fallow land, ensuring that the watertable is deep enough to prevent sustained capillary rise to ground level. Although this may be the ruling parameter and result in a design watertable depth deeper than that obtained purely by crop yield considerations, there is still the possibility of transient waterlogging with a saline watertable.

21. There are few data available on yield response to a saline watertable. However it can be concluded that any additional yield reduction caused by a saline high watertable would be dependent on the salinity of the root zone, which in turn would depend on whether or not sufficient leaching occurs under normal irrigation practice. Clearly though it would be extremely difficult to prevent a very saline high watertable influencing the root zone salinity. Sufficient leaching of fallow land would become increasingly difficult. Fig. 3 shows the likely effect of watertable salinity on the generalised yield response to watertable depth relationship.

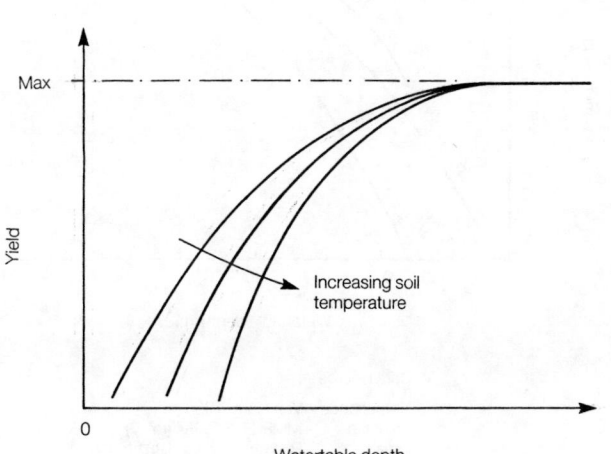

Fig. 2. Effect of soil temperature on yield response to watertable depth relationship

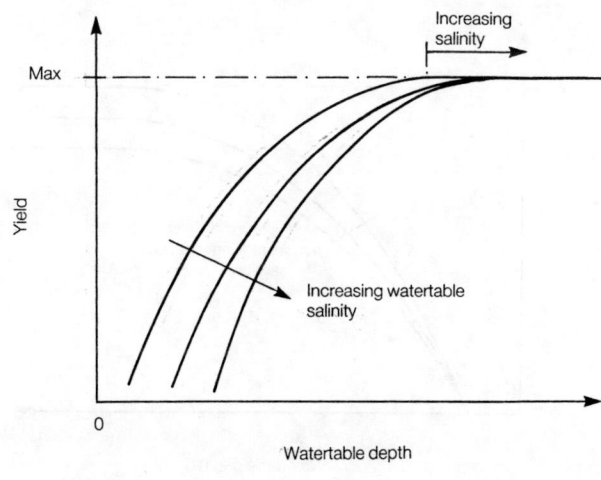

Fig. 3. Effect of watertable salinity on yield response to watertable depth relationship

22. Recent field measurements in Pakistan (ref. 7) tend to confirm the assumed relationship shown in Fig. 3. Measurements were taken to obtain a yield response to watertable depth relationship for cotton. From these measurements it was found that where the watertable salinity exceeded 15 mmhos/cm a soil horizon of high salinity occurred above the watertable. The presence of this highly saline soil layer within the root zone had a limiting effect on root development and hence on yield; however as only 9% of the plots sampled had watertable salinities greater than 15 mmhos/cm, insufficient data points were obtained to produce a significant relationship. For watertable salinities below 15 mmhos/cm there was no relationship between watertable salinity and cotton yields indicating that leaching was effective up to this salinity level.

23. Christopher and Tekrony (ref. 6) have produced a relationship for the combined effect of watertable depth and salinity of the root zone on yields (Fig. 4). This relationship is based on the assumption that the yield reduction due to root zone salinity can be superimposed on the yield response to watertable depth relationship, resulting in a constant percentage additional reduction. As Christopher and Tekrony point out, there are no data to support this relationship; however even if the form of the relationship is correct, including root zone salinity in the relationship is not as useful as watertable salinity. The two are, of course, related by a third factor, leaching efficiency. Perhaps, therefore, the best way to include the effect of watertable salinity would be to produce response curves as shown in Fig. 3 for several levels of water management.

24. There are no data on the effect on yields of transient waterlogging by a saline watertable. However, a saline watertable rising into the root zone will inevitably affect root zone salinity and hence yields. It may well be that repeated transient waterlogging events of short duration are more harmful (in terms of salinity) than a single event of longer period.

EXAMPLES OF AVAILABLE DATA

Constant watertable depth

25. In order to highlight the variability in currently available data for yield response to watertable depth, a comparison has been made of some of the relationships for wheat and cotton. For wheat three different response curves are presented in Fig. 5. As can be seen there is a large variation between the three response curves, with a yield range of between 20% and 70% of maximum yield for a 0.5 m deep watertable.

26. The curve representing the Netherlands data comes from measurements carried out in 1958 by Van Horn (ref. 3) and gives the lowest yield reduction in response to watertable depth. This is mainly due to the low winter soil temperatures in the Netherlands but, in addition, the fresh watertable will also have influenced the relationship.

27. The Pakistan data come from measurements carried out in 1983 (ref. 7) and, as would be expected, the increased soil temperatures in Pakistan have resulted in a greater yield reduction, for a given watertable depth, than that given by the Netherlands relationship. However, the Pakistan relationship is based on a single watertable depth measurement (at harvest) whereas the watertable depth over the growing period will almost certainly have fluctuated and consequently influenced the relationship.

28. The much lower yields, for a given watertable depth, indicated by the Indian data (ref. 8) are harder to explain, especially as the data come from controlled lysimeter experiments. In reality the Pakistan and Indian relationships should have been fairly close as the climates are similar and both data sets include the effect of a slightly saline watertable (about 4 mmhos/cm). Other influencing parameters such as soil type and crop variety would not be expected to have such a significant influence.

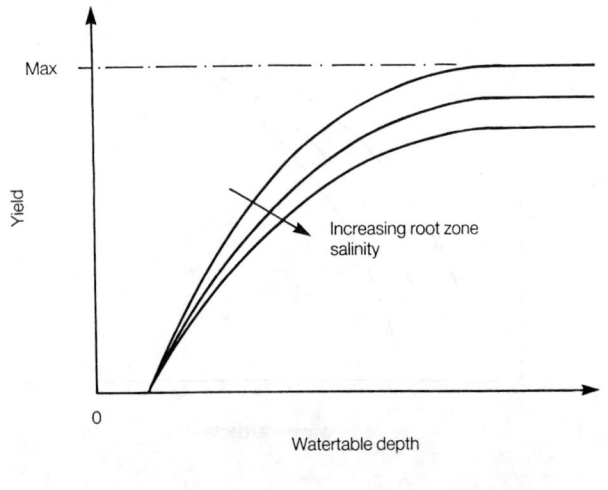

Source – Christopher and Tekrony, 1982 (ref. 6)

Fig. 4. Effect of root zone salinity on yield response to watertable depth relationship

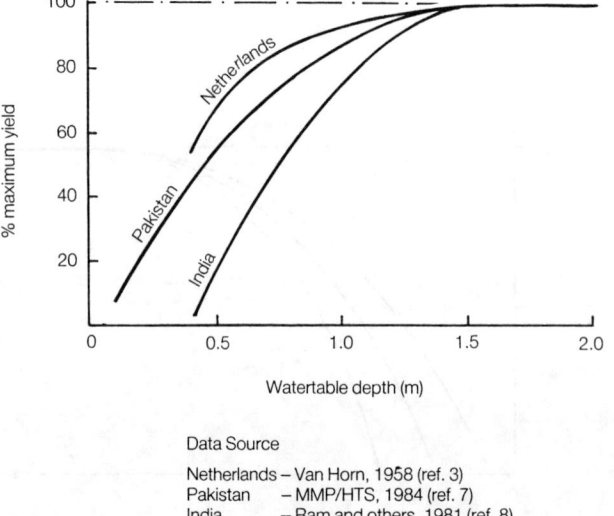

Data Source

Netherlands – Van Horn, 1958 (ref. 3)
Pakistan   – MMP/HTS, 1984 (ref. 7)
India   – Ram and others, 1981 (ref. 8)

Fig. 5. Comparison of yield response to watertable depth relationships for wheat

29. All three sets of data indicate that maximum yield occurs with a watertable depth of 1.5 m showing surprising consistency and, to some extent, confirming the view that differing climatic conditions will not greatly influence the position of maximum yield. However the Pakistan data (ref. 7) indicate that once the watertable depth reaches 1.0 m the scatter of data points makes establishing this maximum difficult. Cannell and Belford (ref.9) from their lysimeter experiments with winter wheat in England found only a small yield reduction (1%) between a watertable depth of 0.90 m and 0.50 m indicating that maximum yield occurred at about 0.90 m. The overall conclusion would therefore be that the maximum yield for wheat occurred with a depth to watertable of between 1.0 m and 1.5 m.

30. Fewer data are available on yield response to watertable depth for cotton. Two response curves are presented in Fig. 6, the Egyptian data come from lysimeter experiments with constant watertables having a salinity of 16 mmhos/cm (ref. 10). The Pakistan data (ref. 7) come from field measurements for fluctuating watertables with salinities generally less than 6 mmhos/cm. The watertable levels used in plotting the Pakistan data are the seasonal means.

31. Both relationships show maximum yield occurring at about 1.8 m confirming, to some extent, the view that salinity of the watertable will not greatly influence the position of maximum yield. However, the expected effect of the higher watertable salinity for the Egyptian data causing higher yield depressions, is not reflected in the two response curves. This is probably due to the fact that the Egyptian data come from controlled lysimeter experiments and the Pakistan data come from actual field measurements.

32. From the relationships for wheat and cotton presented in Figs. 5 and 6 it is clear that available data on yield response to watertable depth are very variable. This is hardly surprising, bearing in mind the different locations and methodology involved in collecting the data and a general incompatibility of the key parameters such as climate, soil type and salinity of the watertable.

Transient waterlogging

33. For transient waterlogging, data are very scarce, the only comprehensive experiments on wheat being those carried out by Cannell and Belford (ref. 9 & 11). The results of these experiments, which involved raising the water level in lysimeters up to the soil surface for a predetermined number of days, showed that the most critical period of crop growth was the pre-emergence stage, when waterlogging for a period greater than 6 days killed most of the crop, but less than 4 days had very little effect on yield. These results indicate how sensitive the crop is to small changes in the duration of waterlogging at critical periods.

34. The only experiments known to have been carried out on the transient waterlogging of cotton are those of Reicosky and Meyer (ref.12) utilising a sloping, repacked slab of soil. However, the data obtained from these experiments are inconclusive except in indicating cotton's tenacity and ability to acclimatise to waterlogging events.

35. The only attempt to date to produce guidelines on the effect of transient waterlogging on a variety of crops has been by Sieben (ref. 5) who developed the SEW30 index. This index is a cumulative measure of daily watertable depths shallower than 0.3 m, a concept based on the results of field measurements carried out in the Netherlands, where 0.3 m was identified as the critical level for winter crops. The higher the index the greater the yield reduction, with significant yield reduction only occurring when the index is greater than 100. In the international context such an index is of little value as it takes no account of variations in climate, soil type and stage of crop growth during which waterlogging occurs. Also each index can be derived in several ways, for example an index of 100 could represent either a watertable at 0.20 m for 10 days during the growing period or a watertable at 0.05 m for 4 days during the growing period, the index giving no indication whether the waterlogging events are independent or consecutive.

FUTURE DEVELOPMENTS

36. The purpose of the preceding discussion has been to highlight the variability of available data on yield response to watertable depth, the lack of data on transient waterlogging and the problem in adopting specific research data for developing general drainage criteria. Drainage engineers could, of course, continue to use existing guidelines which usually ignore the effects of climate, watertable salinity, and crop species and make no mention of the effects of transient waterlogging. With such a design approach horizontal drainage systems would continue to function adequately, but for reasons of engineering conservatism rather than a full understanding of the interrelation between a specific crop's yield and the position of the watertable over the growing period.

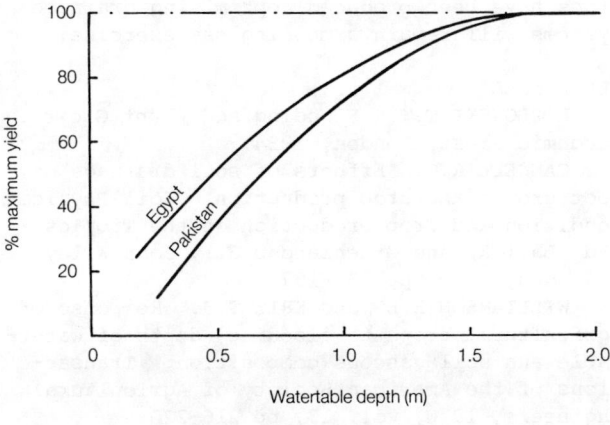

Data Source
Egypt   – Moustafa and others, 1975 (ref. 10)
Pakistan – MMP/HTS, 1984 (ref. 7)

Fig. 6. Comparison of yield response to watertable depth relationship for cotton

37. Adopting existing guidelines for the design of vertical drainage can, however, lead to problems especially if the current trend for deeper horizontal subsurface drains is translated into a 'the deeper the watertable the better' approach. It may well be that if the watertable is fresh the crop relies on water from the capillary fringe to supplement deficient irrigation applications. There seems little point, in these circumstances, in lowering the watertable such that the capillary fringe is out of reach of the crop. To do this would only result in unnecessary recirculation of water.

38. What is needed, therefore, is a comprehensive set of guidelines on yield response to watertable depth and transient waterlogging. Part of the database for such guidelines already exists; a literature search carried out in 1983 of the Commonwealth Agricultural Bureau abstracts, using the keywords wheat, sugar, maize, cotton, watertable and waterlog, listed 264 publications for the period 1972 to 1982. Although many of these publications are concerned with the physiological effects of waterlogging and therefore not directly relevant, some will inevitably contain usable data.

39. Careful scrutiny of the literature will be necessary to determine the extent of usable data and to highlight deficient areas. These areas will undoubtedly be the effects of transient waterlogging and watertable salinity as only ten of the 1972 to 1982 publications deal with transient waterlogging and none with the specific effects of a saline watertable. Once the database has been established it will be necessary for drainage engineers to encourage research establishments to fill in the gaps; but before any research work can be initiated it is essential to define the format of the end product (i.e. the guidelines).

40. Clearly the guidelines need to cover both yield response to watertable depth and yield response to transient waterlogging for a number of crop species. The first section dealing with watertable depth should include the effects of all the ruling parameters which, in order of importance, are as follows:
   a) reliant/unreliant on the watertable;
   b) soil temperature (climate);
   c) watertable salinity;
   d) soil type;
   e) crop variety.

Crop variety has been included because varietal differences within some crop species have been reported. In addition to these five parameters water management would need to be considered as it will influence the effect of watertable salinity.

41. The format of this first section of the guidelines should therefore be a series of yield response to watertable depth relationships for each selected crop species, covering various climatic regions (temperate, arid etc.). The effect of watertable salinity, soil type and, where applicable, crop variety on the basic response curves should also be included. In order to prevent the guidelines becoming unnecessarily complex some simplification of data presentation will be required for the effects of watertable salinity and soil type. It is envisaged that these parameters would be presented as broad categories, for example, a light textured soil with moderately saline watertable. The main response curves would be for a medium textured soil with a non-saline watertable and either a table of factors or additional response curves should be included to enable the effects of other combinations to be assessed.

42. The second section of the guidelines would cover transient waterlogging and should be fully compatible with the first section so that additional yield reduction factors can be determined. The format of the response curves presented should be 'yield against number of days waterlogged', with curves for 25%, 50%, 75% and 100% of root zone flooded. The effects of watertable salinity and soil type will need to be included and more importantly, the stage of crop growth at which the transient waterlogging occurs.

43. As an example of the use of the guidelines outlined, it should be possible to estimate the reduction in yield of cotton growing in Pakistan in a medium textured soil with a watertable at 0.6 m and watertable salinity of 20 mmhos/cm. It should also be possible to estimate the additional reduction in yield caused when the watertable rises to ground level for two days during flowering. Clearly such guidelines would be invaluable to drainage engineers, enabling a more rational approach to drainage design to be made and providing a firm basis on which to assess drainage benefits. For vertical drainage such guidelines would allow wellfield operation to be better defined by making use, at last, of the inherent flexibility in depth to controlled watertable.

44. Finally, future developments in drainage will undoubtedly see an increase in the use of computerised drainage models. Increasingly models such as DRAINMOD (ref. 13) are being developed to assess the performance of existing drainage systems and to optimise the design of new systems. For such models to be of any real value to drainage engineers they need to be able to assess the effects of watertable levels and fluctuations on yield. At present they cannot do this and until the necessary guidelines have been produced, optimising drainage systems will remain a meaningless exercise.

REFERENCES
1. KOZLOWSKI T.T. Flooding and Plant Growth. Academic Press, London, 1984.
2. CANNELL R.Q. Effects of soil drainage on root growth and crop production. Soil Physical Condition and Crop Production in the Tropics (Ed. Layl R. and Greenland D.J.), John Wiley and Sons, 1979, pp 183-197.
3. WILLIAMSON R.E. and KRIZ G.J. Response of agricultural crops to flooding, depth of watertable and soil gaseous composition. Transactions of the American Society of Agricultural Engineers, 1970, vol. 13, pp 216-220.
4. JACKSON M.B. Plant and crop responses to waterlogging of the soil. Aspects of Applied Biology 4, 1983, pp 99-116.
5. WESSELING J. Crop growth in wet soils. Drainage for Agriculture, American Society of Agronomy, 1974, Monograph 17, pp 30-32.

6. CHRISTOPHER J.N. and TEKRONY R.G. Benefits related to watertable and salinity control. American Society of Agricultural Engineers, 1982.

7. SIR M. MACDONALD AND PARTNERS/HUNTING TECHNICAL SERVICES. Left bank outfall drain, stage 1 project preparation, annex 2. Water and Power Development Authority, Pakistan, 1984.

8. RAM H., NATH J. and KHANNA S.S. Effect of depth and composition of underground water and mulching on salt accumulation and yield of maize and wheat. Journal of the Indian Society of Soil Science, 1981, pp 518-525.

9. CANNELL R.Q. and BELFORD R.K. Crop growth after transient waterlogging. Proceedings of Fourth National Drainage Symposium, American Society of Agricultural Engineers, 1982, pp 163-170.

10. MOUSTAFA A.T.A., IBRAHIM M.E. and BAKHATI H.K. Effect of depth of saline watertable on cotton yield and its consumptive use. Ministry of Agriculture, Agricultural Research Review, Egypt, 1975, pp 21-24.

11. CANNELL R.Q., BELFORD R.K., GALES K., DENNIS C.W. and PREW R.D. Effects of waterlogging at different stages of development on the growth and yield of winter wheat. Journal of the Science of Food and Agriculture, 1980, pp 117-132.

12. REICOSKY D.C., MEYER W.S., SCHAEFER N.L. and SIDES R.D. Cotton response to short-term waterlogging imposed with a watertable gradient facility. Agricultural Water Management, Elsevier Science Publishers B.V., Amsterdam, 1985, pp 127-143.

13. SKAGGS R.W. Field evaluation of a water management simulation model. Transactions of the American Society of Agricultural Engineers, 1982, pp 666-674.

Discussion on Papers 13 and 14

MR C. M. BOLT, Hexagon Development Services, UK
With regard to horizontal tile drainage, it is valuable to reuse tile effluent for irrigation after the root zone and the volume above drains have been leached free of salts.

MR W. C. HULSBOS
As the flow pattern of the groundwater moving to the drains in homogeneous profiles is mainly through the subsoil, below drain level, the drainage effluent remains saline for a long period after the root zone and the soil above the drains have been leached free of salts. However, I agree that it is valuable to reuse the drainage effluent after the salt content has dropped to an acceptable level, either alone or mixed with surface water.

MR K. HEFNY, Research Institute for Ground Water, Egypt
Mr Stoner has described the case of rice fields where water seeps from high fields to lower fields. There was a case in the Nile Valley where new land agriculture had been developed in high lands on the fringes of the Nile in upper Egypt for about 25 000 ha along the valley.

The horizontal intensifying drains were dry, but the low old lands suffer from water logging and salinity. Water balance studies carried by the Research Institute of Egypt and IWACO of The Netherlands, using models, showed that a feasible solution was vertical drainage in both high and low lands. The salinity of the water is low in both areas and pumped water could also be used for irrigation in conjunction with Nile water.

MR W. C. HULSBOS
Vertical drainage can be an attractive solution in those cases where the aquifer conditions are favourable and the quality of the pumped water is suitable for irrigation, but in this case the provision of the new irrigated areas along the valley fringes with drainage, horizontal or vertical, has a lower priority than increasing the water use efficiency of the irrigation system including the use of modern irrigation techniques.

MR R. F. STONER
Recent work that we have done in the Nile Valley confirms that there is great potential for vertical drainage and for conjunctive use of groundwater and surface water. As yet this potential has scarcely been recognized at all in Egypt with the exception of the Research Institute for Groundwater.

MR M. MULLER, National Institute for Physical Planning, Mozambique
Given the problems of maintenance in water and sanitation infrastructure in developing countries, what are the maintenance considerations for soil drainage?

MR W. C. HULSBOS
Under average conditions, i.e. normal soil conditions, the yearly cost of maintenance of open drains is mostly taken at 5% of the investment cost. The cost of maintenance of closed tile systems is generally lower provided that they are installed properly. With pipe drains most maintenance is needed shortly after construction and can be minimized or prevented by careful construction. Under good conditions it is even considered possible to install maintenance-free pipe systems, using modern drainage machines, pipe materials and properly selected envelopes of filter materials.

MR R. BAHAR, World Health Organisation, Switzerland
Water logging in the Punjab creates irrigation and health problems. The cost that the people and government pay in terms of suffering, pain and expenditure for the prevention of malaria is high. Is there a solution to this problem?

Since early 1960 the Government of Pakistan has been spending US $2 per year on average on pesticides for the control of malaria vectors, not including the loss of working days due to death and disability and the cost of treatment for the patients.

MR W. C. HULSBOS
The introduction of irrigation is known to introduce or increase health hazards. In numerous countries the extension of the irrigated area has resulted in a considerable increase not only of malaria but also of many

water-borne diseases, the most important being bilharzia.

Surface drainage of excess irrigation or rain water will no doubt help to improve the health situation, as will subsurface drainage through control of groundwater tables that are too high, but drainage alone cannot solve this problem. Adaptations of the irrigation design may be possible in some cases to prevent low flow velocities in channels. However, the general health situation of the population in the rural areas may be favourably affected by the introduction of irrigation which raises farmers' incomes. Increased health problems in irrigated areas should be studied by engineers and health specialists together. Until now inadequate attention has been paid to these aspects.

MR R. F. STONER
Engineers clearly recognize the need to deal with health problems but so far there is no co-ordinated approach between various government departments on this matter. With regard to the papers presented here all that can be said is that any drainage works carried out for agriculture in the Punjab must improve the health position. There is no doubt that more needs to be done.

MR E. A. BURLEIGH, Atkins Land and Water Management, UK
How can drainage systems be modified to conserve water during periods of low rainfall so that the cropping season or the availability of rainwater for plant growth can be extended without affecting the overall drainage capacity required during the rainy season?

In particular, reference is made to banana cultivation in the Caribbean where open drain channels are closely spaced (12 ft) and low rainfall is experienced for about 4-5 months.

MR W. C. HULSBOS
Drainage through a free-flowing outfall system does not conserve water. If drainage is needed during the wet season and it is necessary to conserve water during an extended period of low rainfall this can only be done by closing the drainage system through appropriate structures. These structures can be relatively simple gates or stop logs as they often have to be operated only twice a year. The number of structures depends on the lay-out of the system and especially on the topography of the area. In large flat areas structures are only needed in the main drains. In some cases it may be possible to let water into the drainage system. The success of water conservation depends on the natural drainage conditions through underground flow.

MR G. NEVILLE, Sir William Halcrow & Partners, UK
I would like some clarification of the cost data presented in Table 4 and the text which gives a range of cost increase with depth. This information relates to increasing diameter with depth. Are data readily available on the cost increase with depth for single-size diameters of pipe drain?

What approximate indicators would be used to give a cost range of the total drainage works, as opposed to the range of values given by Mr Hulsbos of up to US $1500/ha for lateral and collector drains alone, but excluding the associated main drainage system, i.e. a factor to give a rough idea of the total cost over the lateral and collector cost alone.

In the Paper, for composite systems a 'junction box' is shown at each lateral/collector connection (fig. 1). Is this intended to be a man access point to provide access to each and every lateral drain, bearing in mind Mr Hulsbos' point that systems should ideally be designed to be maintenance free?

MR W. C. HULSBOS
The cost increase with depth for the same diameter pipes depends to a large extent on local conditions, especially the subsoil and the water-table. Any available data should therefore be used with caution for other areas than those for which they have been produced. A theoretical analysis, based on present prices, drainage equipment and materials, shows that an increase in the drain depth from 1 m to 2 m for the same diameter approximately doubles the cost. An increase in depth from 2 m to 3 m again doubles the cost. For more information see ref. 1.

The cost of the main drainage system depends to a large extent on the outfall conditions. Are extremely long outfall drains required or is the drainage effluent pumped? There is no factor of general validity to calculate the cost of a total drainage system on the basis of field and collector drains. For conditions in The Netherlands and in Egypt I estimate that the total cost of drainage is about double the cost of field drains including collector drains. However, such a system also discharges irrigation tailwater and excess rain-water. The total cost cannot therefore be attributed to groundwater depth control alone.

A junction box is not intended to be a manhole. Experience has shown that manholes or inspection holes protruding above the surface are not only expensive but also endanger the system as they may be improperly used for waste disposal and disposal of surface water. A buried junction box can have different designs, but should have a provision to enable flushing of the lateral and collector pipes if needed.

MR J. F. ROBSON, Sir M. MacDonald & Partners, UK
Reference has been made to the high cost of drainage in Iraq and Pakistan, and drainage criteria and drain spacing design have been discussed in detail.

Would Mr Hulsbos comment on the adequacy of monitoring of performance of installed systems in view of the extensive assumptions that have to be made in spacing calculations, on soils data which are so highly variable? Have any significant additions been made over say the

PAPERS 13 AND 14: DISCUSSION

past ten years?

MR W. C. HULSBOS
A drainage design method using a formula, adopted criteria and extrapolated field data is no more than a comprehensive calculation model in which carefully selected elements produce a useful result. For as long as drainage has been practiced in The Netherlands it has been monitored, and the system works satisfactorily with the criteria used, which are considered as empirically based design standards. They can only be used in this calculation method. Such long-term experience is not available in Iraq and Pakistan. Monitoring of the new systems is indispensable to check and eventually to adapt the design method and/or criteria. In Egypt, where large drainage systems have been constructed during the last 20 years, a rather extensive monitoring system has been initiated. The results show that neither the design method nor the criterion used but the quality of the construction is at present the most important factor which determines the performance of the systems.

MR V. R. ROWLAND, Pauling PLC, UK
The Paper shows that considerable areas still require drainage.

The expected life of the present systems may well be more than 20 years. However, some of the earlier installations with less developed techniques and materials are proving ineffective after a considerably shorter period and their rehabilitation consequently increases the need and opportunity for further drainage works. This is in line with the conclusions of the Technical Notes.

The overproduction now existing in areas of Europe may lead progressively to the removal of cultivation of marginal lands and the concentration of agriculture and development effort on the most favourable areas. Against the higher yields produced from these better lands the necessary enhancement by improved drainage may be cost effective. Would the Author comment on these possibilities?

MR W. C. HULSBOS
An overproduction of agricultural products decreases the attractiveness for investments in agricultural land not only in Europe but also in many other countries. Many countries, however, are still stimulating agricultural production either because they are not self-supporting or because they want to remain self-supporting with an increasing population. The time when agricultural products will only be grown in places where it can be done at the lowest cost is still far away. Moreover, for many countries drainage of agricultural land is mainly a means to prevent further destruction of areas under production and not an extension of production into marginal areas. In any case it is common practice to decide on a drainage infrastructure on the basis of a feasibility study which decides the economic justification.

MR C. M. BOLT, Hexagon Development Services, UK
Breeding crops to adapt to highly saline conditions may be more cost effective than drainage.

MR W. C. HULSBOS
Although it is known that some crops or varieties can withstand rather high soil salinity or saline irrigation water, the economic value of these varieties is not always clear. The use of saline irrigation water mostly requires additional water, high soils and drainage. However, breeding efforts to obtain salt-resistant varieties are important, although results so far obtained indicate that the possibilities for economic use of saline water or soils for growing salt-adapted varieties are limited.

MR R. F. STONER
Current work at the Plant Breeding Institute (PBI) in Cambridge on salt-tolerant wheat is being extended to field trials in Pakistan. Discussions are also taking place at the PBI about the possibility of breeding a wheat that is tolerant to waterlogged conditions. Certainly there is genetic material available that would make such a proposition feasible.

DR T. W. TANTON, University of Southampton, UK
In Europe a high subsidy from the European Economic Communities for agriculture can make land drainage economically attractive to farmers, but in countries where such subsidies do not exist and yield levels are often low intensive drainage is often uneconomic.

In such cases would it not be better to try a less intensive more economically attractive solution such as initial wider drain spacing in conjunction with either more ploughing or soil restructuring? Such solutions may carry a higher risk than a fully intensive drainage system but the potential benefits may more than justify the increased risk.

MR W. C. HULSBOS
Subsidies do not affect the economic feasibility of drainage. Drainage is only justified if it is considered economically feasible at national level. However, it does influence the financial attractiveness to farmers. In European countries the drainage construction rate depends on the existence and level of subsidies. In most developing countries the governments will usually pay 100% of the cost of the systems. Repayment schedules are sometimes made but I do not know of any case where repayment by the farmers occurs. The installation of a suboptimal system under these conditions is independent of subsidies but may be justified for limited financial resources.

DR Z. SVEHLIK, University of Southampton, UK
It is interesting how little attention has been paid in drainage design to the crop requirements. The design is based on the

PAPERS 13 AND 14: DISCUSSION

principles of groundwater hydraulics without any consideration to the soil moisture regime in the unsaturated zone, water extraction by plants etc.

An integrated approach, considering the interactions between the two zones, would improve the quality of land drainage design.

MR W. C. HULSBOS
A draining system has to be designed for present and assumed future crop systems and not for single crops. Drainage design in practice is usually based on average crop conditions as drainage requirements of most field crops do not differ much. For individual crops in the crop pattern with strongly diverging requirements such as rice and vegetables the design must be adapted. Different criteria have been established for fruit-trees, vegetables and grassland.

The influence of the moisture regime on the drainage design in the unsaturated zone is in practice considered to be taken into account by these drainage criteria. An integrated approach taking into account the flow in the unsaturated zone would theoretically improve the drainage design. Such an approach is used in some mathematical drainage models. It is doubtful, however, whether refining the calculations has great practical value in view of the complexity of the system and the variability of parameters.

DR T. W. TANTON, University of Southampton, UK
A review of the world literature indicates that under trial conditions there is little or no correlation between drain spacings and yield increases attributable to drainage. Although little has been published about trials under saline conditions, the review clearly points to the need to obtain data on the effect of drainage on the yield response of the plants and strongly supports Mr Stoner's conclusions.

MR W. C. HULSBOS
In Europe and in most other areas with a temperate climate, the benefits of drainage are not related to better crop yields but to a reduction in production cost. Drains are usually dry during the growing season, the main function of drainage being to make land accessible earlier for farming operations in springtime, which will also make earlier sowing dates possible.

For irrigation in arid zones there is a relationship between drainage and leaching of salts and between soil salinity and crop yields. More data should especially be collected in these areas and I refer to the comment by Mr Robson on the desirability of increased monitoring efforts of implemented schemes, which I support.

REFERENCE
1. BOUMANS J. H. and SMEDEMA. Derivation of cost minimizing depth for lateral pipe drains. Agricultural water management. Elsevier, Amsterdam, to be published.

M. J. SNELL and R. J. WELLS, Sir M. MacDonald & Partners, Cambridge, UK

TN24. The decay and rehabilitation of irrigation systems

Most irrigation systems gradually change, even with adequate maintenance. as a result of changes in farming practice or water availability. The usual pattern is a gradual deterioration of both the irrigation system infrastructure and the system's operation. From the individual farmer's viewpoint this degradation may appear as an improvement of the system, when he makes changes to canals and structures to improve the water supply to his own farm at the expense of others, but the scheme as a whole and the wider community suffer losses. Eventually factors such as shortage, wastage and inequitable distribution of water result in decreasing agricultural production and make rehabilitation necessary. After rehabilitation the cycle of change, deterioration and further rehabilitation is often repeated.

2. The Authors' experience in Chile, Ethiopia, Indonesia, Iraq, Madagascar, Malawi, Nepal, Sri Lanka and Sudan indicates that the cycle described is typical of schemes in widely differing environments. The reasons for deterioration are outlined and some ways of reducing the frequency of rehabilitation are indicated. It is questioned whether or not such measures are worthwhile.

REASONS FOR THE DETERIORATION OF IRRIGATION SYSTEMS
3. Deterioration generally results from inadequacies during planning, design, construction, commissioning or operation, such as

(a) inadequate or inappropriate design of the physical system
(b) failure to implement effective O & M
(c) failure to anticipate changes in the agro-economic situation
(d) failure to recognize social factors affecting performance.

4. In the process of diagnosis and prescription of rehabilitation, as well as in the planning and design of new schemes, it is important to recognize the interaction and interdependence between the physical system and its human environment. The latter involves the farmers, the scheme management and its employees, and also the wider socio-economic systems that determine crop markets and prices, input costs and availability, and the balance of power between the various groups. There follows a tentative check-list of factors that are often significant in the deterioration process. Many of them are the effect or the intermediate cause of others.

Physical factors
5. Physical factors affecting water management

(a) layout unsuitable for systematic operation
(b) inadequate provision for water measurement
(c) water control structures unsuitable (leaky, difficult to adjust, sensitive to small level fluctuations, sensitive to sediment, badly placed)
(d) canals too large, too small or leaky.

6. Physical factors affecting maintenance

(a) tendency to scour, erosion or siltation
(b) poor materials or construction standards (especially canal banks' freeboard or compaction)
(c) inadequate washing and bathing facilities
(d) inadequate provision for animals to drink, bathe and to cross canals and drains.

7. Physical factors affecting cropping

(a) salinization
(b) waterlogging
(c) erosion of topsoil.

Managerial factors
8. General managerial factors

(a) inadequate delegation and autonomy
(b) unsuitable or inadequate training for managers
(c) lack of contact with farmers
(d) inadequate motivation towards efficient and equitable scheme operation.

9. Financial factors

(a) difficult or non-existent recovery of water charges
(b) shortage of recurrent funds
(c) shortage of foreign exchange.

10. Factors concerning technicians and operators

(a) motivation lacking or inappropriate
(b) inadequate training.

Agricultural, social and economic factors

11. Factors concerning farmers

(a) lack of confidence in the system
(b) lack of discipline
(c) lack of motivation (e.g. due to inequitable distribution of benefits, better opportunities outside agriculture)
(d) shortages (often seasonal) of labour and plant for agricultural operations.

12. Factors concerning the wider socio-economic environment

(a) credit and input supply inadequate or untimely
(b) market changes (may lead to a desire to change cropping patterns)
(c) land tenure problems
(d) religious, caste, gender or other social constraints
(e) rivalry between tribes and villages.

WAYS OF REDUCING THE RATE OF DETERIORATION

13. The planning process, whether for a new scheme or for a scheme to be rehabilitated, needs an adequate and accurate data base covering social, agricultural, economic, financial and engineering factors. In the case of an existing scheme the causes of deterioration should be investigated. Merely repairing obvious defects is rarely successful. Both planning and design should involve the farmers effectively and be matched to their expertise, experience and social structures. Socially cohesive farmers' groups should be encouraged, for instance by adjusting scheme layout to village or clan boundaries wherever possible. The design should suit the method of operation and be sufficiently flexible to cope with reasonable changes in cropping pattern or in operation or farming methods.

14. Structures should be tough enough to discourage wilful damage and unauthorized modification (it helps if design is such that any tampering will be easily visible). Minor standard structures should be designed carefully (downstream erosion on all of thousands of small canal structures in an irrigation scheme is all too common). Special attention should be given to areas prone to silting, scour, seepage and erosion.

15. Proper supervision and quality control during construction are essential. Incomplete as-built drawings lead to difficulties when maintenance or repairs are carried out. Farmer participation in the construction of tertiary canals may help farmers to identify with the system and look after it better in the future, but these benefits may be offset if construction quality is not maintained.

16. A long commissioning period helps to establish practical procedures and train the scheme's management and operators. Clear and simple O & M manuals in the local language help to establish and maintain procedures. Payment for water, although difficult to achieve in many countries, may encourage farmers to look after their minor canals. It can also alleviate budgetary constraints on O & M. An integrated operation authority facilitates cost recovery and the planning and co-ordination of water supplies and physical inputs. Communication between management and farmers is also much easier with a single operating authority.

HOW UNDESIRABLE IS A CYCLE OF DECAY AND REHABILITATION?

17. Unforeseen changes will occur, no matter how well a scheme has been planned and executed, and will usually result in degradation of the system and a need for rehabilitation. Fully effective maintenance is also difficult to finance and implement, so periodic rehabilitation may be easier in practice. It may be preferable economically to annual expenditure on ineffective maintenance.

18. The right balance in some circumstances may be a low intensity of routine maintenance and minor rehabilitation (or upgrading) every five years to accommodate changes and implement major repairs. Full-scale rehabilitation may still be necessary, perhaps every 25 years, as structures wear out and circumstances change.

A. A. GADDAL, MD, A. FENWICK, PhD, and
O. TAMEEM, MD, Blue Nile Health Project,
Sudan

15 Health aspects of water supply and sanitation

The provision of water supplies and sanitation is a priority in the developing world, but different standards of supply have to be set according to available resources. The Blue Nile Health Project has obtained results which clearly show that there is a correlation between lack of water supply and prevalence of schistosomiasis, and also that diarrhoeal disease is reduced when adequate water supplies are installed. The auhorities are urged to press ahead with the provision of safe water supplies and sanitation provided that adequate consultation takes place first to ensure that the proposals are acceptable to the recipients.

Introduction.

1. During the current decade the World Health Organisation has urged the provision of safe water supplies and excreta disposal to as many communities as possible. A clean water supply and effective sanitation are accepted as essentials of life in developed countries and are taken for granted. In poorer countries, such facilities are rare and the result of dependance on raw water is a high prevalence of water associated diseases.

2. Cholera, typhoid and diarrhoeal diseases spread so quickly that under the conditions of poor water and sanitation, the risk of a major epidemic developing is very great. One of the best indicators of an inadequate sanitary situation is an abundance of flies which is a further contributory factor to an unhealthy environment.

3. The main constraints to the WHO objective are availability of water and the high cost of distribution. In any area the facilities to be provided are usually determined by the available resources so that provision of water supplies may mean one village tap, or it may mean house connections with shower facilities and taps in every room. Similarly, the sanitation provided may be a simple pit latrine or a sophisticated water closet.

4. In the Blue Nile Health Project which is a joint WHO / Sudan Government venture which started in 1979, the targets are the control of schistosomiasis, malaria and diarroeal diseases using an integrated approach. The comprehensive strategy upon which control is based includes the following components :-

a) improvement of domestic water supply and sanitation.

b) health education and community participation.

c) the control of the vector mosquitoes and snails.

d) provision of diagnostic facilities and chemotherapy.

5. The results already obtained by the Blue Nile Health Project clearly demonstrate how the provision of safe water supplies and adequate latrines can reduce the prevalence of infection of our water associated diseases.

Schistosomiasis and water supplies.

6. Data collected on the prevalence of schistosomiasis in the villages and camps in one area of the Gezira clearly demonstrates a correlation between water consumption and prevalence. (Figure 1.). The three camps with no safe water showed a prevalence of over 65% S.mansoni whereas the villages with piped water supplies indicated a reduced prevalence falling to about 40% when the water provided was 70 litres per capita per day or above.

7. A second example is provided by annual survey results from a Rahad Scheme village in which the prevalence of schistosomiasis increased during a year when the water filtration plant failed to function. The prevalences of S.mansoni from 1982 - 1985 were respectively :-

10.6%   13.7%   12.6% and   40.8%

Diarrhoeal diseases and water supply.

8. WHO have concluded after a survey in seven countries that a statistically significant reduction in diarrhoea can result from the provision of safe water supplies to communities which have reasonable sanitation.

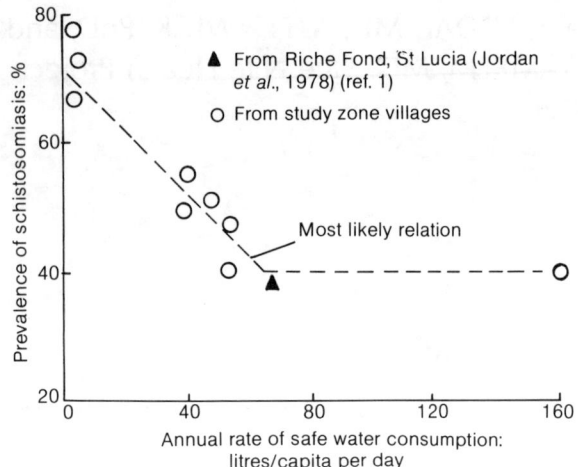

Figure 1. Prevalence of S.mansoni in 9 Gezira villages plotted against the annual mean safe water consumption.

9. In the Gezira Scheme, a survey of the children under 5 years suggested that in villages with an efficient water supply the prevalence of diarrhoea was reduced. The data was collected from three villages in which the per capita water consumption was 57, 55 and 38 litres per capita. The respective diarrhoea prevalences were 48%, 52% and 67%.

10. During the collection of base-line data in 1981 before intervention the diarrhoea-associated mortality rate was 49 per thousand in our study area, while after intervention the rate was down to 2.7 per thousand. The reduction was due to an improvement in water supplies and sanitation, the training of health workers and the provision of Oral Rehydration Salts.

11. The Project has almost completed a research study aimed at demonstrating more scientifically the impact of safe water supplies and sanitation on diarrhoeal diseases in 4 Gezira villages all previously with no water supply or latrines. Two of the villages (A and B) have been provided with shallow bore wells and handpumps. (Figure 2.) The results show a dramatic reduction in the prevalence of diarrhoea cases in the villages provided with water. (Table 1.)

The B.N.H.P. plan of action.

A - Water Supplies.

12. By 1980 about 75% of the villages in the Gezira Scheme in Sudan had been provided with a clean water supply which consists of a bore-hole and a pump (electric or diesel) to raise the water to a tank which then allows gravity distribution to stand-pipes in the village or to house connections. In areas where there is no ground water, then canal water is filtered by slow sand filter and treated with chlorine before being pumped to a raised tank. Unfortunately many of the systems are old and in need of rehabilitation. major

Figure 2. Showing villagers collecting water from their new shallow bore well using a handpump

Table 1. The effect of water supply on the prevalence of diarrhoea in children.

Children with Diarrhoea (%)

Village	Prevalence Survey 1.	Prevalence Survey 2
A	19	13
B	46	3
C	30	25
D	28	26

faults occur in the distribution lines and the pumps are in need of repair or replacement. Also the design capacity is now inadequate to meet the increased demand caused by population expansion in the villages. The slow sand filters have deteriorated and require more maintenance than they receive. The practice of chlorination has disappeared.

13. In these tenant villages The B.N.H.P. strategy calls for an improvement in the existing water supplies and an increase in the number of household connections.

14. The villages without water fall into two categories: -

smaller tenant villages (250 - 500 people)

labourer community camps (60 - 300)

15. The B.N.H.P. has a priority objective to provide the remaining tenant villages with a safe water supply. The sinking of a deep bore hole is too expensive for a small community and therefore the supply will consist of a shallow bore-hole and a hand pump to lift the water. There is usually a seam of water at a reasonable level (say 20 m.) which is adequate for such a small number of people. Three villages have been supplied with water

during 1985, and the results have been acceptable.

16. The labourer community camps are known as 'unregistered villages' because historically they have sprung up on ground not designated for building, and they have never previously been offered any facilities from the government or the scheme. The residents are all immigrants from either Western Sudan or even from Tchad or Nigeria, and they take labouring jobs during the year and are employed by the farmers for cotton picking and other harvesting tasks. They have no water supply other than the irrigation canals and no sanitation. The prevalence of schistosomiasis in these people is very high indeed (Ref. 2) and they play a major role in the transmission of schistosomiasis to all residents of the Gezira. The Project will provide all these camps with small roughening filters which removes turbidity and 95% of all bacteria from the canal water. The filtered water is lifted from the enclosed holding tanks by hand pumps. The villagers themselves are trained to clean the sand filters and to maintain their own supply. Six of these roughening filter systems have been completed and they are proving to be most successful. It is expected that this system will be spread not only throughout the Gezira, but also extended to cover all the 'Haffirs' (rain water reservoirs) in the Sudan.

B. Sanitation.

17. The tenants farmers in the Gezira are relatively prosperous and their houses in the tenant villages are by now mostly brick. The B.N.H.P. is embarking on a health education programme designed to educate the people into more hygienic ways, and to convince them to install pit latrines into their compounds. In order to achieve this, the project has produced a design which aims to minimise the problem of flies at reasonable cost. In a project factory concrete slabs to cover pit latrines are produced. The hole in the slabs has a flap which prevents the escape of flies and odour from the pit. Provided the slab itself is kept clean then the fly population is kept to a minimum. The slabs are available at only 10% of cost price, to all villagers who dig a pit in their compound. The slabs are subsidised by the World Food Programme and distribution centres are being set up throughout the project area. The response to date has varied according to the efforts of the health committees in each village, but in villages where the committee have themselves been convinced of the value of the project, the response is excellent.

18. Perhaps the best design for pit latrines is one in which the latrine is in a room and the pit has an outside chimney with the top of the pipe covered with mesh. The flies may breed in the pit but their only visible exit is through the chimney from which they cannot escape, and they fall back into the pit. This design is considered to be too expensive at present to recommend for use in Gezira.

Recommendations.

19. There is no doubt that socio-economic development is the key to improved health, and in particular that the provision of water supply and sanitation will lead to a reduction in the prevalence and severity of water-associated diseases such as typhoid, cholera and schistosomiasis. The point I wish to stress is that when new development projects such as irrigation schemes are being planned, the welfare of the residents should be given a higher priority than has ever previously been the case.

20. Specifically therefore please :-

a - Ensure that all villages are provided with a water supply which is adequate in quantity and quality to supply not only the planned population, but also to allow for inevitable expansion of the population.

b - Ensure that there is provision for the disposal of village rubbish in general and human excreta in particular.

21. Finally may I give some guidelines for these recommendations and offer some advice of mistakes to avoid :-

a - It is accepted that there will never be enough money in the early stages of a scheme to provide the standard of water supply that developed countries expect. However the standard must be high enough to prevent a disaster caused by a polluted piped supply.

b - The supply should be designed so that in future, improvements can be made to increase capacity and improve house connections.

c - The attitudes and opinions of the population to be provided with water should be taken into consideration, and should their views seem to be unreasonable then they should be educated carefully to the value of water and hygiene.

d - The positioning of the water outlets must be given careful consideration, and in particular the distance to the provided water should be less than the polluted alternative.

e - Again a knowledge attitudes and practices study should precede any attempt to force a sanitation supply on a community. Latrines must be suitable to the customs and habits of the people expected to use them.

f - In general communal latrines should be avoided because of the difficulty of maintaining any level of cleanliness.

g - The design of latrine must be tailored to the people, and to local conditions (eg availability of water and type of subsoil).

They must be sophisticated enough to prevent odour and flies. The digging of a hole on someone's property and calling this a latrine is folly.

22. In conclusion then, health workers can benefit from the efforts of the engineers who provide water and sanitation, provided that a minimum standard is reached. If it is not, then epidemics are likely because an inadequate and polluted water supply is worse than no supply at all, and a bad latrine is worse than no latrine at all.

References.

1. JORDAN P. et al. Further observations from St.Lucia on control of S.mansoni transmission by provision of domestic water supplies Bulletin of the World Health Organisation 56 965-973 1978.

2. FENWICK A. et al. Schistosomiasis among labouring communities in the Gezira irrigated area, Sudan. Journal of Tropical Medicine and Hygiene 85 3 - 11 1982

R. BAHAR, BScEng, MSc, Division of Vector Biology and Control, World Health Organisation, Switzerland

16 Irrigation and health

In an irrigation project, the irrigation system and its management affect not only agricultural productivity (income) but also affect well-being (health), of the settled population in and around the project. Generally, planning, design, operation and monitoring of irrigation projects undertaken by agencies with specialized expertise that focus mainly on technical and financial aspects of projects without considering the project's health and environmental impacts. In view of the complexity of the irrigation project, multisectoral and multidisciplinary (holistic and/or comprehensive) approach must be promoted. For implementation of this holistic approach a mechanism must be adopted and institutionalized. For institutionalization a sequence of steps is described for implementation.

1. INTRODUCTION

1. During the past three decades, water development schemes in developing countries of tropical and sub-tropical zones, for irrigation, production of energy, and the satisfaction of water demands of fast growing urban areas have been greatly expanded. The primary objectives of these water development project is to produce more food, fibre and to raise the standard of living of the people. Unwittingly these projects have added a new dimension to existing health problems, especially in prevention and control of vector-borne diseases such as malaria, schistosomiasis and filariasis. The irrigation projects as well as other water resource development projects, while necessary and advantageous, can have drawbacks because they create favourable conditions for the transmission of diseases and for breeding of intermediate hosts and vectors[1] of parasitic or infectious diseases. For example, prior to 1926 when the Sennar Dam was constructed in the Blue Nile River for the production of energy and for irrigation of the Gezira in the Sudan, schistosomiasis did not exist in the area and malaria was not a health problem in this part of the country. At the present time in some of the villages in the Gezira, up to 80% of the school age children are infected with schistosomiasis and malaria is hyper endemic in the area[1][4].

2. With available technology it is possible to considerably reduce the health hazards, if required attention is given to the elimination of potential vector habitats and breeding places during the planning, design and construction phases of an irrigation project along with improvement of water management in the operational system. For the fulfilment of these objectives, it is essential that the responsible staff for the project including technical and managerial staff, i.e. policy makers, planners, designers, constructors, managers, supervisors and operators who should consult and work in close collaboration with health specialists, from the project's inception through to its completion and throughout its operation.

2. Major Health hazards in irrigation projects

3. Health hazards which may occur in an irrigation project are either a result of:
   i. Environmental changes and ecological disruptions which may produce suitable biotopes for disease vectors such as mosquito vectors of malaria and snail intermediate hosts of schistosomiasis, or
   ii. Socio-economic changes which takes place because of human displacement over-crowding, shortage or lack of safe water and sanitary facilities, absence of health services, introduction of new reservoirs of diseases (endemic and/or new) and the possible arrival of immigrants immunologically susceptible to the endemic diseases prevailing in the project area.[6]

2.1 Diseases related to ecological changes in irrigation projects

4. Diseases are either transmitted directly by water (water-borne) like diarrhoeal infections or are water-associated such as the vector-borne diseases. The principal vector-borne diseases which may be associated with irrigation projects are given in Table 1.

5. Among the diseases listed in Table 1, two emerge as posing very serious public health problems i.e. malaria and schistosomiasis.

2.1.1 Malaria

6. Malaria is the most important and widespread water associated vector-borne disease. According to the latest (1983) available figures, of a total world population of 4676 million about two thirds are at risk of malaria, and the total number of malaria cases reported, (excluding Africa) for 1983, was 5.5 millions.[20]

---

[1] The term "Vectors" is used here in the loose sense and includes such intermediate hosts as snails and cyclopes

Table 1. The Principal vector-borne diseases associated with Irrigation Projects in relation to main habitats of the vectors and intermediate hosts

Diseases Group	Diseases	Vectors	Intermediate Hosts	Rivers & Streams	Lakes & Ponds	Irrigation Canals	Drainage Canals	Rice Cultivation	Human Settlement	Tree holes & leaf axils
Arboviral	Yellow Fever	Culicine Mosquitos (Aedes spp)	− (1)						+	+
	Dengue Fever	Culicine Mosquito (Aedes spp)	−						+	+
	Dengue Haemorrhagic Fever	" " (Aedes spp)	−						+	+
	Japanese Encephalitis	Culex spp	" (2)					+	+	
Filarial	Bancroftian	Culicine & Anopheline				+	+	+	+	
	Brugian	Culicine (Mansonia)	−		+	+	+		+	
	Onchocerciasis	Simuliid (Black Fly)	−	+						
	Dracunculiasis (Dracontasis)	−	Copepods (Cyclops)	(3) +						
Protozoal (Malaria)	Malignant (P. falciparum)	Anopheline	−	+	+	+	+	+	+	
	Benign (P. vivax)	"	−	+	+	+	+	+	+	
	Quartain (P. malaria)	"	−	+	+	+	+	+	+	
	Malaria (P. ovalae)	"	−	+	+	+	+	+	+	
Helminths (Schistosomiasis)	(Urinary) (Haematobium)	−	Planorbid (snail)	+	+	+	+	+		
	Japanicum	−	Oncomelanian (snail)	+	+	+	+	+		
	Visceral (Mansoni)	−	Planorbid (snail)	+	+	+	+	+		

1 in nature monkeys are reservoirs of yellow fever
2 birds and pigs are multiplying hosts and reservoirs of J.E.
3 small ponds and step wells

7. Malaria parasites are transmitted from man to man by species of anopheline mosquitos which breed, depending on the species, in different types of water collections. The parasite, Plasmodium spp., must develop in the body of the mosquito before reinfecting man by a subsequent bite and blood feeding by the mosquito. Only the female mosquito feeds on human blood and transmits the disease.

8. Malaria is primarily a disease of rural areas and may often be associated with faulty irrigation, and impoundment of water and/or incorrect agricultural practices. In many instances the kind of crops grown may encourage breeding of the vector mosquito and consequently malaria prevalence and cause epidemics of the disease in the area.

9. On the impact of irrigation project on malaria (Farid M.A. 1975) mentions "In areas using water from rain, rivers or wells for dry farming and irrigation of winter crops malaria may be absent or at low endemicity (hypoendemic) and does not constitute a serious health problem, unless man-made environmental changes such as introduction of perennial irrigation, wet cultivation and population movements set the stage for malaria epidemics and post-epidemic, hyperendemic situations occur. This tragedy which has been repeated in many areas in sub-tropical arid belts, is due to the lack of vision with regards to the serious malaria hazard connected with these undertakings. It is noteworthy to mention that once malaria is entrenched under such circumstances, it is most difficult to control or eradicate. The high malaria endemicity in the Delta region of Egypt, particularly where rice is grown and the failure to interrupt malaria transmission in areas under perennial irrigation in Syria, Iraq and the Khouzestan province in Iran, in spite of eradication campaigns that have been conducted over the last 2 decades bear witness to the fact."[5]

Table 2. Examples of Increased Prevalence of Schistosomiasis
Resulting from Water Resource Development Projects(12)

Country	Project (year completed)	Pre-project prevalence (percent)	Post-project prevalence (percent)	Schistosome species
Egypt	Aswan Dam (first) (1900)	6%	60% (3 yrs. later)	S. haematobium S. mansoni
Sudan	Gezira Scheme (1925)	0%	30-60% (15 yrs. later)	S. mansoni S. haematobium
Tanzania	Arusha Chini (1937)	low	53-86% (30 yrs. later)	S. mansoni
Zambia and Rhodesia	Lake Kariba (1958)	0%	16% adults 69% children (10 yrs. later)	S. mansoni S. haematobium
Ghana	Volta Lake (1966)	low	90% (2 yrs. later)	S. haematobium
Nigeria	Lake Kainji (1969)	low	31% (1 yr. later) 45% (2 yrs. later)	S. haematobium
Iran	Dez Pilot Irrigation Project (1965) (1 & 4)	15%	27% (2 yrs. later)	S. haematobium
Sudan	Rahad Scheme (1977)	0%	12% (7 years later)	S. mansoni S. haematobium

Source: Rosenfield, P.L., and Bower, B., 1979

2.1.2 Schistosomiasis

10. After malaria, schistosomiasis is the most important and widespread water associated vector-borne disease and could be considered the most important health problem in irrigation schemes in Africa, the Middle and Far East. It is estimated between 500 to 600 million people are exposed to the risk of schistosomiasis infection. It is also estimated that over 200 million people in 71 developing tropical and sub-tropical countries, are currently infected by this disease.(21)

11. The disease is transmitted to man through the contact with water infested with the free swimming larvae of the parasite, the "cercaria". The cercariae penetrate the skin of man in the water. In the human body the cercaria develops to a mature worm. After development in the liver, an infected individual discharges the parasite eggs through faeces or urine into the environment. Once eggs come in contact with fresh water, they hatch and an actively swimming larvae, known as a "miracidium", will emerge from the egg shell. This "miracidium" must enter the body of a suitable snail within a few hours after hatching. In the snail the miracidia undergo an essential development state, lasting 4 to 8 weeks, each miracidium may produce and release thousands of cercaria larvae to start the cycle once again.

12. Like malaria, schistosomiasis is also a disease of rural areas, where incorrect agricultural practices and substandard hygiene with inadequate, if any, sanitation facilties produces conditions favourable for intense transmission. Especially in areas involving irrigation projects, disease transmission becomes intense corresponding to the frequent contacts of the resident population with water. Table 2 below gives some example and demonstrates the association of schistosomiasis with water resource development and irrigation projects.

2.1.3 Onchocerciasis

13. Onchocerciasis occurs, in tropical countries of parts of Africa, America and Asia. This water associated disease, also known as river blindness is distributed widely in tropical Africa, part of south and central America; there are also small foci of infection in Yemen and Saudi Arabia. It is estimated that the number of persons infected globally is in the range of 25-30 million.(10)

14. Onchocerciasis is transmitted by the blackfly, Simulium spp. which breeds and grows during its larval development in fast-flowing water The adult fly through biting transmits the disease from man to man. The worm parasite grows in man's body and multiplies producing millions of embryos or microfilaria. The adult female worm in the human body may grow up to 70cm in length and lives as long as 15 years. The embryo may cause skin lesions and involve the eyes resulting in blindness.

15. The breeding sites of black flies are mainly rapids, spillways and half submerged branches of the trees in rivers and streams. These sites could be also man made. i.e. construcion of small diversion dams, for irrigation purposes, often will create spillways over the dams and the rapids below dam sites which are suitable for breeding of black flies.

16. Especially in Africa the bites of flies, infection and blindness have forced the inhabitants to leave the fertile land along the rivers and settle in nearby plateau far from sources of water where farming conditions are not as suitable. In addition, many of the inhabitants are so debilitated by the disease that they can hardly grow enough food to meet their bare demands even though they are not blind.(11)

2.2 Diseases related to socioeconomical changes in irrigation projects

17. As a result of population movement and formation of new settlements in and around of irrigation projects the local conditions and

style of life, even temporarily, will effect the state of health of the population.(6) The newly arrived population will face shortage of housing, overcrowding, lack or absence of sanitary facilities including safe water. Under such circumstances communicable diseases, such as water-borne, respiratory and skin infectious will spread among the newly arrived population. In addition, loss of traditional economic activities, failure in adaption of resettled population, social insecurity and insufficient family income, often temporarily, with its consequences on education, hygienic, and nutritional conditions can adversely effect the health of the population.(6)

18. The provision of housing and of safe water, the establishment of schools, and adequate health services and the development of an infrastracture for the control of endemic diseases, in and around the project area, will lead to reduction or elimination of specific health hazards and improvement of the general health situation.

3. Health Planning in Irrigation Projects

19. In most cases vector-borne diseases in and around irrigation schemes, are man-induced and executing agencies unwittingly alter and modify the vector habitats in such a way that vector-borne diseases are intensified. Hence, responsible agencies should endeavour to prevent the adverse effects on health and incorporate preventive measures from the very early stages of any irrigation project. Therefore, health planning in irrigation projects must be promoted and implemented as a part of any water resource development project. The whole planning process should take into consideration the objective and scope of the project. The Ministry of Health has a moral and constitutional responsibility for the health of the people, regardless of other sectors i.e., agriculture, irrigation, labour, public work, natural resources etc., and invested activities concerned with health in their area of work. According to the policy and strategy of health for all by the year 2000 recommended by WHO for health planning in irrigation projects, the eight elements of Primary Health Care (PHC) should be considered and provided for. These eight elements are: Health education, adequate food and nutrition, safe water and sanitation, maternal and child health services, including family planning, immunization prevention and control of endemic diseases, treatment of common diseases, and the provision of adequate drugs.(13)

20. In view of possible adverse effects on health and environment in planning irrigation projects, the PHC elements totally or partially must be provided for by the executing agencies, depending on the extent of the availability of such services by the health sector.

21. There are four principal components which are instrumental in successful planning and incorporation of health safeguards in an irrigation projects namely:
i. Availability of a sound policy and adequate legistlative support for incorpoaration of health and environmental safeguards in irrigation projects.
ii. Availability and/or development of appropriate technology for health and environmental safeguards in irrigation projects.
iii. Coordination and collaboration between sectors and organizations involved in irrigation projects.
iv. Community awareness of health hazards and their participation in incorporating health and environmental safeguards in irrigation projects.

22. Coordination and inter-relation between these four components are instrumental to successful planning, design, organizational structure, management and monitoring of any development project including irrigation projects.

3.1 Development of a sound policy and adequate legislative support for incorporation of health and environmental safeguards in irrigation projects

23. Although the necessity of vector-borne disease prevention and control may be stated in various documents, they are often not reflected specifically in irrigation projects. The responsibility, structure and financial allocations for incorporation of the health and environmental safeguards, must be defined. The policy makers in different levels, i.e., international, national and local levels, for protection of the public and of the environment, should for example, consider the health impact of environmental changes as a result of an irrigation project. They should define the responsibilities, liabilities and inter-relations of health sectors, financing agencies, executing agencies including private organizations (consultant and/or contracting agencies) and "consumers" or communities. The responsibilities and liabilities for incorporating health and environmental safeguards for each of the above sectors in each level and in different stages of the project, i.e., planning, design, construction, operation, monitoring and evaluation, must be described. The development of technology for this purpose is of great importance and must be considered in the policy formulation.

24. The inter-disciplinary character of vector management in irrigation projects will require manpower development with knowledge in a variety of basic science and engineering skills. Therefore, manpower development should be considered as a major component of policy formulation in the majority of countries.

25. In most cases, it is necessary that the relevant health policies, partially or totally, be implemented through specific legislation. Legislation must be regarded more as a guiding principal in the performance of a task, than an instrument of enforcement.

3.2 Technology

26. Technology or "tools" for the prevention and control of vector-borne diseases in irrigation schemes exist. Integrated vector control has been in practice since the latter part of the nineteenth century, when the role of vectors in transmission of diseases was conclusively demonstrated. Integrated vector

control can be considered as the utilization of all the appropriate technological and management techniques, to bring about an effective degree of vector suppression in a cost-effective manner.(17) Components of integrated vector control are:
- environmental management measures
- biological control measures
- chemical control

27. The diagram (Fig.3) demonstrates in detail the components of integrated vector control.(16)

28. In terms of integrated vector-borne disease control, the treatment of cases through existing health infrastructure, could also be added to the above 3 components.

29. Of the three components mentioned for integrated vector control the environmental management component is the direct responsibility of designers, engineers, irrigation specialists and agronomists in the irrigation project.

3.2.1 Environmental Management Methods for Vector Control

30. Among the ecological factors affecting vector propagation, the reduction or elimination of open water surfaces and the control of vegetation growth seem to be most susceptible to engineering manipulations. Therefore, the environmental management methods for vector control which have been proved effective are mostly aimed at these two factors. These methods have been described in detail in a WHO manual on environmental management for mosquito control(16). Table 3 below gives appropriate environmental management measures effective for control of each vector.(14).

31. Kuo C. 1984(7) more specifically prepared a list of the environmental management methods applicable to each type of breeding place in an irrigation scheme, which is given below:

(a) Methods applicable to reservoir projects
- Proper pre-impoundment preparation of the reservoir site, in particular, selective clearing of trees and vegetation in marginal areas.
- Deepening and filling to eliminate shallow margins.
- Suitable drainage of reservoir margins
- Provision of the necessary water control structures and equipment for water level fluctuation
- Water level fluctuation
- Effective maintenance of shorelines and reservoir margins by grading and vegetation removal
- Effective maintenance of marginal drainage system

(b) Methods applicable to settlement or resettlement of population
- Suitable siting and provision of adequate of sanitary facilities
- Protection of the work-force during construction by suitable siting of living quarters and office areas, provision of adequate sanitary facilities, application of chemical control measures, and medical screening and chemoprophylaxis procedures.

(c) Methods applicable to irrigation systems
- Use of closed conduits or lined canals for water conveyance as far as possible
- Use of a "safer" type of irrigation (such as sprinkler, trickling or dripping irrigation) as far as possible
- Design of canal gradients and cross sections to ensure suitable velocities for vector control
- Good alignment of canals and avoidance of sharp curves
- Proper land forming and grading
- Inclusion of a drainage componenet in any irrigation system
- Intermittent irrigation and periodic drying of canals and fields
- Effective canal maintenance, in particular vegetation removal
- Good irrigation practices to avoid over-irrigation
- Channel flushing

(d) Methods applicable to drainage projects
- Use of buried drains instead of open ditches in as far as possible
- Lining of ditches or of the ditch invert, where possible, for open ditches.
- Good alignment of ditches and avoidance sharp curves
- Effective maintenance of the scheme
- Ditch flushing as required wherever possible.

32. It is important to point out that most of the environmental management methods for vector control listed above are consistent with good engineering practices. A case in point is effective canal maintenance which is required more for engineering purposes than for vector control benefits. Economic gains may result from at least one method i.e. the deepening and filling operation which could reclaim land for productive purposes. The use of closed conduits or lined canals for water conveyance in irrigation is probably a controversial issue, but in some cases where fertile land is limited it will release land for farming i.e. Egypt.(2) Many take it as "too expensive" and reject it without a project-cost comparison or an economic study. It must be noted that closed

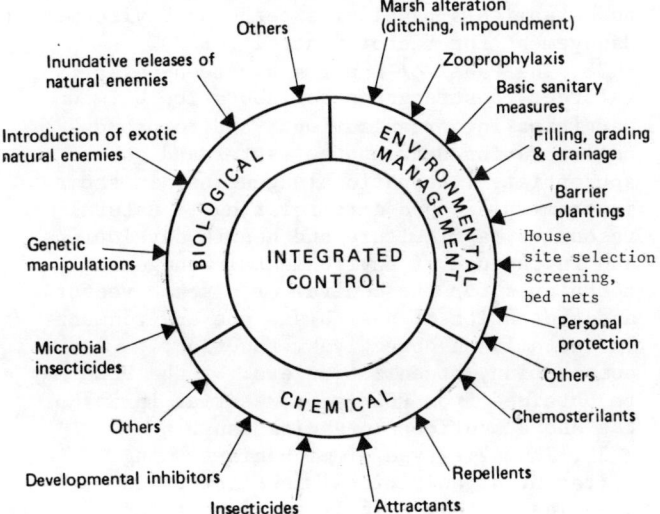

Fig. 3

Table 3. Environmental Management for Vector Control

Vector or intermediate host	Diseases transmitted	Environmental modification						Environmental manipulation						Modification or manipulation of human habitation or behaviour				
		Drainage (all types)	Total earth filling	Deepening and filling	Land grading	Velocity alteration	Impoundment	Clearing and burning of terrestrial vegetation	Shading or exposure to sunlight	Water level fluctuation	Sluicing for flushing	Aquatic vegetation control	Salinity regulation	Water supply and sewerage	Screening and bednets	Refuse collection and disposal	Land-use restriction	Improved housing
Anopheles mosquitos	Malaria	++	++	++	++	+	–	+	+	+	+	+	++	+	+	+	+	+
Aquatic snails	Schistosomiasis	+	+	++	+	+	–	–	–	+	+	+	+	++	–	–	+	–
Culex and *Aedes* mosquitos	Filariasis; viral and other diseases	++	+	+	+	+	–	+	+	+	+	+	+	+	+	+	+	+
Blackflies	Onchocerciasis	–	–	–	–	+	++ a	–	–	–	+	–	–	+	–	–	+	–
Houseflies	Infantile diarrhoea	–	–	–	–	–	–	–	–	–	–	–	–	++	+	++	–	+
Tsetse flies	African trypanosomiasis	–	–	–	–	–	–	++	–	–	–	–	–	–	–	–	+	–
Triatomid bugs	Chagas' disease	–	–	–	–	–	–	+	–	–	–	–	–	–	+	–	–	++
Rat fleas	Plague	–	–	–	–	–	–	–	–	–	–	–	–	–	–	++	–	++
Cyclops	Dracontiasis	–	–	–	–	–	–	–	–	–	–	–	–	++	–	–	–	–

Key:
- – = Little or no directly demonstrated value, or not applicable.
- + = Partially effective (some species).
- ++ = Primarily effective (most species).

a Small dams = adverse effect; large dams = good effect.

conduits or lined canals will considerably cut down the water loss due to seepage and thus, all other conditions being equal, these can be of a smaller size (cross section) than unlined canals, to carry the required quantity of water for irrigating the same acreage of land. The increased smoothness of the conduit or the lined canal surface will also permit a further reduction in the cross section. The land saved in the right-of-way, the reduced maintenance requirements, the reduced water loss and other relevant factors should all be quantified and taken into consideration in a fair cost analysis.

33. The selection of environmental management measures for vector control suitable for a given situation to be incorporated into the project design, will be made by the project engineer, based on findings and recommendations of the health impact study and in consultation with the epidemiologist and biologist/entomologist as and when required. Project engineers should therefore have sufficient basic knowledge of the health implications of water management and the available environmetal management methods for vector control. Unfortunately, most civil engineers have not been adequately informed about these subjects during their university studies and should be encouraged to undertake additional training.

3.3 Coordination and collaboration between organizations involved in irrigation projects

34. The establishment of a mechanism for coordination and collaboration among the different sectors, i.e. irrigation, agriculture, public works, consumers (farmers) and the health sector is of prime importance for the achievement of the project's objectives, including incorporation of health planning in the irrigation project. This mechanism must be so designed to ensure coordination of the work from the highest level (policy making) to the field level (project).

35. Depending on its political and administrative systems and organizational patterns each country should create its coordinating body and mechanisms. Some governments may find it useful to establish or strengthen, if it already exists a national health council (NHC). This national health council should be composed of persons representing a wide range of interests in the field of health, political, economic and social affairs, as well as the population at large, including the rural population. It might be useful also to create such council at subordinate levels, depending on the size and administrative system of the country.[18]

36. Thus, the NHC by itself or a subcommittee to this body, could initiate and coordinate the incorporation of health and environmental safeguards in irrigation projects.

37. Fortunately, at the international level a coordinating body has been established since 1981. An arrangement has been made between the World Health Organization, the Food and Agriculture Organization of the United Nations and the United Nations Environmental Programme to establish a panel of experts on Environmental Management for Vector Control (PEEM).

38. The Panel of Experts has been established in order to strengthen collaboration between participating organizations, and to promote collaboration between the latter and other appropriate international agencies, in their programmes and projects relating to natural resources, agriculture and health development and in the use of environmental management techniques for the control of disease vectors and protection of health and the environment. The details on objectives, function, composition, organization etc. of the PEEM may be obtained from PEEM's Secretariat in any of the above mentioned organizations.

39. The same type of mechanisms among different organizations, institutes and agencies, involved in irrigation projects at the national and local level for coordination of the work should be established.

3.4 Community awareness of health hazards and their participation in the incorporation of health and environmental safeguards in irrigation projects

40. The farmers are the ultimate consumers of the outcome of an irrigation project; that is, of "water and land". "The farmer", as an individual and the farmer's family and rural communites as a group, should know what they are buying and what price they have to pay for it. Undoubtedly, because of economic incentives farmers are enthusiastic and cooperative in implementation of irrigation projects. They must also be made aware of environmental and health hazards of badly planned irrigation projects which will be a threat to their well-being, working capacity and consequently their income. In view of economical incentive, farmers will be more cooperative in community action in irrigation projects. This is a good opportunity and a suitable point of entry to introduce the health component of the project. Moreover, because of the multisectoral and multidisciplinary nature of irrigation projects, the service of primary health workers, agricultural extension agents, social workers and teachers should be enrolled in making aware and educating the community on project's health hazards and what a community can do about it.(19)

41. For implementation of any community health programme there is a need for:
- sound planning at project level with the participation of community leaders
- organized effort on the parts of communities are required to sustain community participation and involvement in this connection, the village council, the farmer's union, the farmer's cooperative all have vital roles to play in the mobilization of the resources for promoting and mobilizing community participation. In this respect, the formation of community health committees would be useful in achieving better community participation.
- Detailed knowlede of socioeconomic and cultural aspects of the community is also useful in formulating, motivating and educating processes necessary for promoting community participation.
- Dissemination of information, specially by providing feedback on the pattern, prevalence, morbidity and mortality of water-associated vector-borne diseases which is a powerful tool to increase awareness of the community of the need for prevention.

42. Governments, institutions, members of the health professions, as well as all agencies involved in health and development will therefore have to take measures to enlighten the public in health matters so as to ensure that people can participate individually and collectively, as part of their right and duty, in the planning, implementation and control of activities for their health and related social development in the irrigation projects.

4. Summary and Conclusions

43. It is recognized that in irrigation projects, the irrigation system and its management effect not only agricultural productivity (income) but, also effect the well-being (health), social and economic relationships of the population, and their potential for further development.

44. In spite of the above planning, design, operation and monitoring of irrigation projects, generally undertaken by agencies and individuals with specialized expertise that focus mainly on technical and financial aspects of irrigation projects and there is rarely any understanding of the health and environmental impact of these types of projects.

45. In view of the previously described complexity of a water resource development in general and of irrigation projects in particular, multisectoral and multidisciplinary (holistic and/or comprehensive) approach must be encouraged. For implementation of this holistic approach a mechanism must be adopted and institutionalized. For institutionalization the following sequence of steps can be flexibly applied:
- As a part of national health policy a comprehensive policy for incorporation of health and environmental safeguards into irrigation projects at the national level must be formulated and adopted.
- Each country should identify, organize and develop an effective set of mechanisms, according to their needs and resources, for implementation of the policy including supportive legislative where necessary and useful.
- All the agencies and institutes involved in irrigation projects must be identified and a linkage must be established among them to strengthen or ensure intersectoral collaboration for achieving the project objectives. A national health council would be a suitable body to promote and strengthen the inter-sectoral collaboration.
- The communities settled in and around the project area are going to be affected economically (positively) and health impact (negatively) by the irrigation system. They, thus should be involved from the beginning as part of their rights and duties, in process of planning and implementation of the project.

46. Health sectors, as well as all agencies involved in water resource development projects, should take measures to inform the community in health and environmental hazards of the project so as to ensure their participation. In promoting community participation the greatest need is to organize mechanisms, at project level, to give people opportunities and capacities to share in decision making and providing support for those activities that they consider most important to them.
- Although the technology for health safeguards, related to vector-borne disease in irrigation projects exists i.e. environmental management, biological control, chemical control and treatment of the cases, the search for new and better, and more cost-effective methods should be encouraged through the intersectoral cooperation.
- With the interdisciplinary character of vector-borne disease management and control in irrigation schemes, there is a need for variety

of science, engineering and managerial skills. However, this does not imply the engagement of extra manpower is required but that environmental management should be added to the curricula of professional training of all categories of staff engaged in irrigation projects, or alternatively through organizing short in-service training courses where skills can be transferred and reoriented to cope with the prevalent problems in the projects. The role of training institutes should be defined and their collaboration obtained. Manpower development needs to be well-coordinated with the overall planning and development programme of the country. A network of relevant educational institutions are already available in many countries and regions of the world, mainly through associations of professional schools or through professional associations, linkage must be established with these associations to promote inclusion of health aspects of water resource development in their curricula.
- Certain activities require special international coordinating mechanisms to facilitate institutional linkage around specific subjects. Some of the activities are listed below:
  i. The promotion and coordination of research for the development of technologies for health and environmental safeguards in irrigation projects.
  ii. The promotion and coordination of manpower development for these types of projects.
  iii. The bringing together periodically project and institutional staff who are working on similar problems in order to exchange information and views to enhance the development of new technology and approaches for incorporation of health and environmental safeguards in irrigation projects. Through such gatherings, intersectoral cooperation could also be promoted
  iv. The dissemination of information through preparation of guidelines, technical reports, newsletters etc.

47. PEEM is currently trying to carry out the functions listed above at an international level and to facilitate institutional linkages.

48. To sum up, in planning irrigation projects, attention should be paid to their possible adverse effects on the health of the people. It is possible to some extent to forecast the health implications of a given irrigation project and to implement corrective measures through sound engineering practices. In view of the manifest lack of essential institutional arrangements for incorporation of health and environmental safeguards into irrigation projects, taking into consideration, prevailing sociopolitical and administrative set-ups, a mechanism must be established to bring together all the sectors involved and stimulate intersectoral coordination and collaboration both at the sector and project level. For this purpose an organizational structure with the necessary planning and managerial capacity is required. As the problem is a man-made one, man alone can solve it. The key to solving this problem is mutual understanding and cooperation among those who apply their skills in various sciences, those who apply them in engineering and those who occupy and utilize the irrigated land.

REFERENCES
1. AMIN M.A. Problems and effects of Schistosomiasis in Irrigation Schemes in the Sudan. In: Worthington, E.B. ed. pp. 407-411 (1977) Oxford Pergamon Press.
2. BAHAR R. Integrated vector control in the Blue Nile Health Project presented in WHO Expert Committee on Vector Control "Integrated vector control". Geneva 7-13 December (1982) (unpublished) pp. 1-9, and Review of Experience of the health impacts of water resource development projects in EMRO. EM/SEM. VBC/11 (1978) (unpublished) pp. 1-4.
3. BAHAR R. and KUO C. Prevention and control of water-associated diseases in connection with rural water supplies. Unpublished WHO document EM/SEM VBC/13.2, (1978) pp. 1-5.
4. Blue Nile Health Project. Annual report (1984), Wad Madani, Sudan, pp. 12-21.
5. FARID M.A. Irrigation and Malaria in Arid Lands. In: Worthington, E.B. ed., OP.cit. (1977), p. 416.
6. HUNTER J.M. et al. Disease prevention and control in water development schemes. WHO-PDP/80.1 (1980), p. 1.
7. KUO C. Health factor in water management. Proceedings of 12th Congress of ICID, Volume I(A) Q38R:G7 Fort Collins (1984), pp. 1067-1068.
8. LINCKLAEN ARRIENS W.T., OOMEN J.M.V., DE WOLF J. Elements of ideal institutional arrangements conducive to incorporation of health and environmental safeguards in water resource development projects. Document EPO/PE/WP/84.1 PEEM, WHO, pp. 1-47.
9. MCJUNKING F.E. Water and human health, U.S. Agency for International Development, Washington D.C. 1983. pp. 39-45.
10. PANT, C.P. Global importance of vector-borne diseases and their control. Presented WHO/DANIDA course on vector control. Shanghai, China 12-31 August (1985) (unpublished). p. 9.
11. RAFATJAH H. The Impact of Irrigation and Drainage on Public Health, proceedings ICID Special Meeting R.6 Moscow 1975, p. 105.
12. ROSENFIELD P.L. and BOWER B. Management Strategies for Mitigating Adverse Health Impacts of Water Resource Development Projects. Prog.Wat.Tech. Vol. 11, Nos. 1/2, Pergamon Press Ltd. (1979), p. 287.
13. Formulating strategies for health for all by the year 2000. WHO Health for All series No. 2 (1979), pp. 1-85.
14. Environmental management for vector control. Fourth report of WHO Expert Committee on vector biology and control. WHO technical report series No. 649 (1979), p. 16.
15. Biological control of vectors of diseases. Sixth report of the WHO Expert Committee on vector biology and control. WHO technical report series No. 679 (1982), pp. 1-39.
16. Manual on environmental management for mosquito control, with special emphasis on malaria vectors. WHO offset publication No. 66 (1982), pp. 1-283.

17. Integrated vector control. Seventh report of the WHO Expert Committee on vector biology and control. WHO technical report series No. 688 (1983), pp. 1-72.

18. Report of Interregional Consultation on National Health Development Networks (NHDN). Colombo, Sri Lanka, 15-19 November 1982. WHO SHS/83.2 (1983), pp. 13-15.

19. Malaria control as part of primary health care Report of a WHO study group. WHO technical report series No. 712 (1983), p. 11.

20. World Malaria Situation 1983, Malaria Action Programme, World Health Statistics quart. 38(1985), pp. 193-211.

21. Schistosomiasis Control: A Primary Health Care Approach (Unpublished document) WHO/SCHISTO/83.71, p. 1.

DR G. WANCHAI, Provisional Waterworks Authority, Bangkok, Thailand

17 Social aspects of water supply and sanitation

1. INTRODUCTION
Water supply and sanitation are generally regarded as social infrastructures. It is easy technically to design and standardize water supply and sanitation systems. However, to make them acceptable to and operable by the users is not a simple task, since knowledge, values, beliefs and practices vary from place to place and usually depend on social class, ethnic group, education background, tradition, etc. Moreover projects in developing countries, especially in small urban and rural areas, greatly depend upon cooperation from local leaders as well as from residents. The planners should, therefore, pay particular attention to the social and political structures of the community.

It is the intention of this paper to show the basic socio-cultural factors that must be accounted for during the project preparation stage to make the project successful. Furthermore, it shows some of the important benefits one can expect down the end of the Project Cycle when the project is operated. Since the author is working for a water supply agency in a developing country, most examples used herein reflect the experience in that area.

2. REQUIREMENTS FOR SUCCESS OF A PROJECT
As mentioned earlier, this paper emphasizes activities in small urban and rural areas, since the social aspects, particularly the socio-cultural factors, seem to be more pronounced than in large urban areas. These factors are the basic requirements if a project is to be successful and can primarily be grouped into three categories, which are: community needs, community participation and technology transfer. The discussions below elaborate on why and how these factors should be taken into account when designing a project.

2.1. Community needs
It was observed that the failure of most community projects is usually because they do not satisfy the real needs of the communities. The following points should be attended to in evaluating the needs.

2.1.1 Demography. The data on population size, average family size, growth rate and mobility as well as migration pattern should be accurately gathered, since they are the baseline data used in forecasting demands.

2.1.2 Health situation. Major health problems, especially those related to water/sanitation should be identified.

2.1.3 Occupation. What are the major occupations of the residents and is there any seasonality of employment?

2.1.4 Level of interest. The community interest in improving water supply/sanitation facilities should be compared to other potential improvements. Commitments from the leaders should also be obtained.

2.1.5 Physical structures. Types and conditions of dwellings and the layout should be recorded and sketched. This information will help in determining the water production or sanitation capacity and in designing the network.

2.1.6 Water use pattern and practices. Preferred sources of water and quantity used should be identified. Existing water supply facilities, if any, should be investigated and popularity determined.

Not recognising any one of those factors may cause failure. For example, for years the Thai authorities have invested hundreds of millions of baht in an attempt to provide villages in the northeast with "clean" deep-well water. To their surprise, the villagers never use this water for drinking as they regard it unhealthy, even to the extent of being poisonous. They would rather drink "milky" water from shallow wells that they are used to, even though they have to walk kilometres to fetch the water. This situation reflects provision of something to the communities that they do not need, right or wrong, and it will be neglected. Had the authorities tried to come up with a different approach, the program might have been successful.

2.2. Community participation
Once the needs have been identified, the mechanism of fulfilling the needs has to be carefully considered. To continue with the above example, should there be a need for water wells, community participation is the next vital step towards success of the project. Only through this process can the "sense of belonging" of the residents be developed. Often times when there is a minor difficulty with the well or the pump, it is abandoned, since the people feel it is the responsibility of the government, or the provider, to repair it and not theirs. They would rather wait without water than try to fix it. Had they been involved since the beginning, be it through labor, money, or material,

World Water '86. Thomas Telford Limited, London, 1986

they would feel that the well and the equipment belong to them and would try their best to maintain them. Usually, the following points have to be looked at when planning for community participation.

   2.2.1 Organization.
   (a) major local organizations and their memberships, so that they can be properly persuaded.
   (b) major local political or social factions which might affect participation.
   (c) important characteristics that would determine the acceptability and influence of outsiders.
   2.2.2 Local technology and resources available.
   (a) local availability of building materials.
   (b) availability of skilled and unskilled labors.
   2.2.3 Local attitudes, beliefs and associated practices.
   (a) general beliefs, important taboos, sharing habits, etc.
   (b) general household cleanliness
   2.2.4 Willingness and ability to pay.
   (a) household income
   (b) household expenditure.
   (c) borrowing and savings customs.
   (d) ownership of land and house.
   (e) availability of alternative sources.

The willingness and ability to pay may, by some authors, be treated separately. However, the author feels it is an important factor that determines whether or not the residents are willing to participate in the project.

2.3. Technology transfer

This step assures that the system installed will be properly operated and maintained. It happens quite often that complicated automation systems are used and no one in the community really understands how to operate them, or to make a minor repair in case of a breakdown. At least one operator should be trained to do the job. Moreover, the following points should be considered.

   2.3.1 Selection of technology. The designs should meet the needs of the users and yet be simple enough for local residents to understand. Of course, this will largely depend also on the education activities in the community and their potentials to be trained, which vary from community to community, country to country.
   2.3.2 Diffusion of knowledge and training. Usually this can be done through an appropriate organization structure such as a selected committee, or cooperative, or even mass media. Selected community members may be trained in facility maintenance, keeping accounts, collecting fees, etc.
   2.3.3 Motivation for adoption. Periodic follow-up visits should be established to discuss any problems that may have arisen and also to motivate the community to care for the facilities. Feedbacks can also be obtained from these visits and can be used in improving future projects.

3. BENEFITS DERIVED

Looking down the Project Cycle, one finds that once a water supply/sanitation project is operated, there is a string of benefits that can be expected. Only few will be cited here.

3.1. Increase in productivity

Direct increases can be observed both in urban and rural areas where water supply condition has been improved, since it is an essential factor for the development of industries. Indirect productivity may be obtained more in rural areas where people can save time that they normally have to spend on hauling water over a long distance. It was observed in Thailand that "women prefer raising vegetables and weaving, because it is not so boring and it lets them use their brains." It is clear that economic benefits will occur.

3.2. Improved public health and environment

Safe water for drinking and washing and appropriate sanitation practices reduce the spread of diarrhoeal diseases which are the most important single group of diseases throughout the developing world. As mentioned earlier, the effect of safe water supply or sanitation on community health depends on the extent to which the community makes use of them. Therefore, a program designed to change public attitudes has to be carried out in parallel.

Improved water supply and sanitation not only improve public health, but also the environment. Sufficient water supply encourages the use of water-sealed privies as well as permits household and public gardens, which beautify the community.

3.3. Improved political situation

There have been indications that reasonably successful economic development programs can bring about loyalty from villagers and, in turn, serve as counter insurgency measures. Since water supply and sanitation are basic factors underlying the success of these programs, the Provincial Waterworks Authority of Thailand, fully realising their importance, has launched many water supply projects along the northeastern borders. The results have been quite satisfactory. Villagers have noticeably better living and, therefore, it is less likely that they will turn against the government.

3.4. Reduced migration

Rapid migration into big cities could be a problem in developing countries, because it strains the social and economic infrastructure of the cities. Good water supply/sanitation facilities could be an important factor that reduces the "push" component of migration.

The Thai government has a strong intention to quickly develop regional cities so that migration into the capital, Bangkok, is minimised. It is the responsibility of the Provincial Waterworks Authority, then, to speed up its investments in those cities to ensure that at least they have access to enough clean water.

4. CONCLUSION

It has been illustrated through socio-cultural and socio-economic factors how to make a project successful and what can be expected from a successful project.

Requirements for success include accurate identification of community needs; good community participation; and full technology transfer.

Among the benefits derived from a successful project are: increase in productivity, directly or indirectly; improvements of public health and environments; improved political situation through reduced insurgency; and reduced migration into big cities.

REFERENCES

1. Dworkin, D. and Pillsbury, B., 1980. The Potable Water Project in Rural Thailand. USAID Project Impact Evaluation Project No. 3.
2. Glennie, Colin, 1982. A Model for the Development of a Self-help Water Supply Program. TAG/WP/01.
3. Kalbermatten, J. M., Julius, D. and Gunnerson, C. G., 1980. Appropriate Technology for Water Supply and Sanitation: A summary of Technical and Economic Options.
4. Provincial Waterworks Authority, 1986. Draft Sixth National Social and Economic Development Plan (1987-1991).
5. Simpson-Herbert, M., 1983. Method for Gathering Socio-cultural Data for Water Supply and Sanitation Projects. TAG Technical Note No. 1, UNDP Interregional Project INT/81/047.
6. Village Water Supply, 1976. A World Bank Paper.

Discussion on Papers 15-17

DR A. M. CAIRNCROSS, London School of Hygiene and Tropical Medicine, UK
With regard to Dr Ghooprasert's Paper on the social aspects of water supply, I would like to underline the importance of a subject mentioned in his presentation - that of local institutions. Much of the literature on community participation in water supply and sanitation neglects the fact that communities participate through their institutions. All human communities have some form of organization through which decisions are made and collective endeavours are planned, and this is what distinguishes a community from a simple group of people.

If rural communities are to participate in the construction or maintenance of water supplies or any other social infrastructure, their institutions must play a major role and will often need strengthening. What is needed is no less than the first step to setting up local government. There is often much to be learned, in this context, from the experience of traditional village institutions, such as those responsible for building and maintaining the rickety looking but practical bamboo channel water supplies of some mountain villages in Thailand.

A major problem confronting such bodies is how to collect cash contributions for water supply maintenance. This is relatively easy where people have house connections which can be disconnected if they default with their water rate payments, but where communal wells and stand-pipes are involved it is much more difficult, as it is harder to apply sanctions against those who do not pay up.

I recently read of a USAID-funded rural water project in Thailand, where the hand pumps of the first phase were neglected and often unused, while the house connections of the second phase were paid for by the villagers and widely popular. Does this reflect a user preference, or the possibility of more effective cost recovery?

Turning to the paper by Dr Gaddal and his colleagues, I endorse their emphasis on the quantity of water used. The quantities they mentioned are very large for people carrying water from pumps and stand-pipes: what percentage of the population had water piped to their houses?

As for the diarrhoea figures they present, my first reaction (as one who has also tried to measure the impact of water supply on diarrhoeal disease) was one of envy at the dramatic impact that they are able to report. However, are influences other than water supply also at work here?

Dr Bahar's Paper gives a comprehensive list of possible engineering interventions for vector control, and he is right to stress the need for training. I would like to support his call for more research. Some of the engineering interventions are more feasible than others; it is difficult for the engineer to increase flow velocities, for instance, when the land is flat. Some of them depend on particular species of vector, and may be excellent in some areas but disastrous in others. In addition, there are various measures involving changes in human behaviour which need to be tested and publicized. What are the thoughts of Dr Gaddal and his colleagues on some of these other interventions?

DR W. GHOOPRASERT
In reply to Mr Cairncross I would like to refer back to my Paper where it says that the Thai women would rather spend time on something else more productive than just hauling water. Perhaps it was this characteristic that made house connections more popular than mere public taps. Furthermore, if affordability is considered, it is found that the fees are not beyond the range that the people can pay. Good evidence is electricity meters that can be found at almost every house in the rural areas nowadays, through the rural electrification programme. Therefore, cost recovery schemes are not impossible in Thailand.

DR A. A. EL GADDAL
With regard to the percentage of the population with piped water, I did not mention that some villages have piped water to their houses. All villagers with a deep bore well can connect water to their houses through pipes whenever there is plenty of water. About 20% of the villagers in the Gezira have piped water connected to their houses. This number is increasing rapidly.

As for the diarrhoea figures, there are other factors, the most important of which is that of overstaffing during the construction of the well itself. The people themselves in the villages with the wells were informed that no cases of diarrhoea will take place if they only use water

PAPERS 15-17: DISCUSSION

from the well.

MR R. BAHAR
I agree that some engineering interventions are more feasible than others under specific environmental, geographical and topographical conditions and also depend on the prevalence of the particular vector species. This implies that for each irrigation project adequate studies and surveys must be carried out before, during and after the implementation of the project. On the basis of these studies the most suitable intervention must be recommended by the health specialist to planners, designers and operators of the project for their consideration and implementation. This procedure supports the statement in paragraph 2 of the Paper which reads, '... policy makers, planners, designers, constructors, managers, supervisors and operators who should consult and work in close collaboration with health specialists, from the project's inception through to its completion and throughout its operation'.

DR P. BOLTON, Hydraulics Research Ltd, UK
I should like to ask Dr Gaddal whether the pit latrines being installed under the Blue Nile health project are lined and whether there have been any problems because of their location in the black cotton clays.

It is disappointing to note that there are fewer irrigation engineers at this session than there have been at previous sessions on the irrigation theme. If the reason is not that they are apathetic towards health issues but that they cannot foresee anything being said which is of practical relevance then there are implications for the training of engineers. Engineers find it difficult to incorporate provision for health risks unless the parameters are defined quantitatively. To take as an example the control of aquatic snails by design of canals with suitably high velocities, it may be physically unrealistic to increase the mean velocity from 0.25 m/s to the 0.6 m/s recommended as necessary to dislodge snails completely but it might, at a cost, be possible to increase it to 0.3 m/s. The engineer may be asked the magnitude of the costs and benefits, whether it is realistic to quantify such effects at present or in the future or whether the human and biological systems involve such complex interactions that such interrelationships cannot be analysed in this way.

DR A. A. EL GADDAL
The pit latrines are lined because, as mentioned, the soil is a clay soil. This is quite expensive; we are trying other kinds without the lining but the concrete slab is placed on the side of the pit with a slanting pipe to the pit - this is still experimental. Fig. 1 shows the experimental pit latrine without a lining.

MR R. BAHAR
Cost effectiveness of environmental management

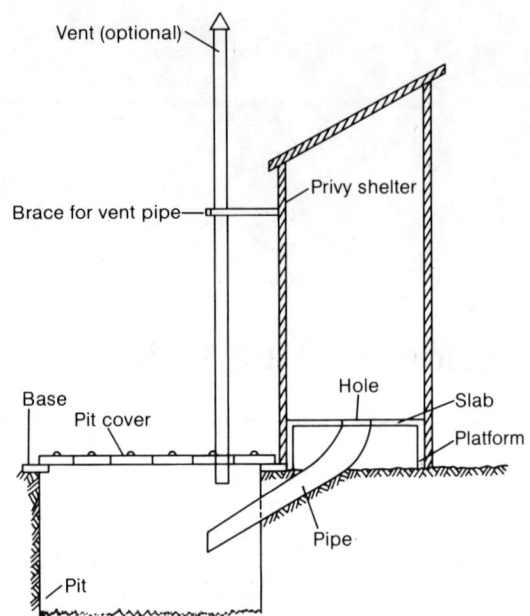

Fig. 1. Experimental pit latrine (side view)

measures applied by engineers for the prevention and control of vector-borne diseases is a major factor in its preference if the time frame of the analysis of cost and effectiveness is taken into consideration. Environmental management measures mostly produce long-term effects and thus, despite their initial high cost, may over a period of time be more cost effective than measures with short-lived effectiveness such as frequently repeated pesticide application.

Furthermore, to calculate the cost effectiveness of an irrigation project, the agency responsible should consider not only the impact of the project on agricultural productivity or an improvement in irrigation efficiency but other components as well. These other components include health, environment and the socio-economic status of the population affected by the implementation of the project.

Mills and Bradley (ref. 1) have discussed in detail a method to assess and evaluate cost effectiveness in vector control programmes. However, the parameters defined still require refinement and further study.

PROFESSOR M. J. HAMLIN, University of Birmingham, UK
The substance of the papers by Dr Gaddal and Dr Bahar were first presented by Professor MacDonald of the London School of Hygiene and Tropical Medicine in the James Forrest Lecture to the Institution of Civil Engineers in 1949 (ref. 2). The cost of engineering solutions although high must surely be outweighed by the disbenefit of the disease which has followed so many irrigation schemes constructed over the past 40 years. How can the disbenefit of disease be properly taken into account in cost-benefit analysis?

MR M. MULLER, National Institute for Physical Planning, Mozambique
In schistosomiasis control, the engineers are not the only professionals who should be

involved. Planners who determine infrastructure standards and the location of human settlements are arguably as important a target group.

MR R. BAHAR
It is true that in irrigation projects for disease vector control the engineers are not the only decision makers and that planners are also involved. Planning, implementation and operation of irrigation projects is interdisciplinary work, and policy makers, planners, engineers, builders, managers, supervisors and operators are involved, each in one or several stages of the project, and they must be aware of the health hazards of irrigation projects and the necessity of incorporating preventive measures in minimizing the possible health hazards.

IR E. ROOSMA, Provincial Waterworks of North Holland, The Netherlands
I totally agree with a progressively integrated approach to cope with the problems. The progressive assessment of water supply and irrigation engineering - formerly the only aspects dealt with - together with the aspects of health care and socio-economic factors is an important development of the approach.

The provision of food to all the people, water for all purposes, housing etc., in terms of quality as well as quantity is the main issue.

Referring to a statement of Dr Bahar that the problems to deal with are mainly man-made, this - in another sense - is the main cause of all the problems on earth. The fast-growing population - up to 5000 million people now - makes it very difficult to find the proper solutions and the possibilities for the needs in time. Mr Brown in his keynote address showed the disappointments mainly due to the unacceptable population growth. It is an almost impossible task to satisfy the needs of the quickly growing population.

The only way to escape from the downward spiral and vicious circle that mankind has entered is an emphasis on family planning all over the globe. It is better to have a smaller quantity of people and a higher quality of life.

This demands an even more integrated approach than already stated. Only with full integration at global, country, regional and local level of the various disciplines, including politics, can the problems be solved on a long-term basis.

MR C. B. BUCKLEY, Welsh Water, UK
The intricate nature of many of the problems which relate to the provision of a safe water supply and a suitable source of irrigation water leads to the requirement for a corporate approach to the provision and management of such schemes. There are public health scientists as well as public health engineers; planners and economists must also contribute to schemes at an early stage. It was disappointing to hear Professor Bradley refer to the 'tame geologist'. The answer received only equates to the quality of the question. It is more appropriate and beneficial to have a range of disciplines available on the corporate team, so that each discipline can contribute without the potential and in some cases real disadvantage of viewing the problem through the eyes of another discipline. This can lead to a distorted view and a less than satisfactory solution to the problem, in partly technical and economic terms.

MR J. HENNESSY, Sir Alexander Gibb & Partners, UK
Irrigation planning and development is a multidisciplinary activity and the substantial health/social component of World Water '86 is therefore particularly welcome. For many years past in the UK a number of ad hoc health/irrigation aspects meetings have been held but improved dissemination is needed for benefits to be achieved. Accordingly the British National Committee of the International Commission on Irrigation and Drainage and the Institution of Civil Engineers are trying with the Institution of Public Health Engineers to arrange (at least) an annual joint meeting with the medical specialists where such vital matters may continue to be addressed. Otherwise the British National Committee of the International Commission on Irrigation and Drainage is always ready to receive reciprocal approaches from the health/social sectors.

MR R. BAHAR
The recommendation to strengthen intersectoral co-operation in the field of multi-disciplinary irrigation projects is one of the principal messages which was delivered in the Paper. It is encouraging to note that the feeling is mutual between irrigation engineers and health specialists. In this connection I support Mr Hennessy's proposal for the organization of a periodic joint meeting between civil engineers, public health engineers and other health specialists to promote interdisciplinary and intersectoral co-operation in the field of irrigation projects.

At an international level the World Health Organisation has always participated in meetings and congresses organized by the International Commission on Irrigation and Drainage. The same type of collaboration between different organization, institutes and agencies involved in irrigation projects must be initiated at a national level.

MR R. CADWALLADER, Overseas Development Administration, UK
I appreciate the health problem caused by engineering projects as I suffered from schistosomiasis for 9 years from Lake Kariba without being aware what was wrong with me.

Having heard what has been said about the problems of schistosomiasis at Lake Volta and the environmental ways of reducing the problem I propose that the studies to be carried out on a power project for Ghana now being considered by the Overseas Development Administration should include a consideration of the possible health benefits of different reservoir operating regimes.

PAPERS 15-17: DISCUSSION

MR A. CHURCHILL, World Bank, USA
In practice can any health benefits be demonstrated from engineering and health interventions? Social variables - education and income - may be more important.

MR A. M. CAIRNCROSS, London School of Hygiene and Tropical Medicine, UK
Mr Churchill's disheartening conclusion should not lead us to think that water supply brings no benefit to health, and there are several reasons why.

First, the negative findings of many published studies reflect more than anything the enormous difficulty of measuring a health impact and attributing it to a specific intervention. Most water supply health impact studies are riddled with methodological flaws (ref. 3), and there are very few in which an epidemiologist would put much faith. To give just one example, in many of the studies reaching negative findings, no-one checked the degree to which the water supplies were functioning and being used.

Secondly many of these studies were carried out under a misconception that health benefits would arise principally from improved water quality, rather than from the increased use of water for hygiene which better access to water makes possible. The emphasis on quality arises partly from the historic and traumatic epidemics of water-borne disease which have sometimes been caused by faulty water supplies, but there is an increasing amount of evidence to suggest that in contrast with those epidemics, the endemic, paediatric diarrhoeas of poor communities in developing countries are more water-washed than water-borne, and more closely related to water quantity than quality.

Many of the studies of diarrhoeal disease to which Mr Churchill refers were carried out in circumstances where people's access to water had not been improved, or not improved sufficiently to cause an improvement in their hygiene, so that no impact on water-washed disease was to be expected in the first place.

Thirdly, there are health impacts which until recently no-one had looked for, such as on trachoma, the blinding eye disease that is common in many arid areas. Several recent studies have shown that the water supply has a powerful impact in controlling this infection (refs 4 and 5).

Finally, Mr Churchill did not mention another benefit of water supply which he has studied closely. His analysis, which I hope the World Bank will soon publish, has shown how, in many practical contexts, investment in water supply is more than justified by the value of the time saved from the chore of collecting water. At least one published study has shown that, if mothers have more time for child care, they will have healthier, better-nourished children (ref. 6). The saving in women's time which water supply can permit means a substantial transformation in people's lives, and if good health is defined as 'being able to finish a day without feeling exhausted', it can be considered a health benefit in itself.

MR R. BAHAR
Mr Cairncross has adequately covered the points raised by Professor Hamlin and Mr Churchill, but I should like to add refs 7-10 to his list.

REFERENCES
1. MILLS A. J. and BRADLEY D. J. Method to assess and evaluate cost-effectiveness in vector control. Environmental Management for Vector Control, 6th Annual Meeting of the Joint World Health Organisation/Food and Agriculture Organisation/United Nations Environmental Program Panel of Experts. 1986, PMO/PE/WP/86.8, unpublished.
2. MACDONALD G. The interdependence of medical science and engineering, with special reference to public health. Journal of the Institution of Civil Engineers, 1949, vol. 32, Oct., 398-416.
3. BLUM D. and FEACHEM R. Measuring the impact of water supply and sanitation investments on diarrhoeal diseases: problems of methodology. International Journal of Epidemiology, 1983, vol. 12, No. 3, 357-365.
4. CAIRNCROSS S. and CLIFF J. Water use and health in Mueda, Mozambique. Transactions of the Royal Society of Tropical Medicine and Hygiene, to be published.
5. KEYRAN-LARIJANI E. et al. Epidemiology of trachoma in the lower Shire valley, Malawi. Archives of Opthalmology, to be published.
6. POPKIN B. and SOLON F. Income, time, the working mother and child nutriture. Journal of Tropical Paediatrics and Environmental Child Health, 1976, vol. 22, 155-156.
7. AMIN M. A. Problems and effects of schistosomiasis in irrigation schemes in the Sudan (ed. E. B. Worthington). Pergamon, Oxford, 1977, 407-411.
8. Annual report on the Blue Nile health project, Wad Madani, Sudan. 1984, 12-21.
9. McJUNKING F. E. Water and human health. US Agency for International Development, Washington, DC, 1983, 39-45.
10. ROSENFIELD P. L. and BOWER B. Management strategies for mitigating adverse health impacts of water resource development projects. Progress in Water Technology, 1979, vol. 11, No. 1/2, 287.

S. KOOTTATEP, S. KARNCHANAWONG,
S. WATTANAJIRA and V. KITJANAPANICH,
Department of Environmental Engineering,
Chiangmai University, Thailand

TN25. The natural disinfection of rural water supply by solar irradiation

Safe drinking water means disease-free water and can be obtained through disinfection. Conventional disinfection (chlorination, ozonation etc.) is not appropriate in rural areas because of its complexity. Natural disinfection by solar irradiation may be an alternative. Disinfection by solar irradiation occurs in solar stills due to a rise in temperature and/or ultraviolet irradiation (ref. 1). Recent reports from Israel (ref. 2) and Beirut (ref. 3) state the possibility of using sunlight to disinfect water. However, it is still necessary to develop more knowledge on natural disinfection. This study investigates the effect of solar irradiation on disinfection of drinking water from laboratory scale models with a view to implementation in the rural areas of Thailand.

EXPERIMENT AND METHODOLOGY
2. Three solar plates were constructed to investigate the possibility of using solar disinfection for rural water supply. The first plate was constructed using copper tube 19 m long with a diameter of 9 mm. The second and third plates were constructed using glass tubes with a diameter of 10 mm and total length of 23.2 m, and ordinary steel water conduit 11.7 m long with a diameter of 12 mm. Copper tube and steel conduit were bent as a zigzag curve on steel plates which were insulated by glass fibre. Glass tubes were connected by using silicone rubber tube to form a continuous tube on a similar plate. Water flow through solar plates by gravity with a constant head and flow rate could be adjusted by ordinary water valve. Treated water was kept in a traditional clay jar and stored for drinking.
3. Water from three different villages was collected and transferred to the laboratory for experiments. Water was put into two 150 litre containers: the upper one was a raw water reservoir and the lower one was a constant head water reservoir. A number of experiments were performed by using well water from the villages which was moderately polluted by faecal coliform and by using well water plus domestic waste. Three detention times of each unit and each type of raw water were studied. Temperatures were recorded by using a data-logger at the inlet point, the outlet point and in the storage jar. Raw and treated water samples were taken for bacteriological analysis. Total coliform, faecal coliform and standard plate count were measured using a standard method procedure (ref. 4). Water flow rate was measured by using a stop-watch and glass cylinder.

RESULTS AND DISCUSSION
Disinfection by solar irradiation
4. The results show that copper tube plate shows the most significant disinfection effect. From 11 a.m. to 3 p.m. on a clear day, for a detention time longer than 15 minutes, the temperature of the water was always higher than $60^{\circ}C$. The maximum temperature ever obtained by this unit was about $70^{\circ}C$ even at 8 minutes' detention time. However, some overcast could reduce the water temperature to a certain extent. At temperatures higher than $60^{\circ}C$ and with well water, all faecal coliform could be killed all the time, but when waste water was added, all faecal coliform could be removed only 80% of the time. The results show that water polluted by domestic waste could not be disinfected efficiently by solar irradiation, but natural water such as well water in which the number of faecal coliform is not so high could be disinfected efficiently.
5. An attempt to use the ultraviolet effect on disinfection was performed by using glass tube as the solar plate. The maximum temperature obtained was about $50^{\circ}C$. The results show that, with a detention time longer than 20 minutes and natural well water all faecal coliform could be removed between noon and 3 p.m. on a clear day. With a detention time shorter than 20 minutes it is not possible to obtain satisfactory disinfection. However, the long-term effect of dirt collected on the solar plate will decrease disinfection.
6. Steel conduit solar plate did not show effective disinfection during the experiments. This was because the heat absorption was not adequate. The maximum temperature obtained was about $58^{\circ}C$ and the average temperature $50^{\circ}C$.

Cooling of heated water by earthenware jar
7. When treated water was stored in a plastic tank, its temperature was about $50-55^{\circ}C$. When it was stored in a traditional earthenware jar, the temperature reduced dramatically. The maximum temperature in the jar was about $35^{\circ}C$ at 4 p.m.; the water could be cooled down to $22^{\circ}C$ by midnight and to $16^{\circ}C$ by 6 a.m. on the following day.

Cost estimation
8. A cost estimation was performed for

copper solar plate and a detention time of 20 minutes, assuming the unit to be used for a family of six with a water consumption of two litres per person per day, the unit being operated only three hours a day (noon to 3 p.m.). The unit was assumed to consist of 20 m of copper tube, one steel plate with glass fibre, 4 m of steel frame, one glass plate, one bamboo holding structure, one 50 litre container and one clay jar. The total estimated material cost was about 1000 bahts per set.* The average family incomes of the three villages surveyed were about 11 169 bahts/year, 26 411 bahts/year and 40 266 bahts/year with average family expenses of 8405 bahts/year, 23 398 bahts/year and 29 376 bahts/year respectively.

CONCLUSIONS

9. It is believed that solar disinfection could be used to obtain safe drinking water in rural areas in developing countries. The system should be used on dry sunny days using well water as raw water.

ACKNOWLEDGEMENT

10. The project was financed by the International Development Research Centre of Canada.

REFERENCES

1. KOOTTATEP S. and C. TEYOON P. Solar still: drinking water for the rural. Paper presented at Regional Workshop on Limnology and Water Resources Management in Development Countries of Asia and Pacific, Kuala Lumpur, Malaysia, 1982.
2. GROSS L. World environment report, p. 85, 30 May 1984.
3. BINGHAM A. World Water, 1985, Apr., 22-23.
4. APHA. Standard methods for the examination of water and waste water, 1975.

---

*37 bahts ≃ £1 sterling.

P. BOLTON, MA, PhD, MICE, and
A. J. DRAPER, BA, MSc, MICE, Hydraulics Research Ltd, Wallingford, UK

TN26. Investigation of the effectiveness of engineering methods for schistosomiasis control in a small holder irrigation scheme in Zimbabwe

There is a clear link between the expansion of irrigated land and increased transmission of the parasitic disease schistosomiasis in many parts of Africa. Enhanced transmission results both from an increase in water contact by the human population and from an increase in the number of water bodies which form suitable habitats for the snail vectors of the disease.

2. Altlhough considerable progress has been made in capabilities to treat the disease and to control the aquatic snails with molluscicides, it is still widely held that a major contribution to the control of schistosomiasis could be made through the careful design and operation of irrigation projects, particularly for non-commercial schemes. The literature contains many suggestions about suitable control measures but few examples exist where they have been systematicaly applied and assessed.

3. A pilot project funded by the UK Overseas Development Administration is under way in Zimbabwe in which the Overseas Development Unit of Hydraulics Research is collaborating with two Zimbabwean agencies: the Ministry of Lands, Agriculture and Rural Resettlement and the Blair Research Laboratory (Ministry of Health). A 600 ha irrigation scheme near Masvingo, which will accommodate approximately 400 small holder families, is being designed and constructed according to criteria selected for the control of schistosomiasis. Certain constraints on the options for control arise from the fact that this is not a new scheme but rather the rehabilitation of former irrigated commercial farms. However, since these are being subdivided into small holdings, most of the in-field works will be new. Irrigation will begin on the first 60 ha in mid 1986.

4. The establishment of snail colonies and the transmission of disease will be monitored for several years. If the control measures prove successful, a set of guidelines will be produced for the design and operation of similar schemes elsewhere in Zimbabwe.

5. The engineering methods available for the control of schistosomiasis may be divided into those which relate specifically to the irrigation system and those which relate to public health. The first category includes measures which generally have as their objective the elimination of snail habitats, whereas the second includes measures which are directed at reducing contamination of water bodies and reducing human water contact. Taking first the snail control measures, an outline of those selected as suitable for the pilot project in Zimbabwe is given below.

CANAL LINING

6. The lining of canals down to field level with concrete is accepted practice in small holder schemes in Zimbabwe to conserve water and assist distribution. Well constructed and maintained concrete canals discourage snails by eliminating vegetation, allowing high water velocities and allowing rapid and complete drying when not in use. Concrete canals also facilitate the use of syphons for irrigation, resulting in better water management and reduced water contact by the farmers. Again, syphon irrigation is already the established practice in many schemes in Zimbabwe.

HYDRAULIC STRUCTURES

7. Where possible weirs, stilling basins, inverted syphons and other types of structure which retain water have been eliminated in order to produce a free-draining system. In addition, a new canal off-take structure is being tested to replace the standard sluice gate which usually results in water seeping down canals when they are not in use.

STORAGE RESERVOIRS

8. It is usually preferable to avoid reservoirs close to human activity but in this case the system which is being rehabilitated relies on night storage dams close to each irrigation block. An assessment is therefore being made of the effects of different reservoir operating policies on snail colonization. In particular, certain key reservoirs will be divided so that one part of the storage can be closed down and drained during periods of reduced demand, at the same time allowing maximum water level fluctuations to be achieved in the other. Where irrigation blocks are supplied by several night storage dams such subdivision is unnecessary provided that bypass canals are constructed to allow dams to be used or drained at will. There are potental dangers in drying out simple bunded storage reservoirs due to the development of seepage cracks, but in the pilot project this does not appear to be a major problem because the clay content of the bunds is relatively low.

WATER SCHEDULE

9. The key to the success of the project will, it is believed, lie in selecting a system of water scheduling, acceptable to the farmers, which requires only small parts of the distribution system to be in operation at any time, so allowing parts of the canal system to dry out regularly. To achieve this, the irrigation layout, canal sizes and structures and reservoir capacities must all be designed and sized to allow rotation of supply rather than continuous flow.

CONCLUSION

10. These measures have not introduced substantial cost increases over conventional small holder schemes in Zimbabwe. However, increased care is demanded in the design, construction and maintenance of the scheme and in operating it according to the planned irrigation schedule. This requires close co-operation between the different groups and agencies involved in establishing and operating the project. Thus the practicalities of implementing the measures as much as the measures themselves are being assessed in the project.

11. The public health measures for schistosomiasis control are no less important than those specific to the irrigation system. Contamination of water bodies will be reduced by ensuring that there are simple pit latrines in the villages for each household as well as some latrines in the fields. Human contact with potential transmission sites will be curtailed through the provision of safe alternative water supplies (from groundwater) and through the careful location of villages so that inhabitants are closer to a safe supply than to any potentially contaminated alternative. Properly drained laundry slabs are also being constructed. Public participation and education are vital components of such control measures. In planning the public health measures, the project is benefiting from the considerable knowledge already available within Zimbabwe.

R. F. LOVERIDGE, J. E. BUTLER, BSc, PhD, and D. A. BONE, Portsmouth Polytechnic, UK

TN27. Hydroponic use of treated sewage effluent and waste waters for crop irrigation

Portsmouth Polytechnic Water and Waste Research Group has since 1976 been involved in projects designed to improve the treatment, reuse and utilization of sewage. These have included novel irrigation systems using hydroponics. The aim is to stimulate agriculture, especially in areas of low or poor quality water resource and to reduce or prevent the pollution potential of treated sewage effluent discharges.

HYDROPONIC TECHNIQUES

2. Water culture has been practised since Roman times with the term hydroponics literally meaning 'water-working'. Better understanding of nutrition and modern materials has allowed outstanding developments with these techniques. The main benefit of hydroponics is the extremely efficient use of the water element, with usually only plant processes of respiration and transpiration accounting for any water losses. This contrasts with most conventional soil irrigation watering to many times that actually needed by the crop. The nutrient film technique (NFT) was used for protected crops such as tomatoes (ref. 1), but subsequently was modified to a simpler system (ref. 2). Gravel bed hydroponics (GBH) were developed to enable root and field crops to be grown (ref. 3).

NUTRIENT FILM TECHNIQUE

3. Devised by Dr Alan Cooper, the nutrient film technique consists of a number of sloping watertight gullies (gradient around 1 in 70); a nutrient solution is pumped to the top of the channels. In a thin film it flows down, past the roots of plants standing in the gullies, to a catchment area tank at the bottom. The solution is then pumped to the top again. There is provision in the catchment tank for the monitoring and replenishment of water and nutrients and for the correction of acidity. The system can be automated and is a closed system.

4. The drawback found when using treated sewage effluent to replace the water in the NFT was that some elements were concentrated in a closed recirculating system, which in turn produced toxic effects and locked up essential plant nutrients making them unavailable to the crop. To overcome this the basic NFT layout was used but made simpler and non-recirculating, fed in at one end and discharged or reused at the other. A wide range of crops has been grown in this once-through NFT although the principal experimental crop has been tomatoes. This was for comparative purposes and to capitalize on the wealth of data on growth, nutrition and disorders available.

GRAVEL-BED HYDROPONICS

5. A system based on the NFT had great potential but a field system was needed which could be constructed out of doors and produce a wider range of crops, including root crops. Watertight channels, deeper than for the NFT to around 12.5 cm, sloping at 1 in 70, were constructed and filled with siliceous rock. The depth of sewage effluent was maintained as a thin film a few millimetres deep.

6. To maximize nutrient removal and reduce phytotoxic build-up of elements, the effluent was again used on a once-through basis. Sugar beet was the principal research crop, having a large tap root and versatility in use, producing sugar, fodder, alcohol and bioenergy products. Support and aeration seem adequate and a number of useful crops have been grown in this system: producers of food fodder, pharmaceuticals and oils.

HEALTH ASPECTS OF HYDROPONIC SEWAGE EFFLUENT USE

7. With any method of sewage effluent or waste water reuse a prime consideration must be the health aspects. Waste waters and effluents vary greatly in their composition and have a potential for virus, bacteria and heavy metal contamination or concentration (refs 4 and 5). Many of these hazards can be avoided by primary and secondary treatment (ref. 6) by using the physical barrier of plant tissues to prevent adsorption and translocation to aerial portions (ref. 7), choosing a crop which involves processing to kill pathogens or in a system of indirect human use.

NUTRITIONAL BALANCE

8. Crop nutritional needs vary and to maximize the use of the nutrients inherent in waste waters deficiencies need to be corrected. To prevent further pollution potential foliar feeding and root block supplements have been employed as well as pH adjustments using acid to optimize dissolved nutrient availability to the crop. These methods can be very cost effective.

FUTURE DEVELOPMENTS

9. Work in the UK has been augmented by collaborative studies in the USA on crop-based primary and secondary effluent reuse and treatment systems.

ACKNOWLEDGEMENTS

10. The Authors would like to thank Portsmouth Polytechnic, the Southern Water Authority, the Science and Engineering Research Council, the Ministry of Agriculture, and in the USA, Florida International University and the National Science Foundation, for their contributions towards this research.

REFERENCES

1. COOPER A.J. The ABC or NFT. Grower Books, 1979.
2. LOVERIDGE R.F. Crop production in sewage effluent. Proceedings 6th International Congress on Soilless Culture, 1984, 329-338.
3. BONE D.A. et al. Sugar from sewage. British Sugar Beet Review, 1981, vol. 49, No. 4, 42-44.
4. WORLD HEALTH ORGANISATION. Reuse of effluents: methods of waste water treatment and health safeguards. Technical Report Series, 1973, No. 517.
5. WORLD HEALTH ORGANIZATION. Human viruses in water, wastewater and soil. Technical Report Series, 1979, Report 639.
6. NELLOR M.H. et al. Health effects study final report. County Sanitation Districts of Los Angeles County, 1984.
7. WARD R.L. and MAHLER R.J. Uptake of bacteriophage f2 through the plant roots. Applied Environmental Microbiology, 1982.

J. M. HEALEY, BSc(Econ), PhD, and
J. B. WILMSHURST, BA(Econ), Overseas
Development Administration, UK

18 Financing development projects

The water sector in less developed countries is characterised by monopoly supply arrangements, usually in the public sector, and "non-tradeable" outputs. Financial viability and efficiency have proved difficult to achieve. Subsidies generally need to be limited although they may be justified for rural water supplies, on health and public welfare grounds. Aid donors now require clear financial objectives to be set for managers of water undertakings and this is generally desirable even though most finance for the water sector is generated domestically.

INTRODUCTION
1. This paper reviews the major financial issues facing a developing country government and an aid agency involved in the development of water resources. It lays no claim to originality but represents a summary of the state of the art in making financial decisions in this sector. A consistent theme is a concern with effective resource allocation and most of the reflections are prompted by the authors experience of aid donor involvement in water projects.
2. The main stages or phases of decision-making reviewed are:-
A. PUBLIC INVESTMENT IN THE WATER SECTOR: ALLOCATION ISSUES
B. DESIGN ISSUES
C. PRICING POLICIES
D. FINANCIAL OBJECTIVES OF WATER UNDERTAKINGS
E. FOREIGN SOURCES OF FINANCE FOR INVESTMENT: THE VIEWS OF DONORS

A. INVESTMENT IN THE WATER SECTOR: ALLOCATION
Economic Characteristics of the Water Sector
3. Most irrigation and drainage investment results in increased agricultural output and higher land values and hence the benefits permit the authorities to charge for the services. They can expect the investment to be self-financing. However practical problems do arise. In practice much major irrigation investment in Asia has been underutilized because of administrative or political problems and hence it has not always been as economically or financially productive as originally expected. The benefits may be widely diffused and not fully recognised by the beneficiaries. Economic viability may not be synonymous with "willingness to pay" and project sponsors may have collection difficulties. Benefits may accrue over a very long time span but the investment may have to be financed on short maturities. These problems are particularly acute for large-scale schemes but they should be surmountable in smaller schemes. For example, in tubewell irrigation the beneficiaries are readily identified and the costs of well-drilling, pumping equipment and running costs can be attributed and recovered.
4. Water supply and sanitation schemes are more a response to people's basic needs than an input for greater output. The main economic benefits are reduction in water-related disease and time saved in collecting water (Ref.1). Water supply and sanitation schemes can be financially viable - and there is a strong case for approaching them on this basis in urban areas - but for these schemes financing problems are often severe. Sponsors often believe that consumers (particularly households) resist paying for such benefits because they cannot identify ways to increase their incomes commensurately with costs. They may not have the capacity to pay and this raises ethical problems. The benefits of better health and more leisure are not obviously greater for those who can afford to pay than for those who cannot, so willingness to pay is not always the appropriate allocation criterion.

Economic Appraisal Issues and Resource Allocation
5. Economic appraisal methodology is now fairly standardised. It entails the valuation of whole life costs and benefits based on their "opportunity cost" or maximum value in their alternative uses. The stream of costs and benefits are discounted to their present values and compared. The discount rate used should reflect the value society places on future income/costs as compared with present income/costs and is normally thought to be revealed in interest rates. A central principle is that the discount rate should be the same across all sectors - there is no justification for, say, a lower rate for water supply and sanitation projects than for projects in the manufacturing sector (Ref. 2).
6. Contained within this outline are a number of problems. The whole process depends on forecasting so that major uncertainties are immediately introduced. Opportunity cost valuations are relatively uncontroversial

World Water '86. Thomas Telford Limited, London, 1986

for goods which are imported or exported (tradeables) but goods and services which are "non-tradeable" internationally present difficult problems when they are domestically traded in markets which are far from free. The water sector is a classic example of non-tradeable outputs and some of the controversies inherent in the willingness to pay approach have already been mentioned. Some goods and services are not exchanged in any markets: how should benefits such as disease prevention and time saved be valued? Capital markets are often fragmented and controlled so that interest rates may not be a simple indicator of society's assessment of time preference; and they also contain elements attributable to inflation and risk.

7. This recitation of the difficulties of valuing the inputs and outputs for investment does not imply that the cost/benefit framework approach to investment decisions should be abandoned. However, it has now become clear that despite all the refinements which have been made in economic appraisal methods for developing countries, the Achilles heel is the degree of uncertainty of outcome. Uncertainty means those likely outcomes for which there is no objective probability distribution so that no measures of central tendency can be deployed in the analysis. A satisfactory procedure for dealing with a situation with a high degree of novelty and hence uncertainty has yet to be devised. Its main elements are likely to make financing decisions more difficult because they will require initial financial commitments to situations where choice will be deferred, exploration will take place, and maximum flexibility of options will be desirable. Frankness about the most uncertain elements in a project will need to have precedence over the spurious accuracy of many past project proposals for financing.

8. Where marketed products are closely linked to water projects, as in the case of agricultural output from irrigation and drainage schemes, and projects are self-financing, economic appraisal and cost/benefit analysis are at least as useful as alternative criteria. The problems noted above (paragraph 6) remain but common solutions can be adopted across sectors so that, for example, the rate of return estimated for a mining project should be comparable to that for an irrigation project. There are too many potential sources of error for appraisal results to ever be accepted as more than relatively crude estimates and for this reason sensitivity testing is an essential part of any appraisal. But subject to that, economic appraisals should provide some guidance to government on both project priorities within a sector and resource allocations between sectors. If there is a persistent tendency for appraisals to show a lower (or higher) rate of return in one sector than in others then there must be a strong suggestion that investment resources should be reallocated. Similar results from systematic ex-post evaluation of investment should convey an even stronger message.

9. Serious problems arise when the output of the investment is not marketed and financial viability is ruled out. Cost-effectiveness analysis is helpful in this situation using whole life costings. This technique is commonly applied to different means and hence costs of achieving a given objective but in the context of water supply and sanitation projects in less developed countries (ldcs) this can mean adoption of common valuation assumptions and wider application. It can permit comparison of different projects against some common yard-stick (e.g. cost per household connected to mains water, cost per hour of travel saved in collecting water) to allow ranking of different projects as well as different options within the same project.

10. Cost-effectiveness analysis cannot tell us how much to allocate to water supply and sanitation overall, however, because any yard-sticks adopted have little relevance to other sectors. Even if values are attached to water supplied (by costing at the rates consumers pay in urban areas, for example) they cannot have the standing of valuations derived from sectors where output actually is marketed. Governments' decisions on the total resource allocation to water supply and sanitation must therefore be based on wider considerations than economic appraisal.

11. Irrigation and drainage can be regarded as essentially economic activities and we have suggested above that the total level of resources devoted to them over time should be substantially determined by their "profitability", or economic rate of return, as compared with other sectors. Political judgements are needed but these are more likely to relate to optimisation than to strategic resource allocation questions. For example, India has had two decades of heavy investment in major schemes and is currently concentrating on securing more efficient utilization of existing capacity while Pakistan is continuing with major schemes (at a cost of £500 million the Left Bank Outfall Drain in Sind is large by any standards). Sudan and Nigeria also have major schemes planned but many other countries in Africa are giving priority to quick-yielding schemes to alleviate pressing economic problems even though major irrigation projects can be economically attractive for the longer-term. In water supply and sanitation the level of costs which can realistically be recovered from consumers has an important bearing on resources allocated to the sector (see section D below) and this reason alone argues for minimisation of subsidies. In the long run there is no reason why the government should determine the total resource allocation to water supply and sanitation - private sector supply is under consideration in Britain, for example - but in the meantime governments' concern to meet the basic needs of their populations will determine the final share of these services in national resources. This is a largely political issue.

B. DESIGN ISSUES

12. Given the objective the main issues in design revolve around standards - capacity, accessibility and quality - and technical

options. Even in self-financing projects these issues are not easily determined because competition is not a common feature in the water sector. Typically there is a monopoly supplier, whether public or private, who is charged with design and cost-recovery. Consumers are seldom offered a choice. They must accept the service offered or refuse it and cannot trade-off price against quality as they do in most other purchases (food, clothes, travel, etc).

13. To ensure that the design of water technology reflects local economic conditions requires early awareness by the designers of the true (opportunity) costs of the various inputs. In the past, economic appraisal has often been brought to bear after the technology has already been fully determined and when it was too late to make significant modification. In many developing countries exchange rates have been chronically overvalued (a situation which is only beginning to change now). The result has been to artificially cheapen the cost of imported equipment and hence bias technology away from greater use of local goods and services. Domestic interest rates have frequently been kept too low in poor countries and have induced excessively capital-intensive technologies and opportunities for more labour-intensive methods may have been missed.

14. By now this is a familiar theme and there is no need to labour the two sides of it. Conventional engineering standards are not chosen lightly. They are the result of detailed research over many years and can be regarded as providing a short-cut to the costly testing which would be needed at individual sites in their absence. Moreover equipment manufacturers use these standards to provide "off-the-shelf" packages cheaply. The alternative view, now widely recognised, is that the resource endowments and needs of developed countries are different from those of ldcs. Standards derived from the former are unlikely to be appropriate in the latter and iterative processes are likely to rank technical solutions differently.

15. A final comment on design issues is that site-specific characteristics can be more important than general principles. Two examples from ODA experience illustrate this. An irrigation project undertaken with British aid in the late 1960s in the Middle East relied on centralised high speed diesel generator sets. Five years after project completion all four generator sets were inoperative due to poor maintenance and a lack of spare parts. Lower technology well-head diesel pumps had to be installed, even though this design is technically inferior, because it is better suited to local conditions. In the mid 1970s an alternative approach was adopted for water supply to villages in Lesotho. Three technical options were available; gravity fed spring supplies, wind pumped ground water or diesel pumped. The first method was more capital-intensive but it was chosen (after applying an opportunity cost discount rate) because it could be maintained by the villagers themselves. The other options needed more spare parts and skilled maintenance.

C. PRICING POLICY

16. Prices have a crucial role in market economies. In a pure free market system (which does not exist anywhere) questions of overall resource allocation, choice of project, optimal design, etc are settled automatically by profit maximisers responding to price signals. The important departures from this system in the water sector have been noted above: monopoly supply conditions and output which cannot be easily valued and priced. The significance of the external benefits from water supply and sanitation schemes - improved health in entire communities - is an important additional aspect in ldcs. Nevertheless the prices charged for the output of the sector are important determinants of all the main financing questions.

17. The economists' traditional rule for a monopoly public sector undertaking is that charges should cover the marginal cost of the water supplied. In this way there will be economically efficient allocation of resources to these water undertakings. If users are undercharged there will be excessive demand for water and too much investment compared to other sectors where users do pay the marginal cost of the service. In the long run water capacity can be adjusted to meet the level of demand and the rule requires that charges should cover long run marginal cost, i.e. the cost of supplying the marginal unit of water taking account of the operating costs and the estimated incremental capital cost of meeting this level of demand. In the short run, water capacity is fixed and the only relevant costs are those which vary with its degree of utilization. Charges should in principle be set to cover these short run marginal costs. This may mean quite high charges when fixed capacity is strained by high demand and needs to be rationed. The use of short run marginal cost principles is relevant for the day to day running of an enterprise while the long run marginal cost rule is relevant to the planning of investment and financial targets for the undertakings (see section D below).

18. The main pricing issues are:-
   a. how to determine marginal costs and set tariffs to parallel these; and
   b. how to determine what proportion of costs should be borne by subsidised consumers.

19. The reasons for charging consumers on marginal cost principles include equity as well as allocative efficiency. But charging on this basis itself imposes costs - meters, billing, revenue collection, etc. These costs are justified only if the financial advantages they bring are likely to be greater than the costs incurred, and this is more likely when recurrent costs are proportionately high or peak consumption can be deterred. To take a British example, the costs of domestic water supplies in the UK are dominated by capital costs and hitherto the economies which might be achieved by metering have been estimated to be less than metering costs (Ref. 3). As a result

domestic water supply and sewerage is still financed largely by charges related to rates on property rather than use of the services although this question is currently (1985/86) thought to merit further study.

20. There is a further argument against marginal cost chargings of individual consumers in rural water supply and sanitation schemes in ldcs because this may deter consumption and hence frustrate the health objectives of such projects. A pragmatic yardstick recommended (Ref. 1) is:-

"A level of per capita consumption should be specified which may be regarded as "adequate" for health purposes (e.g. a "basic need" of say 20 litres per day). Consumption above this threshold level is then classed as "excess" water, and therefore the benefits of metering are assessed as the savings in expenditure on water production only down to this threshold ...".

Similarly connection charges, the obvious means of recouping costs when capital charges are proportionately high, may need to be kept low in order to maximise uptake. Periodic fixed charges over a period of years may be preferable.

21. Even this approach can be difficult when provision is via standpipes, a solution consistent with our arguments above for appropriate standards. Standpipes rule out variable charges to individual households so that the choices for the water authority are either free provision of water or charging the entire community on some rough and ready basis. The latter solution was chosen in the Lesotho Village scheme discussed above. Households were required to pay at least 2 rand (about £1) to the water authority as a connection charge and at least 60% of households had to pay before the village was connected to piped supplies (standpipes). This is little more than a token charge, well below the cost of supply, but at least it established the principle that piped water is not a free good.

22. Pricing and cost recovery is generally simpler for irrigation and drainage schemes, but there can be significant differences between different types of scheme. The ODA is presently involved in two schemes in this sub-sector in South Asia with very different characteristics. The Bangladesh Deep Tubewells Project is a fairly large project costing £79m, and seeks to provide deep tubewell irrigation to village co-operatives. The drilling of the boreholes and provision of the pumping equipment will be supervised by the Bangladesh Agricultural Development Corporation (BADC), but the village co-ops will pay for the materials and services. To assist payment, credit will be made available to co-ops at market rates (14% per annum in 1982) with nine years maturity. The maintenance of the pumps will largely be left to the co-ops, but BADC will provide training and extension services.

23. In this scheme, cost recovery is fairly straightforward, as the co-operatives can be charged directly for the drilling service, pumping equipment, and fuel costs. However, in the Left Bank Outfall Drain project in Pakistan, where the major costs are for the construction of drainage channels and pipes, the costs will have to be met by the Pakistani Government, and a system of charges used to recover these costs. The drainage charges will be designed to cover operation and maintenance, and will commence at 25% of total operation and maintenance costs in the first year, rising every subsequent two years to ensure fuller cost recovery. The farmers will also be required to pay 25% of the material costs of the project and provide all the unskilled labour needed. This charging system has to reconcile the farmers' ability to pay with total cost recovery, in a project where the connection between costs and benefits may not be as obvious to the farmers as in a smaller-scale tubewell irrigation project.

24. These considerations do not justify national tariffs. The cost of water, drainage and sanitation varies from location to location and institutions should reflect this in their charges as far as possible. Obviously there are difficulties in distinguishing costs within an interconnected system but cost differentials between different towns and villages, and between different irrigation schemes, should be identifiable. In the past donors have tended to leave such details to ldc governments and institutions in aided projects but there is considerable evidence that this has been misguided. Administrative and technical weaknesses in local institutions, combined with vested interests (for example, the political power of local landlords), have resulted in tariffs which have an inappropriate structure and are too low. The implication is that "policy dialogue" should precede aid agreements and that technical assistance to overcome financial weakness should accompany it.

Subsidies

25. The thrust of the previous section is to suggest that charges in general should not be subsidized to the user. This rule is necessary for efficient allocation of resources. There is a further reason especially for public sector monopoly undertakings. There must be pressure to ensure efficient performance. If subsidies are implicit and not transparent there is little reason for the managers of undertakings to minimize waste and seek the most economical way of operating. They have no clear financial norms to pursue and to be judged against.

26. In practice subsidies are widespread in the water sector, with charges in water supply and sanitation schemes usually not covering even operating costs and full costs frequently not covered in irrigation and drainage schemes. Inflation has been a significant contributor to this undesirable state of affairs. Governments usually control prices in the water sector and they resist increasing prices in line with inflation for a variety of reasons: because water is regarded as a basic need and social considerations suggest that consumers should be protected; or because control of some prices is seen as an anti-

inflationary measure; or because water services have a high profile and all increases in their prices result in adverse reaction from urban consumers (domestic water) or politically powerful rural land-owners (irrigation water).

27. Even though most subsidies in the water sector are unjustified and excessive there are situations where they can be supported. It has been noted above that the services are often provided under monopoly conditions and this militates against efficiency. Subsidies may be justified to protect consumers, in the absence of alternative sources of supply, until cost control mechanisms are devised (see section D below). Another rationale for subsidies is that the economies of scale and external benefits (such as reduction in communicable disease) require subsidy of the service rather than, for example, welfare payments to individuals (even if these are a realistic possibility) on economic efficiency grounds.

28. The available evidence suggests that households cannot be expected to pay more than 4%-5% of their income (including income in kind) for water supply and sanitation. Given this the burden is thrown on designers to adopt standards which meet general objectives but stretch resources so that maximum coverage is obtained. One mechanism for achieving this is to insist that revenue covers operation and maintenance costs as a minimum: if this is not possible the scheme cannot be afforded.

D. FINANCIAL OBJECTIVES

29. One of the main problems which has emerged acutely in the developing world in the last 10 years is the failure of public undertakings to: (a) set themselves satisfactory financial objectives and establish adequate budgeting and control systems; (b) to recover their costs and make enough surplus to help finance new investment; and (c) to set and achieve satisfactory performance in terms of efficiency, management, manning etc. Undertakings in the water sector are no exception to this rule and indeed tend to lead the field!

30. The rules for pricing discussed in Section C above have implications for financial objectives as also do the project appraisal rules on new investment. The need is therefore to have a quantified objective, in financial terms, which applies to the whole enterprise and which can be a norm against which the performance of the enterprise can be monitored by central government and outsiders. This is an important requirement when a lot of parastatal undertakings have grown up with weak management and no clear direction on what they should be doing.

31. One way to reconcile marginal cost pricing rules (paragraph 17 above) and financial objectives is to express charges which cover long-run marginal costs in terms of a rate of return on total capital assets. LRMC includes the cost of new investment charged at the opportunity cost of capital in the public sector, the likely costs of operating the optimum amount of new investment (which is determined by expected demand on the basis of full cost pricing) and the cost of operating existing capacity. Properly calculated this yields a unique return on total assets employed that a public body should achieve. This will not be the same for all undertakings since it will depend inter alia on the current value of the existing stock of capital assets.

32. Despite its rationality there are a number of problems with this approach. First, it is not an easy system to understand and requires considerable analytical capacity to set it up. Second, much depends on whether valid and accurate (current) valuations of the capital stock are available otherwise the targets may be very misleading. Replacement values of capital assets are essential while very often only historical ones exist (if at all). Third, these financial targets on their own do not set the undertakings any explicit physical or productivity performance goals. The financial targets tend to be set on the basis of existing productivity levels but the undertakings may be inefficiently managed, overmanned and have poor practices and incentive systems. The financial targets need to be complemented by separate performance targets and monitored. It will also be appreciated that LRMC at opportunity costs in ldcs can depart significantly from nominal cost calculations. The financial target needs to reflect this.

33. In the UK public sector (including the water industry) there is an integrated set of financial targets, performance aims, external finance limits (in the sense of funds obtained from outside the undertaking) and required rates of return on new investments. Few ldcs have the administrative resources to operate an equivalent system and simpler disciplines are needed. One element in this is full charging for capital. Even if it is agreed that the service needs to be subsidised it makes no sense to do this by way of cheap (or free) capital since this will introduce the distortions noted above: it will encourage the use of capital as compared with other resources (notably labour) and it will favour investment against recurrent costs (for example, new invesstment as compared with repair and maintenance of existing facilities).

34. The performance aims now established in the UK public sector are one way to approach efficiency and value-for-money at the institutional level. These were introduced in the early 1980s and require water authorities to contain operating costs - defined as total costs before depreciation, exceptional items and interest - within cash limits. These imply productivity increases because they are supported by a battery of 25 objectives relating to quality of service and unit costs are set to rise below the rate of inflation. These performance aims are needed because a monopoly supplier can conform to all the economic principles of allocative efficiency outlined above yet remain inefficient by passing on costs to consumers.

35. These general financing principles apply to both self-financing projects and those that need to be subsidised. The corollary is that subsidies need to be provided to the latter as

transparent, block grants. This can seem bureaucratic: for example, debt service due to the government is paid and then returned to the water authority via the block grant. The advantages are those outlined above - resource costs are fully apparent at all levels - and government has tighter control of its finances. The burden of annual subventions is a spur to greater efficiency and in turn should assist institutions in replicating successful projects. We have noted above that the level of these annual subventions is a political decision but tariffs are an important and controversial factor.

36. In developing countries the various difficulties have led to the use of an alternative type of financial objective for public undertakings. This is the proportion of its investment that it should be obliged to finance from its own net earnings by a certain date. This implies adequate charging levels which cover more than operating costs in order to yield a surplus for re-investment. There may need to be supplementary and explicit targets for raising tariffs to ensure movement towards full cost charging as well as performance targets on productivity, management and staffing practices. The weakness of this approach is that it does not provide a guide for the appropriate level of investment. There may be a temptation to under-invest in order to meet the self-financing target rather than raising net earnings and investment to a satisfactory level.

E. FOREIGN SOURCES OF FINANCE

37. Non-concessional sources of foreign funds - notably the banks - generally look for evidence of financial viability in the projects and undertakings to which they make loans. This provides the collateral which is a normal condition of commercial lending (albeit that non-convertibility of most ldc currencies emasculates this security). Most water supply and sanitation projects and institutions are unable to fulfill this requirement and many irrigation and drainage undertakings are in the same position. Even with better management of the latter they would run into further problems with commercial lenders because water assets are typically long-lived. Maximum repayment periods for commercial loans may be of the order of 15 years whereas many water assets have lives of 60 years or more (dams, canals, embankments, underground piping, etc). Moreover the debt service problems of many ldcs today are such that banks are increasingly recognising that even undertakings which are viable in domestic currency terms may be bad risks.

38. Concessionary funds - aid - are therefore the only realistic source of foreign finance for investments in this sector. Currently availability is limited because (a) real aid flows are not growing significantly; and (b) more of these are being used for balance of payments support rather than financing new capacity. As an illustration, about 50% of British bilateral aid is currently for projects and only 4% of this goes to the water sector. The proportion is comparatively low for most other donors also. The majority of the resources available for the water sector are likely to be generated domestically in these circumstances.

39. Where aid is channelled in this direction donors are increasingly concerned with the issues disussed above. Much emphasis is placed on fuller utilisation of existing capacity, rehabilitation of this where required, adequate maintenance, and better cost recovery and appropriate charging for water services. Attempts are being made to help improve the local institutional capabilities and financial management improvements are often made a condition of the provision of finance. Achievement of targets are required before further tranches of finance are released.

REFERENCES
1. OVERSEAS DEVELOPMENT ADMINISTRATION. Manual for the Appraisal of Rural Water Supplies. HMSO 1985.
2. MINISTRY OF OVERSEAS DEVELOPMENT. A Guide to the Economic Appraisal of Projects in Developing Countries: HMSO 1977.
3. NATIONAL WATER COUNCIL. Charging Households for Water. 1980.

The views expressed in this paper are those of the authors and do not necessarily reflect the official policies of the ODA.

G. THARUN, DiplPol, South East Asia Programme Office, Asian Institute of Technology, Bangkok, Thailand

19 Project casework based training for better planning and management of water pollution control in developing countries

After briefly introducing training as medium for change and discussing its relevance for development efforts, water pollution control is analysed as a management problem and subsequent training task. The resulting development of project casework, based on the training model and its structure and function, is described in more detail. Overall assessment of the educational approach as such and of feedback reporting successful application of learned knowledge and skills, underlines its practicality, problem solving potential and high motivating factor as complementary professional learning exercises.

TRAINING AS MEDIUM FOR CHANGE

1. While people, especially at work, are more or less learning all the time, although often in a diffused and unsystematic way they might or might not be conscious of, training is a deliberately planned activity to canalise and activate their learning behaviour to meet specific ends. It is a systematic learning process designed to change people's job performance (ref. 1) to make them better able to carry out certain functions in their work which also means to make them more effective and efficient.

2. Training needs exist when performance standards are not met or newly introduced equipment or processes cannot be coped with sufficiently or when new tasks are assigned which require new knowledge, skills or attitudes. Training, as a systematic process or cycle to change behaviour, usually involves a number of steps which are also characteristic for any systematic decision making process, i.e. to:
   - identify training needs,
   - develop training objectives,
   - plan and design a training programme,
   - implement it,
   - evaluate training results.

The simpler a given job is, the easier it is to carry out these steps. It is for, instance, relatively easy in clearly defined manual work or if this task can be performed within a given organisation, be it a water supply undertaking or effluent treatment plant because staff's capabilities, in comparison to job requirements, can be described because they are known. It is much more difficult or, in real world training business at the international level, even impossible to do when actual job situations are not really known to the organizer of training, but can only be imagined. This is the case in most of our regional or international workshops, seminars and courses, especially with respect to environmental planners, managers and policy makers of different institutions.

3. While sound principles of effective training and its steps should be applied whenever possible, it is not always manageable. More realistic foresight and imaginative power have to help out to enable us to do the right thing with training. If end products of training are correctly defined as adequate performance of a specific job to be achieved at the right time in actual working situation (ref. 2), then certain questions relating to steps at the beginning of a training cycle, as well as afterwards, are to be left to qualified guesswork because exact job situations are usually not known to us and are quite diversified. Furthermore, actual job performance after training cannot readily be observed and measured with our means, but more indirectly and only occasionally. That it still works depends most probably on the right assessment of the overall problem situation in the highly complex field of water pollution control and management for developing countries. This is covered in more detail below.

RELEVANCE OF TRAINING IN THE DEVELOPMENT CONTEXT

4. In recent years, it has increasingly been stressed in national and international development assistance circles, that training plays an important role in the field of development cooperation. Although the funds for professional training have increased over the last decade, their portion of the total development aid cake, whether on national or international scale, still remains minimal if compared with the budgets allotted and channelled through financial or technical assistance schemes. The Personal Cooperation or Human Resources Development programmes might further increase their weight in the development policy considerations of developing and industrialised countries alike as well as of multinational and international bodies. However, some caution might be advisable as it seems that man tends to jump from one craze to another when over-excitement for one earlier development policy approach

fades away and gives way to another one. This latter is then uncritically looked upon as saviour with expectations that can hardly be fulfilled however reasonable and sound certain aspects of the new approach might be.

5. The training scene has been no exception to this. Whether it was the sudden drive to search for partners in developing countries in the mid-seventies or the German administration's desire to shift training venues from Western industrialised countries to the developing regions itself as much as possible or the basic needs as well as the alternate technology approaches, all the then attached importance and urgency to each of these policy shifts led one to wonder how policy makers sometimes tumble from one straw to another. All these turns and twists have more or less to be followed then by the administration and related agencies. One cannot but help to assume that a high degree of insecurity and naivety in professional standing of 'expert' advisors might also be responsible for these rushes. A more unbiased look at the different options seems to indicate that almost each of these has its validity if properly and cautiously applied to suitable circumstances and might play a meaningful role in combination with other measures to reach a certain end. Professional training can certainly become a viable instrument in attaining certain development goals if properly arranged and applied either alone or in concert with additional actions. However, as hardly any human activity can reasonably be perceived as a panacea, training is certainly no exemption to this simple truth which should be obvious enough, but still seems to be overlooked all too often, even (or above all (?)) by professional educators.

6. What matters in the course of human action, especially with regard to development intervention, is to choose and develop an appropriate activity or set of activities to reach or work at least towards common goals. It depends upon the interrelated web of circumstances, targets, available resources and existing constraints and, subsequently related sound judgement to decide whether training at all can provide any meaningful alternative. If this is the case, then the type of training has to be established and eventually, its linkage to other types of activities, e.g. respective R&D studies, consultancy services, political moves, information dissemination, etc. However, the big question is HOW to proceed because there are plentiful courses, seminars and conferences covering numerous subjects. The question remains what is their contribution to actual problem-solving, if any? (ref. 3) It has to be realised that standards are very different, even within professional organisations, for advanced training. This problem is even more complicated because many subject matter-specialists still believe that training organisation and continuing education skills are something almost everybody can acquire quite naturally overnight. This rather widespread mental blockade, which in its most serious form is coupled with academic ignorance and arrogance, makes the joint search for reasonable and relevant training solutions not an easy task.

7. If we perceive our own roles either as direct or indirect change agents and want to activate some improvement in the equilibrium of continuing socio-economic development on the one hand and a sustainable environmental resource base for it on the other, both of which together constitute the foundation for people's well-being, we have to act now with what we have and accomplish what we can (ref. 3). Therefore, professional educators have to be humble and reasonable enough in their approach to address certain development problems with the means at their disposal, namely, adequate training or, in a wider context, human resource development. Given that development problems have to be properly transformed into training problems, yet due to the aforementioned obstacles of a mainly perceptive nature, the way to do it will be debatable. An operational way out could be opting for PRACTICALITY and PROBLEM SOLVING CAPACITY of training as main guidelines for organisation of course work etc. because the programme participants as direct or indirect target groups will make the final judgement over the quality with regard to their professional problems.

8. Even this practice orientation does not offer any easy solution. Just a helping yet essential hand as pragmatic compass to navigate the course development and implementation through the numerous pros and cons of adequate action for appropriate training or what can be perceived as such. This organisational work of informal education and training of professional standards can be likened to incremental ways of muddling through in the cybernetic-systemic sense (ref. 4). Despite all the experience gathered, and expertise derived from and applied to it over the years, no one single training project turns out to be the same because of the countless changing factors involved of which many are, and will always be, beyond our control, especially in special management training courses for environmental planners and decision-makers from several developing countries of a particular sub-region, for instance South East Asia. In the author's case, these constraints can range from certain - often inadequate - administrative procedures in sponsoring agency as well as nominating authorities in developing countries over administrators' mis-conception of necessary quality-input in terms of allocated budget, time, personnel etc. to differing priorities and attitudes of partner institutions and respective communication problems, just to mention a few.

9. The higher the quality demands on the organiser's side, the more intense the necessary coordinating requirements and professional input which make the work itself much more challenging and rewarding. At the same time, it will carry much higher risks with respect to the principally unknown outcome before the exercise is over. The professional participant is the client and as such to be treated like a king with respect to his or her professional requirements and justified expectations. With this underlying

principle coupled with maximum problem solving orientation and practicality - both for the organiser and the ultimate target group - as a working concept for advanced training for developing countries, at least the overall direction in a systemic sense seems to be right if judged from the feedback received so far. Such a guiding principle for senior level training and human resource development is only a general set of formula. At that level, there is no detailed cookbook at hand from which the recipes could blindly be copied.

WATER POLLUTION CONTROL AS MANAGEMENT TRAINING PROBLEM

10. A Thai commercial calls water the source of life and it is for many Thais and other South East Asian people. Water is an essential part of their culture, their mode of living and feeling. Water is the element which permeates their life both as a natural resource and as a hazard. According to its usage it has been classified as an important economic resource and as basic human need (ref. 5). To great deal still considered as a free or common good by man, water has been widely exploited and misused and has only recently reached dangerous proportions in so far as the natural regenerating capacity of water bodies has been threatened by a number of different actions due to their proportion and intensity. Water resources are not only central to life-support systems, but are also essential input in most of our economic activities (ref. 6), be it for transportation, fishing, irrigation, recreation and tourism, water supply, industrial requirements for cooling and effluent diluting purposes, hydropower, etc. The highly interwoven relationship of the hydrosphere and its eco-systems with the other three spheres of the total environment, the atmosphere, the geosphere and the biosphere, makes it also highly vulnerable to pollution and destructive impacts taking place in these other spheres. As is widely known, deforestation leads to serious soil erosion, destruction of groundwater resources, flooding and siltation with further negative effects on fish farming, change in climates etc. Another example is air pollution affecting water quality and thereby marine life and public health often far from the source of pollution. Additionally, big infrastructure development projects such as huge reservoirs and dams can have serious side-effects due to water-borne diseases etc.

11. The intricate network of the hydrosphere within the total environment and its eco-systems with numerous interdependent cause-and-effect relationships on the one hand and manifold use of water resources by man and the highly complex relationships of involved agencies, industries and private households as consumers and polluters of water and other natural resources on the other makes the achievement of necessary solutions to water resource protection, development and wise utilization a foremost management problem. To achieve reasonable results in controlling deteriorating situations due to over-exploitation and depletion of resources as well as over-saturation and destruction of carrier function of eco-systems with regard to waste matter absorption and dilution and the combined impacts of both is an almost insurmountable complex task due to degrees of urgency, extent, intensity and cross-sectional nature of problems.

12. What is needed is not only to reconcile conflicting interests of social groups and nations but also to coordinate efforts of different institutions, organisations and sectors' multidisciplinary expertise, opinions and interests for achievement of common goals. Practical plans have to be worked out and implemented addressing the most serious problems at the right time, at their source, if possible. This means, above all, to be a management task of developing and pulling together reasonable objectives based on sound assessment of problem situation, the work on possible strategic options for action, the selection of most suitable, affordable and acceptable sets of action plan and timely implementation and enforcement and monitoring and evaluation. Figure 1 illustrates the key steps in development of an environmental strategy either as a whole or related to field of medium in question, i.e. water, soil, air etc.

13. Numerous problems arise for involved subject-matter specialists and decision makers on the one hand and concerned laymen on the other, to agree on a course of action. This is partly caused by different type of expertise and respective expert language as well as outlook in relation to commitment and responsibility and their relative unpreparedness to tackle such management problems of cross-sectional, multidisciplinary and inter-institutional nature. No formal education programme has ever prepared them for this type of job, i.e. to synthesize and coordinate all available means for common ends. The higher educational system based on specialisation and division of labour has not left any room for it. This situation is even more aggravated in most developing countries which not only face developments at an accelerated rate but suffer serious shortages in manpower, resources and efficient administrative systems. Where developed countries can still afford the wastage of manpower specialisation which is often inadequate to actual problem solving needs in job practice (ref. 8) because of a slowly grown web of regulations, abundant special knowledge and their general richness, developing nations cannot do the same due to their different and much more severe development-environment-interface situations. Their environmental and development planners, managers and policymakers will be in a better position to cope with their serious water pollution problems under given local circumstances, they have to be prepared for it by adequate training which will enable them to get familiar with the principal elements in sound, i.e. practical, decision making processes and apply them for purpose of problem balancing, management and control. Figure 2 demonstrates these steps and their interrelationship from examination of water needs and water pollution problems over the development of water quality management objectives and alternative control

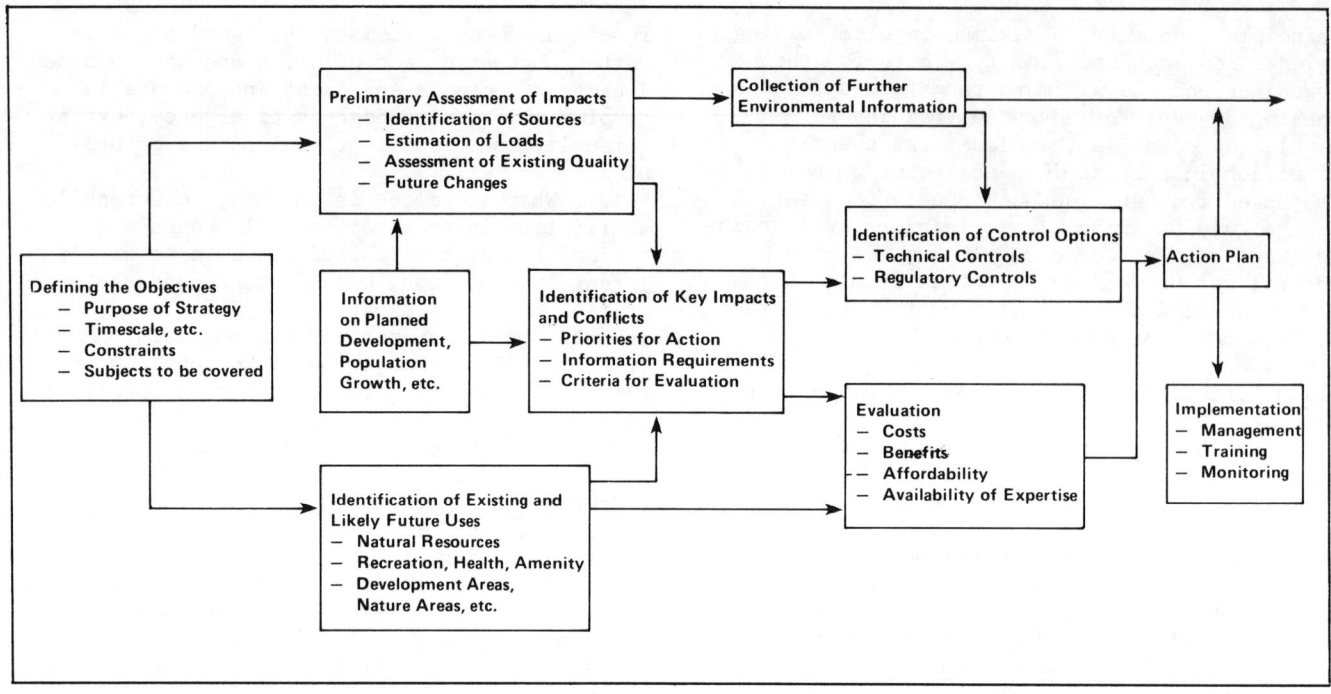

Fig. 1. (Source: ref. 7). Key steps in development of an environmental strategy

approaches to plan preparation, implementation and evaluation.

14. This overall assessment of environmental problems, especially in developing countries, as primarily being a management problem (also ref. 10) of situation analysis, strategy formulation and coordination of available technologies, resources and efforts for plan implementation to improve and better cope with the often detrimental circumstances has led us to search for an adequate and manageable training approach for senior environmental specialists and decision makers in the water pollution control field. This somewhat novel approach will be further explained below.

THE PROJECT CASEWORK BASED TRAINING APPROACH

15. There does not exist any uniform formula about how to proceed with professional training (ref. 11). It is necessary to cope with great amount of uncertainty on the one hand which opens up countless chances for trying new avenues on the other. Although it seems to be much safer and, without doubt, involving less preparation time and effort if one sticks to conventional seminar or course approach, the ploughing of new ground is, despite or even because of its inherent risks to failure, an exciting undertaking. The approach to be described here has been one of those successful trial-and-error adventures into new methodological terrain. Whereas the conventional approach primarily and overwhelmingly consists of lectures and discussions which, in its more updated versions, are complemented by some group work or case study session and excursions, the new approach reversed the whole composition of usual seminar components and the weight they carry. The usually dominant lecture was dethroned and removed from its central place within the seminar and given a still essential, but highly reduced role at the side stage. The case study was turned into the fundamental part of the whole seminar which, in its arrangement, is more a blend of the project method and casework, hence called project casework. This project casework is further sub-divided into three to four sessions whose requirements during progress of assigned group work dictate the additional input of certain other components, e.g. background material, certain technical papers and its presentations and deliberations, participants' country papers as well as field visits and demonstrations. According to this concept, any additional input is arranged in such a way as to accommodate the informative needs for smooth progress of project case work sessions which are conducted in small mixed country groups of five to seven participants each.

16. This programme pattern has proved very suitable in making senior level participants conversant with all significant aspects of environmental management and illustrating their interdependence through the medium of intensive project casework in small groups. Through thorough analysis, particularly of the cost factors involved, following out of group work and subsequent critical discussion of the results, their appreciation of what analytical work must be undertaken if, for instance, wastewater management problems are to be successfully and rationally tackled, was distinctly enhanced. The underlying structure of this seminar approach for intensive training on planning and its inherent decision making processes, while developing a master plan, is outlined in Figure 3. Similar to all complex decision making and management situations, it is an open-ended

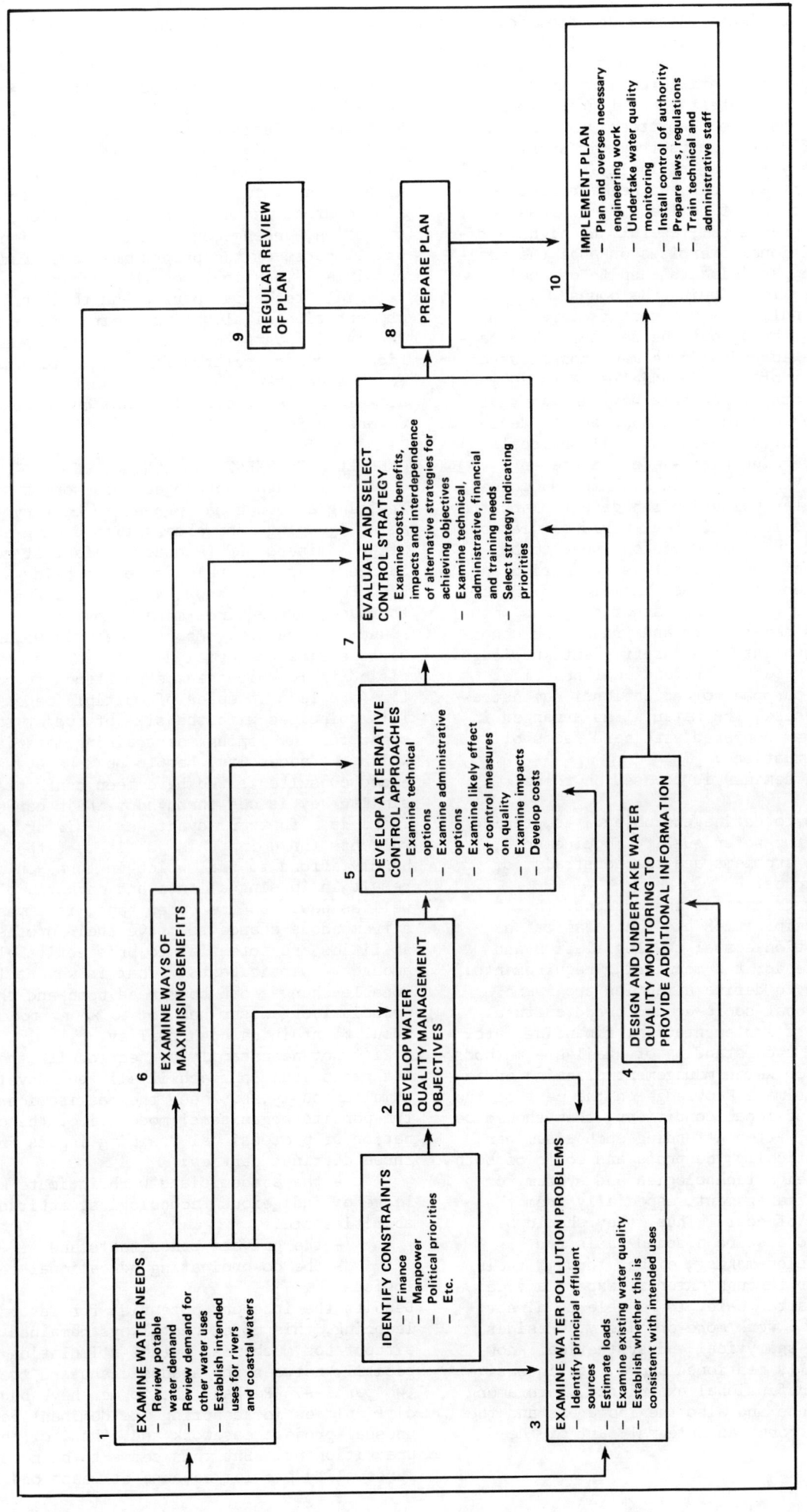

Fig. 2. (Source: ref. 9). The principal elements in the decision making process for water pollution control

educational experiment with enough room for adjustment, changes and inclusion of other elements. These have been the addition of other methodological components, for instance, from role play in the presentation of group work results to respective committees, whose members are formed of participants from different group, not only assess critically the pro's and con's of expert analysis and related recommendations but also the effectiveness of presentation. These practically useful and highly stimulating additional exercises enhance the motivation of the participants inspite of their very heavy workload during the course.

17. The handling of the cost factors is a theme running throughout the seminar. The participants are accordingly be made conscious of the importance of the financial aspects in most decision processes. In this way, the participants will be able to reach appropriate decisions in respect to their national or local problems concerning waste-water management. While working themselves through the succeeding steps of a water pollution management plan, they can experience from this real-world simulation exercise that numerous decisions need to be taken on objectives, priorities for action and how this action may be best implemented. Otherwise, assigned group work cannot progress within given time constraints and prevailing conditions of uncertainty and insufficient knowledge. All of these resemble professional practice, although for purposes of educational demonstration and exercise, the material is arranged in simplified form compared with hard facts of real world situations.

18. Such a seminar is focused on these questions:

    a) How to define the objectives, taking into account the water quality requirements, the pollution problems and the constraint of limited resources.

    b) How to examine the options: the technical possibilities and their capital as well as operation costs; the legislation and enforcement needs; the monitoring requirements.

    c) How to define an action programme suitable to local socio-economic and cultural conditions, its implementation, financing, etc. With a deeper understanding of available methods of analysis for water management planning and a presumably intimate knowledge on the part of the participants of local conditions, they should be in a position, after attending such a seminar, to assess critically the pro's and con's of both local and foreign technologies and system for water quality management, especially from the point of view of cost. Thus, they should be better able to acquire a sound basis for rational decision-making, either through their own efforts or through external expert advice and, in the latter case, to be able to direct and control the work more precisely. Besides imparting the analytical and synthetical knowledge and skills mentioned above, the intensive exchange of professional experience, both among the participants and also the lecturers and the participants, forms an integral part of the seminar.

19. In general, it must be emphasized that at every stage in the realisation of such a seminar the practical bearing and relevance to problem-solving for the participants were looked upon as the most important criteria. With the project casework arranged as the central feature of the seminar (Figures 3 and 4) to enable the participants to deal proficiently with the essential components of a water pollution control master plan, the function, construction, arrangement and optimal orientation of all parts of the programme can be made possible. This includes the preparatory material, technical papers and their brief presentation highlighting their salient points, the participants' contributions giving additional country-related first-hand information for better steering the seminar to their needs as well as the field visit and/or laboratory demonstration.

OVERALL ASSESSMENT AND CONCLUSION

20. Although the project casework based training approach has proved to be very encouraging, judging from observation during and after seminars and feedback so far, it certainly needs careful analysis before being applied, i.e. FOR WHOM, BY WHOM, WHERE, HOW, WHEN? Furthermore, it requires more time than usual for sound preparation. About a year is required for the first material development in any specified field which can be shortened to half the time and less in cases of multiple repetition. This coincides with the size of budget to be spent for developing the training material which, by necessity, has to be done by experienced consultants who have been confronted with similar key issues throughout their career. There is a further advantage in hiring consultants for doing this job in that they can hardly afford to fail which in turn would result in loosing clients and markets, old as well as new. On the other hand, it gives them a tremendous chance to prove their professional abilities and potential to prospective clients among the participants. That is why still reasonable charges can be agreed upon and the principal organiser has not to worry too much about maintaining high quality.

21. Not necessarily connected with the methodological approach itself, but developed simultaneously, has been the realisation of the tri-partite organisers' model, i.e. the combination of professional training inputs from three distinct parties:

    - the academicians with intimate knowledge of indigenous technologies, attitudes, abilities etc.

    - the private consultants and

    - the co-ordinating educational technologists.

Despite the inherent potential for interactive learning, this model has always remained behind without too much real mutual stimulation especially between the academicians and the other two parties. Perhaps the former have had many difficulties in accepting the dominant role of unusual project casework and, thus, of the practitioners. But this seems to be more of a personality problem because at least one of a

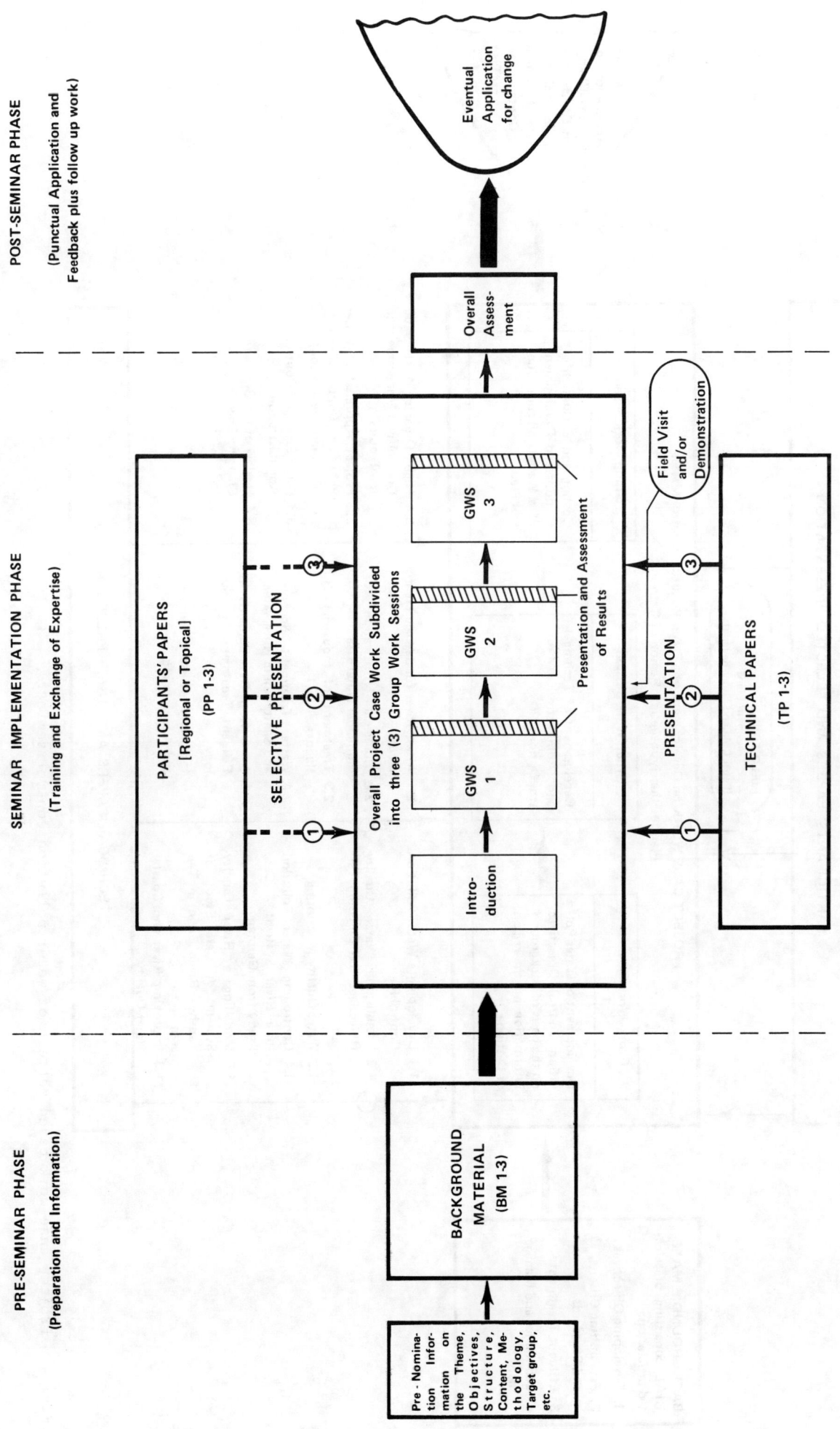

Fig. 3. Conceptual model of project case work based seminar training approach showing the interrelationship of different program inputs

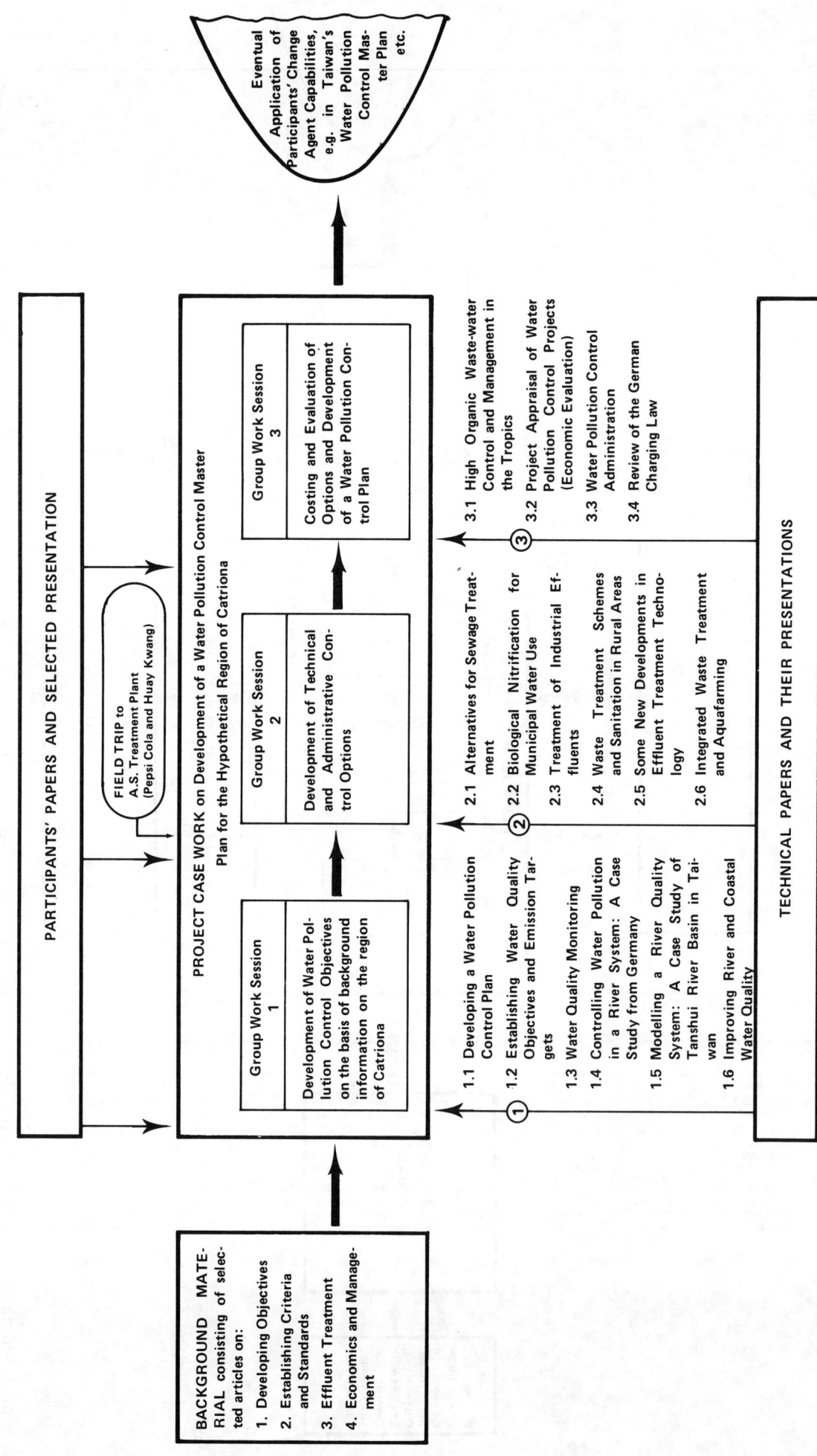

Fig. 4. Project case work based pattern of water pollution control seminar

few attending faculty from other Asian universities has already applied with much success the project casework approach in modified form in his teaching (ref. 12).

22. Even one country as a whole has implemented the concept of the project casework approach on developing a Catriona water pollution control programme with great success by having put the problem of water pollution in Taiwan under control (ref. 13). Altogether it can be stated that the approach as such an educational exercise has been extremely strong in its practice and problem solving orientation. It has generated high motivation, participation and inter-active communication. The simulation of real world planning and decision making situations through the medium of assigned project casework has led to intensive team work in small mixed country groups which has stimulated mutual learning crossing the barriers of academic disciplines, institutions and nations etc. There is still room for improvement, e.g. for better coordination of papers with project casework, selection of participants, thorough applications of educational technology skills etc.

23. Although the great majority of senior participants found the project casework approach the most challenging and interesting part of the one week-long seminars, we have not yet been able to compose a more comprehensive picture of what has really been implemented by how many former participants after returning to their jobs etc. While environmental training in and for developing countries in general receives a poor grade (ref. 14), just the few examples of very positive feedback as mentioned above (see ref. 10, 12, 13) might suffice for our assumption that the approach, as initiated, developed and implemented by us in close collaboration with ERL, carries a remarkable potential for really contributing to better water pollution control and management in developing countries. It also serves as another example of the Carl Duisberg Association's practical contribution to Asian-German cooperation within the framework of German development aid policy and its institution building efforts together with the Asian Institute of Technology.

REFERENCES
1. KLEY W. TRAINING SKILLS. A Manual for Fisheries Trainers. FAO (Tecmedia production), July 1983, p. 24.
2. WHO International Reference Centre for Community Water Supply. Suggested Steps in Development of a National Training Delivery System (Water and Sanitation Sector), Third Draft, Sept. 1978, p. 2.
3. JONES P.H. Educating the Inventor. The Environmentalist, 1(1982), p. 35-37.
4. MALIK F. Strategie des Managements komplexer Systeme. Bern/Stuttgart, 1984, p. 200.
5. ARBHABHIRAMA A. Development Research for Developing Countries. Paper presented at AIT-CDG Regional Workshop on Research and Technology Transfer for Asian Development. AIT, Bangkok, 1-3 Nov. 1983, p. 8f.
6. ESCAP ENVIRONMENT NEWS. Water Crisis in East Asia, Vol. 3, No. 3, July-September 1985, p. 9.
7. BIDWELL R. Environmental Assessment and Management: Problems and Approaches. In: Tharun G., Thanh N.C., Bidwell R. (ed.). Environmental Management for Developing Countries, Vol. 3 Environmental Assessment and Management, CECS No. 3, AIT, Bangkok, 1983, p. 23.
8. BEURET G., WEBB A. Goals of Engineering Education (GEEP). Engineers: Servants or Saviours? CNAA Development Services Publication 2, London, January 1983.
9. BIDWELL R. Development of a Water Pollution Control Plan. The Key Decisions and Choices. In: Tharun G., Thanh N.C., Bidwell R. (ed.). Environmental Management for Developing Countries, Vol. 2 Waste and Water Pollution Control, Evaluation and Decision Making, CECS No. 2, AIT, Bangkok, 1983, p. 192.
10. CHIU M., SHIN P. Project Management for Environmental Protection in Hong Kong. Country Paper prepared for International Symposium on Project Management for Environmental Protection, Pattaya, 24-28 Feb. 1986.
11. The Center for International Environment Information. Environmental Training in Developing Countries. A study of the practices of major national and international development financing agencies, Sept. 1980, p. 17f.
12. Letter to author by Dr. S.S. RACHAGAN, University of Malaya, Kuala Lumpur, 11 Sept. 1983.
13. LEE C.D. Programming and Implementation of Water Quality Management in the Republic of China. Country paper for International Symposium on Project Management Approach for Environmental Protection, Pattaya/Thailand, 24-28 Feb., 1986, p. 2 & 14.
14. KUNZMANN K.R., DERICIOGLU K.T. Environmental Education and Training in and for Developing Countries. IRPUD Reprint 5/85 from: Third World Planning Review, No. 1, Vol. 7, Liverpool 1985.

Discussion on Papers 18 and 19

MR C. BIGGIN, Watson Hawksley, UK
What measures would the Author of Paper 18 suggest to overcome the equity problem of families being unable to pay the charges for water for irrigation (or for domestic supply)?

MR J. M. HEALEY
There is no reason why water should be generally subsidized. Irrigation water has a value for the farmer through enabling greater yields and income which should more than cover the cost of water which is often a small item in his total costs. Drinking water should also cover its cost if possible. There must be limits to subsidization. Someone has to pay. Most less-developed countries governments' budgets are now running large deficits which result in inflation, and internal subsidies divert revenue from the provision of essential public services. There must be concern for the poorest groups but general indiscriminate subsidies to water supply are inadequate for this. They frequently help the more wealthy not the poorest.

MR V. R. ROWLAND, Pauling PLC, UK
In many countries the main sources of wealth benefit a section of the community directly and there is a wide range of incomes, those of the poorest communities being low. In these circumstances without some form of income redistribution such as negative taxation there is no case for subsidy and, if the principle of subsidy is accepted, there is no reason why this should not be applied to the provision of water or even to selected projects for the provision of water.

This leads to the possibility of including social benefits within an economic cost-benefit analysis. Occasionally the value of the time otherwise spent in carrying water is included in such an analysis, but it is a step beyond this to quantify the value of increased life expectancy from drinking relatively pure water.

What are the Author's views on the practicality of any such extension to cost-benefit analysis?

MR J. M. HEALEY
Assessing the benefits of water projects is difficult in practice. Certainly the time savings otherwise spent in carrying water - usually by women - should be quantified wherever possible. This is most practical when the women also cultivate. Valuations of health and life expectancy benefits are difficult to judge and in practice are unquantified objectives for which a cost-effectiveness approach to different project designs is most likely to be used. However, where a project is otherwise marginal or just sub-marginal a consideration of such benefits can tilt the balance of decision even when not precisely quantifiable.

MR A. CHURCHILL, World Bank, USA
Subsidies are difficult to justify on equity grounds as well as the usual efficiency considerations. Somebody has to pay and those that receive the benefits of investments are usually the better off. Equity considerations thus support full cost recovery.

DR T. VIRARAGHAVAN, University of Regina, Canada
I would like to comment on Mr Churchill's observations. I do not advocate subsidies either. How can a realistic cost be put on cost recovery when agriculture is very important and if farmers are driven away from farming? If there is insufficient food production, the government has to import food at high costs, bearing the expenditure through the general tax revenues. It may be easy to keep 'cash crops' out of the 'subsidy' syndrome, but it may be impossible to keep irrigation projects from some subsidy. In many countries especially basic food crops are to be grown in the fields irrigated by that water.

MR N. V. P. SARATHI, Central Water Authority, Mauritius
Subsidies are a highly political issue. In Mauritius 50 million m^3 water for irrigation are supplied. On a marginal basis it costs about 15 Mc/m^3 (1£ = 20 MRs). By law, the planters must be charged 4 Mc/m^3, but they refuse to pay even 2 Mc/m^3; in the process 5 million MRs are lost annually.

As sugar is the main crop of the country the supplies to the planters cannot be cut off. As a result, the domestic consumer is made to subsidize the planter.

Thus it is just a question of 'who pays for what?'

PAPERS 18 AND 19: DISCUSSION

MR G. THARUN
Coping with developing societies' increasing demands for water for different purposes with varying quality standards and quantity levels needs appropriate planning and decision making, co-ordination, co-operation and communication between the providers and the users.

Hence, the provision of usable water and its utilization, its pollution control and reuse etc. are the foremost tasks of management, i.e.

(a) to analyse the (problem) situation (departmental management)
(b) to decide on appropriate objectives
(c) to develop practical strategy options
(d) to select a course of action (departmental management)
(e) to prepare and to implement an action plan
(f) to monitor and control implementation (departmental management)
(g) to evaluate results (and to readjust if and when necessary) (departmental management).

The type of management is either departmental when the overall situation of demand and supply is under satisfactory control, now and in the foreseeable future, and mainly needs proper administration and execution of established guide-lines only or project management if there is a perceived need for substantial change to improve the water supply, usage, recycling etc., including the prevention of anticipated future losses/damage to resources.

Project management is understood as the drawing together of the usually scarce financial, physical and manpower resources in a temporary organization to achieve specified (environmental) goals within a specified time frame under given constraints. This applies especially in development/environmental/interfacial situations.

Development perceived as a planned change to produce a better quality of life requires the application of sound project management principles, i.e.

(a) establish a single point of main responsibility (if possible, also authority for the project)
(b) clearly define the project's direction (objectives, targets), its duration (start and end points), its boundaries (what is included, what is not), its means (financial, material, manpower), its communication and information needs, its external and internal relationships and its constraints
(c) develop an action plan
(d) monitor progress and evaluate results.

The tasks involved in water quality management and development are

(a) time, location and culture specific
(b) cross-sectional
(c) multidisciplinary
(d) inter-institutional
(e) value loaded
(f) complex.

Water resources development is evolving in interwoven phases of a project cycle with varying requirments of input qualities and quantities according to the progress of implementation. However, education and regular professional and vocational training are so compartmentalized and technically specialized that they, normally, do not provide preparation for this complex job of managing such fluid situations as the course of action of the enterprise and the administration of the project.

Whereas developed industrialized societies can still afford the luxury of poorly practice- and problem-solving-oriented education and training, developing countries with often serious manpower, financial and material constraints cannot. Therefore, relevant complementary training is needed by developing countries. Training is a systematic learning process designed to improve people's job performance. Therefore, it is a deliberately planned and executed activity to channel and activate their knowledge to specified targets.

Training is a medium for change. It is a systematic process that evolves in cycles to improve knowledge, abilities and attitudes. It follows the same pattern as any systematic decision-making process, i.e.

(a) to identify (training) needs
(b) to develop (training) objectives
(c) to plan and design for (training) action
(d) to implement the action
(e) to evaluate (training) results (and to readjust activities).

However, its in-built constraint is that it is mainly a medium only, of more indirect effect and depends on many factors that are completely beyond the control of the training manager.

The eventual application of training depends on numerous complex interrelationships between

(a) personality
(b) organization
(c) society.

Training is by no means a cure-all. However, if wisely applied, it can be a powerful tool for long-term conditioning and can enable preparation for appropriate problem solving in the overall evolution, but the higher up the target group the more complex the task becomes (in practice and, subsequently, in training).

There are no recipes for ready-made solutions, especially for senior management training and at international level.

The practical way to muddle through is to aim constantly for relevant training inputs and outputs. General guide-lines for 'relevance' are resemblance to practice and problem solving. This requires experience and seasoned judgement and foresight.

Despite the fundamental operational difficulty arising from high degrees of uncertainty and probabilities, as is also inherent in complex management tasks, training offers remarkable potential for flexibility and innovation due to little or no rigidity which is usually associated with solution manuals.

Advanced training therefore has gained

steadily in importance and volume in the context of development co-operation, but training should be applied only if it can really contribute to problem solving, either alone or combined with other measures.

For the highly complex task of environmental management training, especially in the field of water pollution control, the following project-casework-based training approach has been developed

(a) the entire course is based on and has a stucture of intensive project casework in small groups (mixed country, multidisciplinary and inter-institutional)
(b) thus, the project casework forms the dominant and central theme of the whole course, not just an interesting appendix
(c) the need to progress project casework, to provide overall direction, and to place and sequence all inputs, whether the structured groupwork assignments themselves or additional lectures and discussions, technical resource papers, papers about the participants' countries, field visits and/or process demonstration, background material, role-playing games etc.
(d) active involvement and co-operation of senior level participants to achieve success which, in turn, generates their motivation and commitment and thus leads to success in learning for change
(e) exercises of simulated water pollution control planning and the inherent continuous decision making processes prepare mentally for real action to achieve improvement and to offer insight into the nature of often hidden value judgements by professionals and comparisons between personal judgements
(f) an indispensable precondition is the practical and problem solving orientation which is highly interrelated with the objectives, content, materials, media, methods, target group mix and, last but not least, the resource personnel
(g) thus, project casework materials have to be developed and presented by seasoned practitioners with sufficient 'hands-on' experience and, subsequently, insight into what is really needed in practice etc.

The whole process of organizing and executing such a course is itself quite complex, especially in the international arena, and is a constant act of steering a way through to the anticipated targets. Success and failure are almost as random as in football matches or car journeys, which can take unforeseen and unexpected turns due to sudden developments and influences. This is similar to management practice and is a useful indirect example and experience in life.

Owing to time, manpower and financial constraints no overall assessment of the application of results in practical situations has yet been possible, but, occasionally, encouraging feedback is received, e.g. Taiwan's water pollution control scheme has been developed according to this concept and seems to work successfully. Others have reported that they have gained from training and have been able to apply much of it in their work, but in a more unquantifiable way.

Thus, the project casework approach, if wisely handled, can not only evolve as a highly stimulating and motivating educational exercise with good results but, as the ultimate goal of training work, can also contribute significantly to changes in real world conditions towards the improvement of life.

MR B. LUNOE, Hifab International AS, Norway
With reference to Mr Tharun's paper on training, I consider all aspects related to operation and maintenance of rural water supply and sanitation projects as possibly the most important items to be considered. It is not possible to talk about operation and maintenance without considering training or human resources development.

Insufficient time has been devoted to give the issue the attention that it deserves and ought to have been given. It should be ensured that the issue is given priority at the next and earliest opportunity.

MR G. THARUN
I agree that training could definitely play a vital role in the operation and maintenance of rural water supply and sanitation projects if it is carried out in an appropriate way, i.e. professionally. Those who should best know the requirements and special circumstances of a given project, usually the consultants in charge of it, should play the major role in such training. The main purpose is to secure the optimum practical solution to the problem by training. Thus the project manager should play his role fully in operation and maintenance training and not leave it completely to outsiders. If he discharges his responsibility towards training for his project reasonably well, then there will be a high probability that training programmes will really contribute to the progress of the project. However, training can never be a panacea. It will always be just a medium for change, not change or action in itself. This should never be forgotten with regard to any training programme.

MR I. VICKRIDGE, University of Manchester Institute of Science and Technology, UK
I would like to challenge Mr Wild's definition of appropriate technology as 'using 18th century designs'. Although old designs may be appropriate in certain circumstances, modern 'high' technology may also be appropriate under other circumstances. Good design is always appropriate.

I would like to add 'equity' to the list of criteria for judging whether technology is appropriate as outlined by Mr Williams. If the objective is to improve health, then the provision of a few litres of water to many people is more appropriate than the provision of many litres to a privileged few.

I agree with Mr Wild's comments on simplified components, spare parts, visual aids and after care. Do writers and editors of operation and

PAPERS 18 AND 19: DISCUSSION

maintenance manuals learn something from the authors of do-it-yourself car maintenance manuals which are often well illustrated with diagrams and photographs?

Good operation and maintenance requires good training which can be provided by many agencies and institutions. An integrated approach by the various bodies is required.

One mechanism for technology transfer (as implied by Mr Balfour and Mr Wild in their Technical Notes) is the use of operations, maintenance and training (OMT) contracts.

At present there is no standard form of contract for OMT and several critical questions need to be answered. How long should an OMT contract be? Who should carry out OMT? How can performance in operations and maintenance be evaluated? How can the efficacy of a training programme be assessed? On what basis should an OMT contractor be paid?

The Department of Civil Engineering at the University of Manchester Institute of Science and Technology is embarking on a programme of research designed to answer these questions and to draw up guide-lines for OMT contracts. Has anyone had recent experience of such contracts?

MR L. WILD, Laurie Wild Consultants, UK
I should like to defend my statement that 18th century technology is being provided by much of the water decade's resources. In Europe at that time there were numerous wells with hand pumps, and pit latrines were common. Sometimes the only changes now are the installation of better hand pumps.

One problem that has not been mentioned at the Conference is that liquid from the latrines must go somewhere, and if there is a convenient aquifer it will seep into any well or borehole in the vicinity. This probably accounts for the findings in Mr Lehmusloto's Technical Note which gave details of faecal coliform contamination in wells and boreholes in Kenya. This, incidentally, was a lesson learned in London in the 18th century after it had caused thousands of deaths.

My Technical Note comments that the water decade and this Conference divide into low cost technology in developing countries on the one hand and advanced technology (which often fails after a few years) on the other hand. However, as the low cost approach only marginally answers the needs of the world, what are needed are the most advanced designs now developed to avoid the disasters that we have seen over the past few years. This is even more essential now that most of the developing countries are facing increased populations. My Technical Note provides some guidance as to how modern designs should be directed towards the abilities of the recipients, and it is obvious that those clever enough to develop the equipment should also be able to make its operation and maintenance 'appropriate'.

Where the water decade is failing is that, although many areas are being provided with water for household needs, the people are not receiving means of wealth making - or even self-sufficiency - with irrigation and industry, and so in times of drought they cannot help themselves. This is where the speakers who have expounded the need for the local recipients to pay for the assistance received are ill informed. How can they pay when most have no means of making any money?

The aid agencies must therefore be urged to start the circle - injection of capital to provide the situation whereby the populations themselves can afford to pay for the service and to upgrade their own infrastructure. Agencies must also devote more of their resources to operation and maintenance by 'on-the-job' training as there are too many instances of water and sewerage systems failing after a few years.

People should pay an economic price for their water, but they should first be given the means of making some money. The question to Mr Churchill is not can we afford such aid, but dare we not afford it?

From this side of the Atlantic Ocean if, for example, the European Economic Community's food mountain were abolished there would be plenty of extra money for aid.

MR M. MULLER, National Institute of Physical Planning, Mozambique
Low cost sanitation in Mozambique was included in all major sanitation plans and in the main cities of Maputo and Beira sanitation is being developed on a zonal basis with parallel investments in sewered sanitation.

The production workshops in Mozambique are self-sufficient in local currency. Costs of imported inputs could be met but no foreign exchange allocation is available which is why the programme needs donor support. The local currency surplus is invested in programme expansion.

In general, more attention needs to be paid to the question of marketing. What the experience of Mozambique shows is that sanitation cannot be looked at in isolation from the other 'products' on the market. Rather than a marketing promotion approach what might be preferable is one of consumer protection whereby potential customers are helped to use their resources wisely to meet their needs. This might mean restricting less useful products (taxing Coca-Cola more heavily, banning the sale of medicines that do not work) rather than competing in a free-for-all market-place. These interventions are easier to make in a planned economy but they should also be feasible in mixed economies. The World Bank's investigations into the determinants of effective demand for water and sanitation in the developing countries are awaited with interest.

MR C. R. C. JONES, Sir M. Macdonald & Partners, UK
We have heard about governmental views of financial aspects, training and institution building, but what about the users? Would Dr Healey comment on the idea that beneficiaries from water supply and sanitation schemes identify more strongly with what they have paid for or helped to construct? Does this not make

cost recovery especially relevant to the motivation for devolving maintenance to the users? For example in the irrigation context, a benefit of small-scale schemes is that the users often completely control their own maintenance and patterns of use. Mr Muller has described low cost sanitation experience in Mozambique; it can be inferred that a similar process is in operation there.

MR J. M. HEALEY
I agree that there must be more economical use of water if users have to pay the cost of it. Such a scarce resource must not be wasted. Adequate maintenance has been achieved without cost recovery and financed from the governments' general revenues. However, there is more likely to be best use and adequate maintenance if the users pay and the state authorities responsible for operation and maintenance have the power to raise the revenue that is necessary for this.

MR M. MULLER, National Institute of Physical Planning, Mozambique
One reason for presenting information on water and sanitation investments in Beira is that the city will play a key role during the difficult few years of transition that lie ahead for the region and this information may be of use to those donors, consultants and contractors who work in the area.

More generally and more importantly, however, in many developing countries, Mozambique included, aid is a critical component in project finance. For this reason, it is hoped that aid helps rather than complicates the difficult management tasks that confront water and sanitation sector managers.

To ensure efficient use, it is necessary to be able to discuss openly and frankly the problems implicit in the administration of aid which is what I have tried to do in relation to Beira. The problems are familiar

(a) the added costs of tied aid
(b) the excessive use of feasibility studies and investigating missions
(c) the failure to co-ordinate aid within a national programme
(d) delays in the release of resources allocated.

Although the practical political constraints are recognized, there are many instances where aid could be more effectively used and it must be the objective to ensure this. Effective use, however, will not occur while under constraint from discussion of the problems, although the barriers have been broken in some sessions, notably that about the health components of the Gezira rehabilitation scheme.

The constraints for the various groups are obvious

(a) consultants do not want to upset their clients
(b) aid recipients do not want to offend their donors

(c) donor agencies are looking over their shoulders at their political masters and domestic constituencies
(d) producers of goods and services believe that only good news will sell them.

However, there is one group which should be able to speak without these constraints and that is the professional group, in forums such as this one, organized to debate the issues which confront us in our professional life. It is of general concern that the role as professionals in guiding the best use of nations' scarce resources is recognized. We heard in other sessions of the need for advocacy, the desirability of putting before a wider audience an assessment of the needs and the strategies for meeting them in this vital water sector. It is this point which I wish to emphasize. The way to ensure that our professional role is recognized is to exercise it in meetings such as this.

DR E. ROOSMA, Provincial Water Supply Company of North-Holland, The Netherlands
With regard to the demand of training the responsible drinking-water supply companies in developing countries to make it possible for them to fulfil their tasks properly, in The Netherlands 'twinning' takes place between several Dutch water supply companies who co-operate on a long-term basis with an appropriate company in a developing country to help them in all aspects that are considered to be necessary and deal with their day-to-day problems and more important those in the near and far future. They also give training in technical and administrative and financial management, at high and middle levels, and assist in public relations aspects to produce better attitudes in the customers towards reduced water consumption and better acceptance and appreciation of higher performances. The companies help to solve leakage and non-accountable water problems and help in the operation and maintenance of distribution systems, and in parts of the production facilities including purification installations.

Once the twinning agreement has been signed – involving politicians in the discussions during the period of preparation as well as the aid agency – the long-term co-operation can start. Training is carried out both in The Netherlands at the Dutch water supply company and in the twinning partner's company.

The essential feature is of being partners, who are in close co-operation on a company level and who have sufficient knowledge of the economical/social and cultural backgrounds to work together for a more efficient use of the technical personnel and financial capabilities for producing safe water in the future.

This is an appropriate way to help to solve drinking water problems in developing countries. It might also be a good idea for other water supply companies in the western world.

D. H. TOMPSETT, BSc(Eng), FIEE,
Immediate Business Systems plc, UK

TN28. Meter reading and billing with portable microcomputers

In Paper 18 Dr Healey and Mr Wilmshurst discuss the principles governing decisions by water authorities, their governments and the aid agencies in setting tariffs for water and thereby recovering some agreed target fraction of the costs incurred in water supply projects. They state (para. 19) that the functions of metering, billing and revenue collection themselves impose costs which are justified only if they are outweighed by the financial advantages they bring. It might be added that a modern but simple and efficiently operated metering and computer system incidentally provides a valuable data base for the patterns of water consumption on which to base better informed resource allocation decisions in the future. The value of this facility must therefore itself be assessed.

2. Assuming that a credit metered supply has been provided for a consumer, the water supplier's aim will become that of obtaining periodic readings of the meter in order to calculate a true bill (i.e. one based on actual consumption at the published tariff rate) and then to deliver the bill to the customer. Until these consecutive steps have been achieved the final objective of the supplier - collection of revenue from the consumer - cannot proceed. In classical billing systems, as used by utility companies in most developed countries, the minimum delay between meter reading and bill delivery is 3-4 days (when full account is taken of weekends, public holidays, etc.). In some less developed countries this delay can extend to several weeks. Examples exist of monthly billing arrangements where the meter reader delivers the bill resulting from the previous reading a month earlier.

3. The immediate billing system permits the meter reader to calculate, print and deliver a consumer's bill a few seconds after reading the meter. For this purpose the meter reader carries a pre-programmed hand-portable battery-driven microcomputer together with a lightweight high quality printer holding a roll of bill forms (see Fig. 1). The new meter readings and any special notes or records are entered into the computer and stored in its memory system. This is done either manually by the keyboard or automatically by connection of the machine to an enhanced meter. At the end of the day's work this new data is downloaded into the utility's data processing system so that all consumers' records can be updated. This downloading operation is immediately followed by a reloading one in which the following day's meter reading route and details are inserted into the machine. This route information has been compiled on a host computer and work station located in the district office to which the meter readers report. The host computer, in turn, passes new data to the utility's central computer and receives from it the details of its district's consumers whose meters are next due to be read.

4. The same type of data collection and processing system can also be used with a simpler form of portable microcomputer designed to enable meter readers to collect meter readings but without printing the bill on the the spot. For this reason the two functions of data capture and bill printing can conveniently be handled by two separate units which can be connected by a cable if the printing facility is required (see Fig. 2).

5. The benefits from the application of such machines to meter reading and billing are largely self-evident. With both types of

Fig. 1. Portable billing machine printing electricity bill

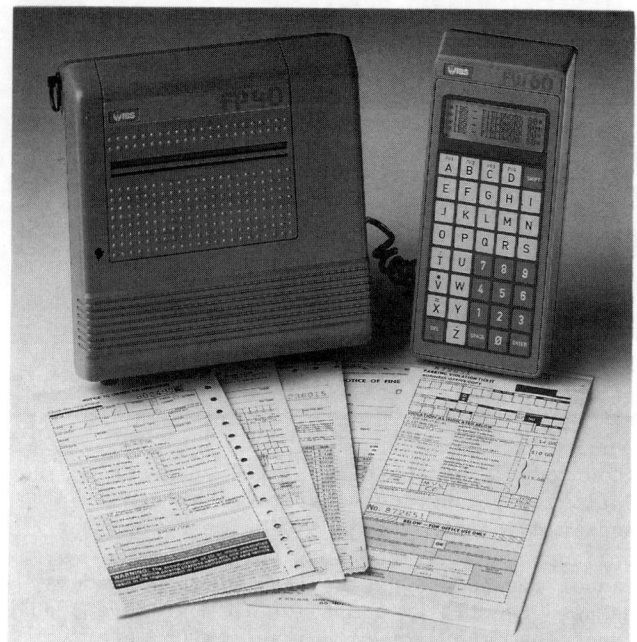

Fig. 2. Combination IBS hand-held printer and data-capture units

system the accuracy of the meter readings is greatly improved. The indicated consumption is compared with an estimated figure and if the reader has made a serious error this is automatically indicated to him on the display screen so that he can recheck the reading. The screen is also used to guide him on the route foreseen for the day's work, although he is able to override this sequence and take customers in another sequence if necessary. Both types of system shorten the delay mentioned in para. 2, although this benefit is naturally considerably less in the data-capture-only case compared with immediate billing. The billing system eliminates the cost of postage or hand delivery of the bills. This is a very substantial benefit, increasingly so if billing is quarterly, bi-monthly or monthly for which 4, 6 or 12 postal charges are saved annually.

6. Many incidental advantages and efficiency improvements have been found to arise from such distributed processing installations.

The district staff are in a far better position than are the central staff to react to the day to day events influencing the meter readers' activities, such as bad weather, accidents, breakdowns, illness and other unpredictable disturbances and exceptions to the theoretical regime being followed. Responsibility for the smooth flow of meter readings to the central computer for consumer billing has always rested on the district. The district staff have been found to be well able to take maximum advantage from the flexibilities built into the meter reading management system. Receptivity of the meter readers and of consumers to the system has been excellent. The availability of extended information to local and central managements has been exploited. It has been fully proven in operational service that meter readers can perform the same number of readings per day while printing and delivering consumers' bills as when merely collecting meter readings. Equipping meter readers with an obviously highly advanced product improves their status within the distributor's staff and in the eyes of the consumers, leading to benefits in both industrial and public relations.

7. Two British electricity boards and two utilities in the USA have fully implemented immediate billing and over 10 million bills per year are being delivered on the spot by the meter readers. Numerous other utilities in the USA are collecting meter readings with portable microcomputers. The technology for both systems is fully proven and adaptations of the software developed for them to suit other utilities and other operational conditions can be readily undertaken. Dr Healey and Mr Wilmshurst conclude their paper by saying that improvements in financial management are often made a condition of the provision of finance. Effective applications of portable microcomputers provide significant moves towards such improvement.

REFERENCES
1. TOMPSETT D.H. Immediate billing. Aqua, 1984, No. 4, Aug., 239-242.
2. TOMPSETT D.H. Water industry applications of portable computers to meter reading and immediate billing. International Conference on Water and Data Processing, Paris, 1986.

M. MULLER, MICE, formerly UDAAS
(National Water Supply and Sanitation
Agency), Maputo, Mozambique

TN29. Beira water and sanitation — the aid component

The city of Beira is a new settlement, less than 100 years old, with a population of 250 000. It owes its economic importance to its port which serves Zimbabwe, Zambia and Malawi as well as the Zambesi valley in Mozambique itself.

2. The water supply for the city is derived from 60 km upstream the Pungue river, beyond the limits of saline intrusion. Built in 1952, the system could produce 24 000 m^3/day, principal capacity determining components being the treatment works and the 50 km transmission main.

3. Sanitation is a problem due to the flat and impermeable terrain. The colonial nucleus is served by a system comprising 15 automatic pumping stations, 52 km of collectors and 17 km of pumping and gravity main - 'the most extensive system ever built at one time in all Portugal' as it was described when inaugurated in 1967. This serves perhaps 20% of the population. The remainder rely on septic tanks, latrines or are without any sanitation provision.

4. The water supply is run by a state company which, with new tariffs, covers all operating costs. Sanitation services are being integrated into the water company and will be self-sufficient when a proposed supplementary tariff is introduced.

NEEDS

5. Mozambique's urban infrastructures serve only a small proportion of their population, so needs are great. Beira's water supply needs major expansion to meet demand from currently under-served communities and to keep up with population growth and port and industrial demand. Present demand is estimated at 41 000 m^3/day and 72 000 m^3/day in 1995.

6. The sewerage system needed urgent mechanical recuperation as its deterioration was resulting in flooded streets and buildings and polluted beaches. Provision was also needed for the unserved majority of citizens.

INTERVENTIONS

7. Since Mozambique is effectively in a state of war, major investment in social infrastructure is virtually impossible to finance from state resources. Agencies in fields such as water supply are therefore heavily dependent on external sources of finance. Availability of finance rather than need tends to determine the investments made.

8. Analysis showed that water supply capacity could be significantly increased by duplication of 7 km of main followed by further stepwise buildups of system capacity. After failed attempts to finance purchase of ductile iron pipe from Europe or steel pipe from Zimbabwe, 7500 m of steel pipe were bought with a Spanish government line of credit. The $3.3 million project was built by local contractors.

9. Attempts to finance either the complete expansion of the system or further partial duplications began in 1981 and are continuing — negotiations with Italian aid authorities are advanced.

10. Since varied packages were needed for presentation to donors, the initial consultant's study based on presentation of a single set of contract documents has not been helpful. Given the timescale involved, however, it would have been unrealistically expensive to retain a consultant to support project preparation and negotiation throughout (ref. 1).

11. Early attempts to recuperate the sewerage system using only local engineering resources failed due to the high imported component needed. In 1983 Dutch aid authorities agreed to fund the purchase of equipment on a year by year basis (a separate basic needs project supports low-cost sanitation activities (ref. 2)). The $1250 000 works have progressed well, if slowly, under local management and apparently compare favourably with similar World Bank efforts in Dar es Salaam (ref. 3). The two locally contracted expatriate engineers involved were able to work usefully in other areas, notably organization and training, whenever work was held up.

PROBLEMS

12. Plant breakdowns caused substantial delays in pipeline construction since the line of credit did not allow purchase of spare parts for local equipment nor major items of new plant. In current negotiations it is evident that the $14 million estimated foreign currency component for the complete expansion of the water supply will be low if participation is limited to Italian contractors. Unit costs presented have been up to twice those estimated and now confirmed by Development Bank tenders for other Mozambique projects.

13. In the sanitation project, the only major problem arose because of donor policy

which, while not formally tying aid, prefers the use of Dutch equipment where possible. Since macerating equipment comprised nearly half the first year's order, it was agreed to buy Dutch rather than substitute existing British-made units. This required difficult and costly structural and pipework changes in the buried pump chambers. These were in fact unnecessary because when the equipment arrived it was invoiced in sterling, the Dutch supplier being only agent for another British manufacturer.

COMMENTS

14. The principal observation is that a poor country like Mozambique is effectively denied the opportunity of making optimum use of the resources available to it. Thus Beira buys second-best pipes (steel rather than ductile iron) from a source (Spain) further away from and more expensive than the obvious one (Zimbabwe). Their installation is hindered because funds can be spent on pipes and fittings but not on plant and spares. Local resources, notably the pipelaying crew, may be redundant if further financing is available only for turnkey projects. In the sewerage system, the decision to demolish and rebuild rather than simply replace equipment was taken for political rather than technical reasons.

15. These are not criticisms of the agencies mentioned since both Holland and Italy have excellent track records in the Mozambican water sector. At issue is the fact that the aid process often leads to second-best solutions and that it is in the interests of all parties involved to avert this.

REFERENCES

1. Sanaqua, Abastecimento de Agua à Cidade da Beira, Reforço do Sistema Existente, Maputo, 1983.
2. JONES F. Operation and maintenance services for Dar es Salaam city sewerage and sanitation. 11th WEDC Conference, Dar es Salaam, 1985.
3. MULLER M. Low-cost sanitation in Mozambique. World Water 86, pp 71-72, Thomas Telford, London, 1987.

L. O. WILD, FICE, MIMechE, FIWES, Laurie Wild Consultants Ltd, UK

TN30. Appropriate technology?

Many of the water projects constructed in developing countries have not performed satisfactorily, or have even failed, soon after commissioning. The failures led to a body of opinion which advocated appropriate technology (i.e. using plant and designs more suitable for the 18th century rather than those required for today), thus robbing the recipients of many modern advances.

2. The poor performances of these projects should not have caused surprise as they followed the transfer of modern designs to non-technologically advanced countries with little thought to the staff available locally.

3. The concept of appropriate technology, however, is wrong as it invariably consists of providing minute quantities of water, which can only continue (albeit slightly better) the beggarly situation of the local population. Cases are even known where supplying a few litres/head per day has caused great problems to the social structure as it does not assist in providing the infrastructure needed to overcome problems such as drought and increasing population.

4. The basis of any infrastructure needed to raise the standard of living is an ample supply of wholesome water and a means of dealing with the resulting waste products. Irrigation and industrial development can then follow, but only by the use of modern designs for water services.

5. The need for appropriate technology should therefore be directed at the designers and manufacturers, who should provide the equipment in such a way that local staff, with the minimum of training, can use it. This should present no problems as anybody with the technical ability to design such equipment should be able to present it so that anyone with intelligence can keep it working.

6. There follow four basic considerations on how to make modern concepts acceptable for use in developing countries. All may increase the overall cost, but this is preferable to failed plant.

(a) Simplified components. Most designs include items that increase efficiency but cause problems to the uninitiated user. For example, consider pumps. Why fit wear rings to centrifugal pump impellers when using a resistant material, for the impeller will ease maintenance problems? Use larger clearances - even use a vortex type pump. Avoid mechanical seals; it is much better to use gland packing etc. The same considerations can be given to most items if the same care is given to subsequent maintenance as to the original design. Operating procedures can be thought of in the same way. Labour is seldom a problem, so avoid remote control complications. Installation of buffer tanks in systems reduces critical control, and sluice and reflux valves are more appropriate than butterflies. For identification, the greater use of colour avoids confusion over manufacturers' names for parts.

(b) Spare parts. It is common practice to supply the minimum of spare parts: only items that wear or are prone to failure. However, it is far easier to fit complete assemblies: bearings in their housings etc. This consideration particularly applies to electrical or electronic gear, engines, etc., where made up parts are simple to fit.

(c) Visual aids. How often does a manual or instruction book leave the impression that the writer has never actually carried out the task described? It is best to show operation and maintenance details visually with the use of video recordings. The cost is minimal and little expertise is required to use modern cameras. Any sequence of operation or fitting instructions can be observed by the local technician who, by using the frame freeze and reversing facilities, can repeatedly watch the task being carried out by an expert. Furthermore, dubbing the tapes in any language is far cheaper than translating the written word.

(d) After-care. The time when consultants and manufacturers completed their task when the water or sewerage works had been constructed should now be past. Various methods can be used to ensure that satisfactory operation continues. These include the requirement that manufacturers should operate and maintain their plant for a period, sometimes including training of local staff, with overall supervision by the consultant or main contractor. This at least ensures that teething troubles are dealt with, but it also has disadvantages. For example, at present, few consultants or manufacturers have any significant experience of operation or day-to-day maintenance. Also, each manufacturer is concerned only with his own item of plant and has no responsibility if another item in the process fails.

7. It is therefore required that one organization is given the responsibility for the subsequent operation, be it the consultant (expanded to include operational staff), a turnkey contractor or a service contractor who specializes in operation, maintenance and training. The period of the contract depends on the complexity of equipment but should be so written that, as the locals become competent at their various levels, they can replace the contractor's staff.

8. Training is recognized as being important, but here again the attitudes of those responsible should change. It is wishful thinking to expect anyone not brought up in a technological society to assimilate quickly the complexities of the equipment; operation and maintenance should therefore be changed to a multitude of set tasks which can be individually taught. One group should be given tuition in bearings, another with glands; one in electric motor windings, and yet another in some aspect of switchgear, etc. Over the years, as competence in each item is achieved, another section can be added. Such methods are not new; during the Second World War aircraft workers and maintainers quickly learnt the simple tasks which, together, resulted in overall satisfactory operation. The real skill was in breaking down the tasks to ensure simplicity but achieving the whole.

9. Modern equipment is required in developing countries to provide the essential water services needed by increasing populations. This leads to the establishment of an infrastructure including industrial development and increased food production by irrigation. It is essential that such plant should be designed so that a minimum of guidance and training is required, to enable local staff to maintain a satisfactory service. All aid packages should therefore include a sum to be expended on setting up a management organization which includes operation, maintenance and a financial structure, to ensure satisfaction for all involved.

General discussion

MR A. HAYES, Ove Arup & Partners, UK
I am a member of the Register of Engineers for Disaster Relief (REDR) and have had three secondments to provide emergency water supplies to refugee camps: three months in Northern Somalia in 1981-82, four months in Uganda in 1983 and three months in Eastern Sudan in 1985. I was seconded to the Save the Children Fund (SCF) in Somalia and Uganda and to Oxfam in the Sudan.

The real tragedy of disasters is the world's inability to prevent them from happening in the first place and to deal with them efficiently when they have happened. In the field the problem is usually tackled by voluntary agencies who often rely on highly motivated although sometimes inexperienced people.

The importance of water supplies, sanitation, roads, shelter and other engineering functions is recognized by most agencies but few of them are equipped to offer practical help. Oxfam and the SCF are notable exceptions. In particular, Oxfam have developed a range of emergency water kits which can be air freighted into the field at short notice. The kits provide basic water and sanitation systems which can be erected rapidly. Excluding military equipment, these Oxfam kits are unique.

However, it is incredible that a disaster on the Ethiopian/Sudanese scale should have been dependent for water supplies on a relatively small charity like Oxfam, albeit acting on behalf of UNHCR. Disaster relief should not be left to volunteers or treated as a vocation. The additional cost of putting experts into the field (not merely to produce reports) can be repaid many times by getting the job done quickly and efficiently. What is more it would save lives.

I would like to see the creation of a national disasters unit aimed specifically at the engineering aspects of relief work with particular emphasis on emergency water supplies, sanitation and transportation. This would require funding, permanent staff and large stocks of equipment. Such a unit would have to be independent of the bureaucracy and vacillation that is usually associated with governmental and international organizations. The unit would then be capable of responding rapidly to an emergency. The capability and expertise already exist in organizations like the REDR, Oxfam, the SCF and the WEDEC.

DR S. STOVELAND, Ministry of Water Development, Kenya
Inadequate financial resources often put the sanitary engineer in a difficult situation whereby he is forced to plan schemes that supply water of poorer quality than that described in the World Health Organisation's standards. It would be most helpful for engineers working in third world countries to have water quality guide-lines which also describe in general the consequences to the community of using water of inferior quality.

MR P. G. WILLIAMS, National Institute for Water Research, South Africa
The work done by organizations such as Oxfam in developing technologies for providing and treating water in disaster areas is very valuable. However, design drawings and so on should be widely published so that information and experiences can be applied more generally.

PROFESSOR M. J. HAMLIN, Department of Civil Engineering, University of Birmingham, UK

Summary of Conference

Summing up at an international meeting is a two-edged sword. One is proud to be asked but, as the time approaches, the magnitude of the task becomes daunting. This is particularly true of a conference with joint sessions, and one in which there is such a wide diversity of papers and technical notes. It would be easy enough to present a 20th paper or to summarize each of the papers presented. However, this is not my task and what I shall do is draw together my perception of what were the more important threads that have run through the meeting.

I shall start with a somewhat evocative quote which is taken from the special D-day edition of World Water.

'On the 10th October about 12 000 people died in the El Asnam earthquake in Algeria; another 250 000 were made homeless. The world was rightly shocked into action and millions of dollars of aid was poured into Algeria. On the same day 30 000 people died in the third world because they had inadequate water or sanitation facilities. Tens of millions of women spent half their working day walking in the hot sun to carry polluted water ....'

The conference has given the opportunity to focus once again on the enormity of the task, to take stock of what has been achieved, and to look at the issues which have yet to be resolved before the goals of the water decade can be achieved. The conference has also included other related topics not concerning the water decade.

Her Royal Highness the Princess Anne has clearly drawn attention to the many human problems which the provision of a proper water supply could solve and this has been referred to by many speakers: the problems of mother and child and of the introduction of new techniques when traditional methods might have been developed to achieve similar results; the difficulties in resolving conflicts between fuel to boil water, and deafforestation, increased erosion and subsequent loss of water resources. She has referred to the contribution of voluntary agencies and the need for training. Finally she has reminded us of our responsibility as professionals who between us, and by sharing our responsibilities with people of other countries, have the task of bringing the decade to a successful conclusion.

In the recent Voluntary Services Organisation television programme a participant said 'If we each do nothing, nothing is done. If we each do a little the problem can be solved.' Dr Brown has also reminded us of the agreement during the 1983 meeting that 'the Decade is not just an impersonal internatonal effort but our Decade - yours, mine and everyone's.'

These then are some of the major topics that have been addressed here

(a) what has been achieved - what remains to be done
(b) the need to incorporate water resource planning and management into the water decade so that aspects of water quality and water quantity are properly linked
(c) the theme of irrigation which was included at the founding conference in Mar del Plata in 1977
(d) the basic problems of rural water supplies, rural sanitation, urban water supply for both domestic and industrial use, and urban sanitation for both domestic and industrial use.

A discussion of training will not be included in this summary. However, this subject permeated almost every session. Authors and discussors alike recognized the need for training at very many levels ranging from master of science courses at university to training women in the maintenance of village water supply systems. The paper by Tharun (Paper 19) emphasized particularly the importance of project-based training.

In considering the first of the four major topics I shall confine myself to Dr Brown's keynote address. Before the conference I had noted that we are standing still in relation to the population without adequate water in both urban and rural water supplies, i.e. new schemes are being installed at a rate which just keeps up with population increase and the drift to the urban areas. With respect to urban and rural sanitation ground is being lost. There are 110 million more people without adequate sanitation than there were in 1983. Infant mortality figures are unacceptably high and Dr Brown suggests that the figures could be reduced by a half with an adequate sanitation programme. Dr Brown introduced the term software investment and urged an increase from about 1% to say 5% which should be spent on human resources development, strengthening institutional performance, health education and other support communication, and also by linking water-related schemes to other relevant development programmes, e.g. in primary health care. The other major point which he raised was the need to have a single water programme not a series of separate conflicting or overlapping programmes put forward by different donor agencies – the need for one sector programme to be effectively linked with other sector programmes and, perhaps above all, to have a strong government department to oversee all aspects of water.

For the future he stressed

(a) better use of human resources
(b) rehabilitation, operation and maintenance
(c) the mobilization of local money
(d) the need for decade campaigns
(e) the re-emphasis of decade concepts.

The second theme is the need to incorporate water resource planning and management into the water decade. His Excellency Dr Kilani introduced this theme into his keynote address and it was taken up by Fish and Shenyi (Paper 1) and in much of the discussion in the irrigation group. As a water resources engineer I welcome this move. Although, in many instances, the development of a rural water supply or sanitation scheme has a negligible impact on the regional resource the concept of regional resources tends to bring together some of the agencies responsible for water under a single authority. This becomes increasingly important as industry and more intensive agricultural practices produce point and non-point pollution, the effects of which may take many years to become apparent, e.g. the case of industrial dumps used decades previously polluting underground and surface water, of which Love Canal in the USA is a prime example. Integrated management also makes it easier to deal with problems like acid run-off (I avoid the term acid rain) and the effects of possible climate changes. People in Britain have become perhaps a little obsessed with integrated river basin management which is very suitable here but is not always exportable. I was pleased therefore to see in the paper by Fish and Shenyi the statement that 'while integrated river basin management of the style employed in the UK is desirable wherever it can be arranged, this is not as essential as it is to have water resources in all respects managed by a single body'.

As a hydrologist I believe that the weather pattern is controlled by the astronomical cycle on which is imposed the enormous effect of noise in the atmosphere. A consideration of the forces in the atmosphere leads me to believe that weather prediction will be limited to a very short lead time, of days rather than months, and I have no confidence in the ability to influence the climate. It will always be necessary to plan, design and manage with that element of uncertainty in all aspects of the hydrological cycle which is inherently due to atmospheric noise.

The introduction of irrigation into the programme of the conference has generated six papers with a seventh paper linking health and irrigation and another paper on the health aspects of village communities within irrigation schemes. The issues raised in this group of papers were as follows.

Firstly, what is known about the hydrology of developing countries and perhaps particularly of semi-arid regions? I have been working with a colleague in Birmingham on a scheme in Sri Lanka which was constructed 1400 years ago and we have been the first people to try both to model and to assess the natural inflow to the irrigation tank.

The second issue concerned irrigation efficiency. How can irrigation be made more efficient both in the conveyance system and in water distribution and application at the farm level? Rangeley and Barnsley (Paper 11) conclude that improvements in management, operation and maintenance will continue for many years and the interweaving of physical and sociological constraints will take time to resolve.

Thirdly, problems of distribution by overhead irrigation generated much discussion particularly with respect to the transfer of systems developed in western countries to less well-developed countries, with some unhappy examples being highlighted.

Fourthly, drainage is now a serious problem in many schemes and two papers addressed this topic. In many countries water-table levels have risen to produce water-logged and saline conditions which make further productive irrigation impossible. The conflict between horizontal drains leading to large drainage canals and vertical drainage using tube wells where pumped water can either be used directly or by blending with water of better quality has still to be resolved and there is probably no general solution. As a measure of the scale of the problem in China alone 17 million ha need to be drained by the year 2000 and £500 million is being spent on the Left Bank drain in Pakistan.

Finally, the paper by Bahar (Paper 16) highlighted once again the dangers of disease associated with irrigation development both to the irrigators and in the paper by Gaddal et al. (Paper 15) to those who live in the associated villages. In 1953-54, I attended courses at the London School of Tropical Medicine and I was taught by Professor MacDonald. Professor MacDonald had highlighted these and similar problems in the James Forrest lecture (ref. 1). This paper together with those by Mr Bahar and Mr Gaddal should be mandatory reading for all engineers.

The final paper, although not in terms of presentation, by Le Moigne (Paper 10) reviewed the whole field of irrigation and, in the summary of his paper, he highlighted the problems which need to be solved

(a) international riparian issues that influence water use
(b) emphasis on rehabilitation of existing schemes through modernization, improved management and better institutional support: in 1985-86 World Bank funds were made available for the rehabilitation of 1.5 million ha compared with 0.5 million ha for new schemes
(c) drainage will become a major activity

(d) new technologies such as control systems for indenting for and scheduling water releases, low pressure sprinklers and drip irrigation will penetrate developing countries

(e) contributions from irrigated agriculture will continue to increase.

In the discussion the issues were concerned with the rate of return, the cost of producing irrigated crops in some developing countries compared with world prices and the use of variable irrigation areas depending on the water resources available in a particular season.

More people attended the sessions on water supply and sanitation than the sessions on irrigation but I commend the papers on irrigation for reading.

No doubt most if not all delegates still regard the water decade primarily in terms of water supply and sanitation (since that is its name). To divide this section into four topics is perhaps pedantic and yet the issues raised, while having many common themes - ability to pay, involvement of consumer groups, etc. - are still seen separately by many authors. The papers divide between high technology and low technology (but presented in a way which I hope recognizes the need for appropriate technology) and a conference on World Water including issues not concerning the water decade should perhaps include this wide spread.

I shall start by considering the common themes.

Many speakers made the point that what a community can afford now should be provided rather than to wait to provide a better solution at a later date.

Equity is very important. The provision of a low level solution for a large proportion of the population should be much more acceptable than a high level solution for a small proportion of the population.

The multidisciplinary teams working within a single authority are likely to produce a more acceptable decision than one imposed by a single disciplinary group and reluctantly I conclude that this observation is probably directed towards civil engineers.

Contrary to the cultural tradition in many western countries women often play an important if not dominant role and this has to be recognized. Dr Brown pointed out that if a rural water supply system breaks down it is the women who have to carry the water so they are the people who should be trained in maintenance.

We should be honest about the true cost of water. Dr Wright has pointed out that water/sanitation is only one of many services to be provided in developing countries. He suggested that the benefits could probably more easily be sold on the basis of individual convenience and privacy in the community than on the grounds of health.

I hope that I report Dr Wright correctly in saying that the bank lends in response to requests and that only 4-5% of the money is loaned for sanitation projects. This he concluded reflects the perceived priority. Sewerage and sewage treatment has for long been disregarded in the western world so it is not surprising that this aspect of water also has a lower priority in developing countries.

What then are some of the particular issues in the four subsections?

As far as rural water supplies are concerned it is perhaps relevant to refer to the Technical Note by Williams and Van Vuuren. Mr Williams suggested that the relevant requirements for rural water supplies were

(a) convenient access
(b) reliable source
(c) adequate quantity
(d) safe quality.

He argued that improvements could be considered as a stepwise process where an improvement of any single factor was a movement towards the ultimate solution. He went on to explain how the notion of team work and technology transfer could be used to achieve all four goals. Technology was specified in terms of being affordable, effective, reliable and acceptable.

Experience of implementing rural water supply in the Philippines included the education of local communities through their existing institutions towards setting up water committees. Rural water systems are financed either by grants or by loans of 90% of the scheme costs. The collection of loans is only about 40% successful and requires programmes which include

(a) a massive education programme
(b) friendly persuasion
(c) periodic follow-ups

to collect all the money due. Problems of the maintenance of rural water treatment plant arise in almost all countries either due to inappropriate technology or inadequate maintenance and this problem can only be solved by suitable training programmes.

The session devoted specifically to rural sanitation was covered by papers on the non-sewered solution for urban sanitation and the use of low cost low energy treatment systems.

In reviewing the contribution to urban water supplies two aspects should be considered. The first relates to water quality and the second to water distribution systems.

In the paper by Helmer and Ozolins (Paper 2) the difference in water quality standards between developed and underdeveloped countries is typified by the chemical versus bacterial water quality dichotomy. In the west the bacterial problem has been largely solved and the carcinogenic effect of trace organisms and other pollutants is of concern. How then should the problems of water quality in developing countries be considered? Is it appropriate to apply World Health Organisation standards which are meaningful only in those countries which have already built or are committed to building the necessary infrastructure? Although a stepwise approach to achieving standards is more rational than the imposition of unattainable standards ab initio for countries with limited resources Helmer and Ozolins nevertheless point out the dangers of adopting this procedure lest the provision of less than desired standards becomes accepted as the norm, thus leading to a false sense of security. Nevertheless the development of standards, whether for urban supplies or rural supplies, must proceed. The standards should be practicable, however, and attainable within a relatively short period of time.

The second aspect of urban water supply is the adequate protection of the reticulation system to prevent leakage and also the ingress of polluted groundwater. The work of the Water Research Centre on leak detection, water mains rehabilitation and sewerage rehabilitation is of significant importance. There are many cities in developing countries which face the same problems of rehabilitation of significant underground assets as are faced in Europe. Although local research is desirable the basic work done by the Water Research Centre probably has wide applicability. Cities where there is a 60% distribution loss and water is available for only eight hours a day can have a 24 hour supply restored if the losses can be cut by half through proper detection and maintenance systems. These are areas which are surely just as important as investment in new schemes and which perhaps in the past have received too little attention.

The provision of urban sanitation is probably the most pressing problem in relation to community health. I am one of the few people who 30 years ago was misguided enough to install a sewerage system in Zambia to replace what was the perfectly adequate provision of aqua privies. I have already referred to the need to provide what a community can afford and there is little doubt that for many communities a non-sewered system is the correct and the most appropriate solution.

Where a sewered system is installed the need for either a conventional works or a series of ponds, aerobic treatment, aerated systems or ponds with aquatic plants must be considered. The papers by Huiswaard and Bruins (Paper 7) and, in a different context, by Thomas and Rachwal (Paper 8) address this problem. In the discussion, Mr Cook from the World Bank referred to a successful scheme in Israel developed jointly with farmers in which the effluent from a small community was used for irrigation. He also referred to several large cities which successfully treat their sewage by non-conventional means. Lest the importance of the adequate treatment of industrial wastes in developing countries is forgotten, I draw attention to the paper by Downing et al. (Paper 6).

I have not done justice to all that has happened at this conference. However, I hope that I have raised some of the more important issues and that this will give encouragement to continue the contribution to the aims of the water decade at international, national and individual level.

REFERENCE
1. MACDONALD G. The interdependence of medical science and engineering, with special reference to public health. Journal of the Institution of Civil Engineers, 1949, vol. 32, Oct., 398-416.

A. CHURCHILL, Director, Energy Department, World Bank, USA

Closing address

These remarks are the observations of a non-expert. I am neither an engineer nor a water supply man by either training or inclination. My remarks are those of an interested and committed observer, who believes strongly in the importance of what is being done but at the same time has the luxury of being able to take a more detached view of the effort. These remarks are meant in this spirit.

First I have seldom encountered a group of professionals that are as dedicated and committed to their work as the group represented here. All the members have a sense of 'righteousness' that is unusual in any profession, except perhaps in the medical profession. This comes from the firm conviction that what you are doing is of great benefit to humanity, and this is undoubtedly right. It is a strength in that it gives both the energy and the drive to overcome the many obstacles, but it is also a weakness. Your conviction of the righteousness of your case often leads to an intolerant attitude towards people who do not share your fervour. It also leads you to be less critical and to be less accepting of the inputs of other disciplines - particularly economists, of which I am one.

I shall illustrate this with my experiences in putting together a strategy paper for the bank on rural water and sanitation investments.

Improving the access to adequate water and sanitation services of about two-thirds of the rural population in the developing countries must be one of the greatest challenges to be faced, yet more than half-way through the International Drinking Water and Sanitation Decade the problem has only begun to be addressed. The record of past investments leaves much to be desired. In Africa, in particular, the record is one of more failures than successes.

When I asked the question why we should be making these investments in the first place, my colleagues were taken aback. To them the answer was obvious - to promote the health and welfare of the population. It was so obvious that to justify investments it simply needed to be stated. Rates of return were obviously high but unquantifiable.

My suspicious nature led me to question the obvious. If the rates of return were so high why were more impressive results not being seen? A closer examination of the evidence suggests that things are far from obvious. It is, for example, difficult to demonstrate in practice the existence of any substantial health benefits from these projects. If they exist, they are too small to be the major justification for these investments.

However, a major benefit tended to be overlooked - the savings in time and energy of the rural population, particularly women, used in hauling water. Even at very low wage rates or values of time these savings tend to be large and sufficient to justify a much larger level of investment than typically was being put in place. One hand pump over the traditional well might make the source 'safer' but added little benefit to the villagers who were faced with haul distances that were just as long, but two or three pumps within the village would dramatically change the length of the haul and would produce large benefits.

Once it becomes possible to quantify the benefits it then becomes possible to make sensible decisions about the size of the investment - the number of pumps or whether or not a piped distribution system should be installed. It becomes possible to make reasonable decisions on the type of technology to be used and so on.

Another assumption that is usually made is that the people are too poor to pay for these services, thus condemning them to dependence on charity. Again, a more dispassionate look at the evidence suggests that this is far from the case. There are some parts of the population in some parts of some countries where this may be the case but for the bulk of the rural populations this is far from true. Improvements in services are possible within the constraints of these people's income. This is a very hopeful situation. It means that it is possible to deliver services without having to rely on charity or the need for large public subsidies. It means that alternative institutional delivery mechanisms that rely less on inefficient public institutions and more on individual and community initiatives can be used.

I have dwelt on the rural water supply issue because it is illustrative of the point that I am trying to make. Much of what is done requires modifications in human behaviour if it is to be successful. Engineering solutions are not solutions unless they are put into this broader framework of social and economic factors. Here it is necessary to listen more to what others have to say, and I mean really listen - not just say, 'of course, it's obvious,' and then go about business as usual.

World Water '86. Thomas Telford Limited, London, 1987

CLOSING ADDRESS: CHURCHILL

Another challenge is sanitary services in the large cities of the developing world. The reality of the situation is that most of these cities - whose total populations are now approaching 2000 million - cannot afford the traditional water-borne sewerage systems. What is to be done? Are there alternatives? I think that there are, but the answers require quite a different perspective to the problem. On-site disposal systems require that the household and its behaviour with respect to what are in all cultures private and intimate functions become the centre of focus.

In our work on this problem, for which we have had the support of the United Nations Development Programme and many bilateral agencies, it has been found that it is this human dimension that has been the most difficult to deal with. The technical solutions are fairly obvious - but to gain their acceptance and use remains a major challenge. One way is to pay more attention to what the customer wants.

There are other examples, of which I shall just mention one more: the problem of adequate maintenance for existing systems continues to be a major issue. In some countries systems are falling apart faster than they can be built. It is seldom a purely technical problem. The political, economic and social factors that have led to the present sorry state need to be examined more thoroughly and solutions that will address the problem in terms of those factors need to be suggested.

This may all be obvious, but when is something to be done about it? Some starts have been made at this conference but there is much more to be done. In front of us is one of the major challenges of this century - to provide adequate water and sanitation services to the large and growing portion of the world's population that lives in the developing countries. Let us just remind ourselves that challenges of this order of magnitude require the efforts of many.

Midland — Maintaining the Flow Worldwide

Midland Bank is proud of its long association with international water and sanitation projects. It is a founder member of the British Water Industries Group, a quasi-governmental organisation created to bring together the diverse segments of the U.K. water industry. Midland Bank Group, one of the world's largest banking organisations, is currently funding the Greater Cairo Wastewater Project and through its specialist team can provide the sophisticated financial packages demanded by the Water Sector.

Contact:
John Sillett, Trade Finance Director,
International Trade & Export Finance Dept.,
Walker House, 87 Queen Victoria Street, London EC4V 4AP
Telephone: 01–260 5516   Telex: 887305

MAY 2 3 1989